W9-AKX-866

MICROELECTRONICS

MICROELECTRONICS

Theory, Design, and Fabrication

EDWARD KEONJIAN, *Editor*

Staff Scientist, Arma Division, American Bosch Arma Corporation
Chairman, Advisory Committee on Microelectronics,
Electronic Industries Association

FOREWORD BY *Jack A. Morton*

Vice President, Bell Telephone Laboratories, Inc.
Fellow, Institute of Electrical and Electronics Engineers

McGRAW-HILL BOOK COMPANY, INC.

New York Toronto London

MICROELECTRONICS

 Library of Congress Catalog Card Number 63-13140

34135

LIST OF CONTRIBUTORS

William N. Carroll (CHAPTER 4)
Senior Engineer, Command Control Center, Federal Systems Division, International Business Machines; Kingston, New York

Stanislaus F. Danko (CHAPTER 3)
Deputy Director, Electronic Parts and Materials Division, Electronic Components Department, The U.S. Army Electronics Research and Development Laboratory; Fort Monmouth, New Jersey

Robert A. Gerhold (CHAPTER 3)
Deputy Chief, Modular Assemblies Branch, Electronic Components Department, The U.S. Army Electronics Research and Development Laboratory; Fort Monmouth, New Jersey

Edward Keonjian (CHAPTER 1)
Staff Scientist, Arma Division, American Bosch Arma Corporation; Chairman, Advisory Committee on Microelectronics, Electronic Industries, Association

Charles J. Kraus (CHAPTER 4)
Senior Engineer, Components Division, International Business Machines; Poughkeepsie, New York

Vincent J. Kublin (CHAPTER 3)
Director, Electronic Parts and Materials Division, Electronic Components Department, The U.S. Army Electronics Research and Development Laboratory; Fort Monmouth, New Jersey

Gordon E. Moore (CHAPTER 5)
Director, Research and Development, Fairchild Semiconductor, Division of Fairchild Camera and Instrument, Inc.; Palo Alto, California

Eugene D. Reed (CHAPTER 6)
Director, Components and Solid State Device Laboratory, Bell Telephone Laboratories, Inc.; Allentown, Pennsylvania

Jacob Riseman (CHAPTER 5)
Senior Physical Chemist, Components Division, International Business Machines; Poughkeepsie, New York

Ian M. Ross (CHAPTER 6)
Director, Semiconductor Device and Electron Tube Laboratory, Bell Telephone Laboratories, Inc.; Allentown, Pennsylvania

Rudolf E. Thun (CHAPTER 5)
Manager, Film Electronics Development, International Business Machines; Poughkeepsie, New York

Edward S. Wajda (CHAPTER 5)
Senior Physicist, Components Division, International Business Machines; Poughkeepsie, New York

J. Torkel Wallmark (CHAPTER 2)
Head, Solid State Device Technology, RCA Laboratories; Princeton, New Jersey

FOREWORD

Our present age is characterized by its exponentially growing complexity—almost any statistical measure one examines demonstrates such behavior. There are more of us, and we have more machines. We need to communicate more often with one another and our machines. The densities of men and machines are higher, and yet we go farther and faster away from each other. As a result, population density, transportation speed, telecommunications volume, and information-processing volume are growing exponentially. There is no indication that this growth is turning over; hence, we may expect the complexity of our existence to increase still further.

Man's attempt to cope with this increasing complexity has been more and more through electronics. In 1939, the electronics industry stood forty-ninth among American industries. Today it is fifth, and the most rapidly growing. It permeates our whole economy and affects almost every aspect of our social-economic life. Through complex computers, transmission methods, and automation, it is becoming a marvelous extension of man's senses and mind. It provides the essential tools man needs to cope with the staggering amount of information he must process to control his complex world. As such, its size and complexity are also growing exponentially.

Early large-scale systems were primarily extensions or joinings of smaller simpler systems. This led naturally to detailed specification of the elemental components by the user. Very often these statements of requirements, without knowledge of the component development possibilities, led to unforeseen, and sometimes insurmountable, compatibility and interconnection difficulties when the components were assembled into systems.

Conversely, the component designer has often developed components only in terms of his own specialized technology and without knowledge of their eventual system function and environment or their compatibility with other specialized components of the future system application.

As system scale and complexity increase, the cacophony of specialist sciences, technologies, and languages may result in dinosaurlike systems; many of which could die shortly after birth from the diseases of inefficiency, cost, and low reliability.

Despite such difficulties, by 1950 electronic man had greatly extended his sensory and information-processing abilities through ever improved and ever smaller electron tubes and associated conventional passive components. By orderly progression, we went from conventional prewar tubes to miniature and subminiature tubes, with corresponding improvements in size and reliability of the passive components.

Along with these component-device developments came new methods of assembly into circuit functions, using printed wiring and modular construction. As one measure of progress, equipment assemblies in 1950 attained the level of 5,000 to 6,000 components per cubic foot.

However, the attendant power density and active device reliability prevented further progress in all but a few cases, where the high cost of power and maintenance could be tolerated though not desired. What was needed was an order of magnitude, or more improvement in power density and reliability, to allow further advance to the more complex electronic systems which, in principle, we knew how to design.

Indeed, this is why the transistor promised such potentially important advances to electronics a decade ago. It was obvious that it was very much smaller and consumed very much less power, and it was thought that semiconductor devices would not degrade as do the cathode or vacuum of an electron tube.

However, early transistors were limited severely in performance, manufacturability, and reliability. Correspondingly, slow realization of transistor potential came during the first seven to eight years.

Since about 1955, power and frequency performance of semiconductors have been extended where today most of the important applications can be covered. Manufacturing techniques have been improved to the point where cost is competitive with that of tubes. We are slowly achieving the anticipated low failure rates, but at higher cost and size than desired, because additional processes and more costly and larger protective enclosures are required.

With these transistor improvements, similar impressive improvements in tubes, and the reductions in size of passive components made possible by the lower voltage, current, and power demands, further miniaturization of electronic apparatus has resulted. But the drive must go on to "make it faster, make it cost less, make it smaller, and make it last longer."

In response to these needs, new techniques of microelectronics promise another large increase in component density. At this early stage in these new developments, great emphasis—and rightly so—is being placed upon component density, since speed and size of complex new equipments are of tremendous importance, particularly in military and space applications.

But component density is only one measure of the effectiveness of new designs for the new large-scale complex systems. Analysis of our new electronic systems discloses that in most cases 60 to 80% of first cost and a correspondingly high percentage of maintenance expense are attributable to the elemental active and associated passive components.

The component designer must, therefore, take account not only of performance and size, but importantly cost and reliability as well. All too frequently, too much early emphasis on just performance and size results in a system being fully developed for manufacture, only to find it is too costly to make and maintain.

Until recently, the primary attack on the problem could rightly be called palliative—applying a lot of science but significant quantities of art to make better elements per se.

Standardization of component geometry and automation to interconnect have been and continue to be useful. They will always be with us, and we must be clever in deciding how much and when we apply them.

Miniaturization up to now has been a great aid in reducing costs of materials and in the assembly of components. But too much of a good thing is possible. The things that were too big to handle have become almost to small to handle and control. Miniaturization of components, *at least as discrete parts*, may be reaching the point of diminishing returns.

Duplication or redundancy aggravates the basic problem of numbers—it is to be used only as a last resort when component reliability is inadequate.

Sanitation, or the use of ultraclean environments and handling procedures, is already standard for tubes and transistors. With our new component designs and techniques, it seems certain that it will pay off in improved yield and reliability.

These measures will continue to have importance. They must be applied aggressively now and in the immediate future to hold the line until more basic programs can succeed. But—by themselves—they are not a sufficient solution for the future.

Basically, the problem is one of defeating the "tyranny of numbers"—reducing the large number of elemental components per system function that must be made, tested, packaged, shipped, unpacked, retested, and interconnected *one at a time.*

The integrated-circuit approach directly attacks the problem of numbers. It recognizes that it is not only the number of elements in a circuit diagram that causes the trouble, but rather the fact that they have been handled as separate elements in their manufacture and assembly into system equipment.

Two major technologies are currently being explored in this area by the industry:

The two-dimensional approach is based upon the precise control of thickness, structure, and chemical composition of thin films of metals, insulators, and semiconductors by cathode sputtering, evaporation, or epitaxial growth.

The three-dimensional approach uses the bulk and junction properties of semiconductors and transistor technology to achieve not only the transistor-diode functions but also the passive-component functions within a single block of semiconductor.

One strength of the thin-film approach is that large groups of high-quality passive components and connections can be made in one batch in a few compatible operations and materials. The chief weakness is that active devices cannot yet be made as part of the same structure. They must be appliqued either as separately capsulated devices or as raw wafers after which the whole circuit is singly encapsulated. This, however, is a transitory situation and will one day be overcome with new techniques now in exploratory development. Meanwhile, our problem is simplified through separation of the technologies, and we can move rapidly in integrating the advances in passive films and active devices currently under way.

In semiconductor integrated circuits, transistor and diode requirements of the material and processes are stressed. Capacitor and resistor functions are satisfactory but somewhat limited for many applications. It seems difficult at present to match the precision, performance, and reliability that can be achieved with passive components made from more suitable materials and techniques. However, here too, progress can be expected as the art is gradually replaced by more scientifically based technology.

The functional-device approach exploits our potential ability to perform electronic-circuit functions by going directly to the physics of solids without being impeded by classical concepts of circuit elements. Increasingly, we can expect the invention and development of physically simple single devices which will replace circuits having large numbers of classical elements.

A few functional devices have been in use for many years, though not dignified by that name. The piezoelectric crystal, as a resonator, is equivalent to an assembly of coils, capacitors, resistors, and connections, but nowhere within the crystal can one identify this part as a coil or that part as a capacitor.

In the semiconductor area, newer examples of functional devices are the *pnpn* and Esaki diodes and the *pnpn* shift register and counter.

With other materials too we have demonstrated the feasibility of performing complex logic and memory functions directly in a monolithic wafer of ferrite, cryogenic, or ferroelectric material, with corresponding reductions in individual component-connection count per function.

Although this functional approach holds the greatest long-term promise of a basic attack on the tyranny of numbers, it is yet based upon invention. Before it can become broadly useful, methods of synthesis to prescribed functions must be developed.

Let me restate the main challenges that confront us today in the problem of component development for new large-scale complex systems.

Our heritage in components is essentially the physical embodiment of the mathematical concepts of circuit theory. With this viewpoint, the system designer has translated his over-all system requirements to those for components, thinking only in terms of classical inductance, capacitance, resistance, tubes, and transistors.

The component designer, adopting this viewpoint, therefore, has been limited in his permissible solutions only to finding new techniques and materials for the classical elements. It is becoming apparent that these conventional concepts are finding increasing difficulty in realizing what we now expect and demand of future electronics.

The variety of new techniques provides exciting opportunities to greatly improve performance, size, cost, and reliability. To capitalize on these possibilities both system and component designers must change their basic viewpoints and start asking different questions.

The system designer must start at a new level of synthesis, specifying his needs only in terms of basic system functions with properly weighted objectives. This, in turn, will stimulate his imagination and effort to higher levels of sophistication in system organization and logical design. Thus the range of possible solutions open to the component designer will be greatly broadened in terms of such functional requirements.

In turn, the component designer must not look only for new and better ways of making and interconnecting elements into functional modules. Without being impeded by circuit-element concepts, he must go directly to the physics of electrons, atoms, phonons, magnons, and photons in seeking basic functions. Developments of semiconductor stepping devices, gyrators, isolators, acoustic delay lines, and masers prove that such things can be done, if only the right questions are asked from the proper viewpoints.

After all, the aim of electronics is not simply to reproduce physically the elegance of classical-circuit mathematics—rather, it is to perform desired electronic-system functions as directly and as simply as possible from the basic structure of matter. If we are to make wise choices from such an expanded range of permissible solutions, we must develop better integrated measures of effectiveness, which will first serve to evaluate and compare these functional techniques, and later serve as detailed design criteria for optimizing the chosen ones.

In this book, your editor and his authors have made a bold and timely attempt to present an integrated exposition of the industries' program on integrated electronics. At this stage of the game, different views and evaluations are not only inevitable but healthy. We have come a long way in the last few years. We are bound together by common objectives, even though we differ as to means. Miniaturization may be a common and necessary condition for success. We all agree it is not sufficient for its own sweet sake. Electronics must produce its increasingly complex functions at lower cost, at higher reliability, without sacrificing performance and flexibility.

Even though it is a first attempt at the difficult job of integrating the wide variety of viewpoints and technologies in microelectronics, this volume has achieved a high professional degree of coherence and objectivity. As such, it should provide us with a valuable and lasting strategy in our fight against the tyranny of numbers in electronics.

J. A. Morton
Vice President
Bell Telephone Laboratories, Inc.

June 1, 1963

PREFACE

We are witnessing an enormous growth of interest in microminiaturization, especially in that area of electronics commonly referred to as "microelectronics." This area has captured the imagination of scientists and engineers who find in it a fascinating source of far-reaching potentialities. Realization of these potentialities will have a massive impact on space exploration, communications, controls, and many areas of scientific endeavor for years to come.

To the users of electronic equipment, microelectronics promises such tangible rewards as improved reliability; drastic reduction in size, weight, power consumption and cost; superior performance, etc. Therefore, the electronic industry has not hesitated to invest considerable sums during the past few years to enhance the state of the art in microelectronics. In 1962, the electronic industry accounted for over $13 billion in sales and ranked about the fifth largest industry in the United States. It has been estimated (*Control Engineering*, April, 1961) that by 1970, its total dollar value will reach the $20 billion mark and that about 20% of all electronic equipment will be microminiaturized. According to *Business Week* (Dec. 8, 1962), industry sales of integrated circuits alone will reach the $200 million mark by 1967. Many types of microelectronic component parts and circuits are on the market today as off-the-shelf items. These remarkable developments, which have resulted from the combined efforts of government and private industry, have already made feasible operational equipment using the principles of microelectronics.

This situation has led to the demand for a book on the theory and techniques of microelectronics which will provide practicing engineers with basic information on this subject. Such a book is particularly necessary in view of the extremely limited sources of systematic treatment of the subject.

Accordingly, this volume has been written to assist those who wish to become thoroughly acquainted with the existing concepts, design, and technology in the field. The book provides the broad basic information, supplemented by a considerable wealth of practical material, necessary for the design and manufacture of microelectronic circuits and equipment. For workers in the various specialized branches of microelectronics, it will provide an over-all perspective of the status of related areas; for those newly entering the field, it will provide a much-needed orientation; and, for the equipment designer beset on all sides by data, claims, and projections, it will provide a measure of clarity to guide him in his plans for current and future applications of the various concepts of microelectronics.

The material in the book is based primarily on the authors' personal experiences in the field. Each has played a leading role in finding new approaches, and in developing new techniques which have contributed substantially to the state of the art. The authors are fully aware that the field of microelectronics is still in a state of flux. Nevertheless, they feel that, at present, enough experience has been accumulated to justify fully the publication of such a book. Any suggestions on the part of readers leading to improvement of the text will be gratefully received and considered.

The organization of the book is as follows: The first two chapters are of an introductory nature. Despite the wide prevalence of microelectronics today, it is a fact that nomenclature and terminology in this field have not kept pace with developments. Therefore, there is still considerable confusion existing as to the meaning of many new terms that were generated by this novel concept. Accordingly, the first chapter contains, in addition to a general *classification of the field, the glossary of microelectronics terms* as developed and recommended in 1962 by the Advisory Committee on Microminiaturization of the Electronic Industries Association. These terms are consistently used throughout the text.

The second chapter sets forth the *basic criteria,* such as value of miniaturization per se, minimum size of components, maximum packaging density, heat problems, reliability, etc., which are common to all existing approaches to microelectronics. These criteria, which have been developed from the standpoint of "system engineering," are instrumental in the establishment of a realistic evaluation of requirements and potentialities of the field.

The first two chapters are followed by chapters which deal with the three broad concepts of microelectronics: discrete-component parts, thin-film technology, and semiconductor integrated circuits. The *discrete-component-parts* concept (Chapter 3) deals with existing microminiature component parts as well as with techniques of their integration. This is chronologically the "oldest" approach, and as such contains considerable valuable data on design, manufacture, and evaluation. The rapidly growing *thin-film technology* is treated in Chapter 4. Here the authors offer a detailed description of fabrication processes for thin-film passive component parts as well as their integration into circuits and subsystems. In addition, technologies for producing magnetic, semiconductor, and cryogenic thin films, which are still in the developmental stage, are also presented in this chapter.

Chapter 5 is devoted to *semiconductor integrated circuits,* which is another fast-growing approach to microelectronics. The progress in this field has been phenomenal. In less than three years, the semiconductor industry has mastered the production of integrated circuits of many different types. Reflecting this progress, Chapter 5 deals with capabilities and limitations of the underlying technology and the device structures, and it uses specific integrated circuits only as examples of the broad capability inherent in this expanding technology. This emphasis has been to point out the unique features associated with integrated circuitry. As in the preceding two chapters, much of the information and data is original, and previously unpublished. It therefore is indispensable to those who are associated with the technology and utilization of semiconductor integrated circuits.

The last chapter (6) on *functional devices* takes the reader beyond "conventional" microelectronics. This elegant functional concept, still in its infancy, holds the greatest long-term promise. After being mastered, it may revolutionize the entire present practice of design and production, providing higher reliability combined with simplicity and low production cost.

Each chapter of the book is replete with references, over 300 in all.

On behalf of my co-authors and myself, I wish to express sincere appreciation for the assistance received from many sources in the preparation of the book. In particular, we want to thank the following organizations, which kindly made their material available for the book and gave moral and technical support in this work.

1. American Bosch Arma Corporation, Arma Division
2. Bell Telephone Laboratories, Inc.
3. Fairchild Camera, Inc., Semiconductor Division
4. International Business Machines, Components Division
5. Radio Corporation of America, David Sarnoff Research Center
6. U.S. Army Electronics Research and Development Laboratory

The Editor is also pleased to acknowledge his indebtedness to C. W. Perelle, president of American Bosch Arma Corporation, and to the management of the Arma Division, in particular to W. L. Larson, M. J. Regan, and I. Harkleroad for their direct support. Sincere appreciation is expressed to Miss Roberta Koehler for her assistance with the manuscript, and to Mrs. Dorothy Granese for conscientious proofreading of the book in its final stage of preparation.

Finally, my own special thanks are due to the dedicated team of authors whose conscientious efforts and cooperation have been responsible for the creation of this volume, which we all earnestly hope will be a valuable contribution to the state of the art of microelectronics.

Edward Keonjian

CONTENTS

CHAPTER 1

NOMENCLATURE, DEFINITIONS, AND CLASSIFICATION

By Edward Keonjian

1-1. INTRODUCTION

In the past microminiaturization in electronics has been largely a practical enterprise guided by experience; however, now fundamental relations in the field are emerging. Though they are still in a state of flux, these relations are beginning to form a pattern: a general theory of microelectronics. Being put on a firmer basis, future developments may come less from cut-and-try methods and more from scientific methods controllable by exact analysis. The achievement of these goals will be facilitated when some commonly agreed criteria have been established and a common language has been developed among researchers, manufacturers, and the users of microelectronics. Accordingly, this chapter introduces a glossary of terms to be used in the book, and various criteria are defined, supplemented with some practical examples.

1-2. NOMENCLATURE AND DEFINITIONS

It should first be made clear what the term *microelectronics* implies since the name appears in many forms—microminiaturization, integrated electronics, microsystems electronics, molecular electronics, etc., and since the term is in itself somewhat misleading. In this book,

> *Microelectronics encompasses the entire body of the electronic art which is connected with, or applied to, the realization of electronic circuits, subsystems, or the entire systems from extremely small electronic parts (devices).*

However, the primary interest in microelectronics stems not from the fact that small size can be achieved, but from the much more important fact that the techniques used should ultimately lead to low cost, high reliability, and, in some cases, improved performance. Small size is, of course, of extreme value in many applications, such as in space or in portable equipment. However, in the overwhelming number of applications, small size is of only peripheral interest, while low cost, high reliability, and improved performance are of paramount importance.

Table 1-1 contains microelectronics nomenclature and definitions for 22 commonly used terms as recommended by the Electronic Industries Association in the latter part of 1962.

In addition to the definitions given in Table 1-1, a few more terms should be defined or expanded, which are widely, although not always discriminately, used. These terms are the following:

1. Component-parts Density

Quantitative engineering discussions of size, volume, and weight efficiency of microelectronic component parts and circuits utilizing different approaches obviously require some units of comparison. One such measure *is component-parts density or packaging density.* As it is defined in Table 1-1, it indicates the number of component parts (or devices) per unit volume, usually the cubic foot (CP/ft^3) or the cubic inch ($CP/in.^3$).* Admittedly, this is not a perfect unit since the component parts are of different size and each application uses different selection of these parts. In addition, the packaging-density figures may be misleading unless it is stated whether they are based on bare component parts only or whether they also include the necessary protection, insulation, encapsulation, and connections. For example, a microelectronic circuit using microminiature diodes in combination with many other microminiature component parts may have ten times less average packaging density than a circuit made up chiefly of diodes alone. Similarly, the packaging density of the entire equipment, which usually includes some bulky parts, may be several times less than the packaging density of the microelectronic circuit alone.

The packaging-density figures also can vary with the type of circuit and the type of equipment, which again emphasizes the necessity of using this term with discrimination, always indicating whether it refers to bare components, to a complete circuit, or to a final equipment, and stating its type. Only under these conditions does component-parts density give a reasonable (although not very accurate) measure of the packaging efficiency. Figure 1-1 shows approximate packaging density for various component parts and miniature equipment. It should, however, be stated that the application of this measure to integrated circuitry (especially to functional circuits) becomes less meaningful. In this case, the terms *equivalent component parts per cubic foot* or *equivalent circuit functions per cubic foot* has been recommended to maintain continuity with discrete component-parts density.

* 1 $CP/in.^3$ = 1,728 CP/ft^3, and 1 CP/cm^3 = 16.4 $CP/in.^3$

TABLE 1-1. DEFINITIONS OF TERMS IN MICROELECTRONICS

Terminology	Definition
1. Microsystems electronics...	That entire body of electronic art which is connected with or applied to the realization of electronic systems from extremely small electronic parts.
2. Electrical element.........	The concept in uncombined form of any of the individual building blocks from which electric circuits are synthesized.
3. Transistance..............	The characteristic of an electric element which controls voltages or currents so as to accomplish gain or switching action in a circuit. Examples of the physical realization of transistance occur in transistors, diodes, saturable reactors, lumistors, and relays.
4. Device....................	The physical realization of an individual electric element in a physically independent body which cannot be further reduced or divided without destroying its stated function. This term is commonly applied to active devices. Examples are transistors, *pnpn* structures, tunnel diodes, and magnetic cores, as well as resistors, capacitors, and inductors. It is not an amplifier, a logic gate, or a notch filter.
5. Component part	See Device. Commonly, the term "component part" is applied to passive devices.
6. Active device.............	A device displaying transistance, i.e., gain or control. Examples of active devices are transistors, diodes, vacuum tubes, ferromagnetic cores, and saturable reactors.
7. Circuit...................	The interconnection of a number of devices in one or more closed paths to perform a desired electrical or electronic function. Examples of simple circuits are high- or low-pass filters, multivibrators, oscillators, and amplifiers.
8. Component...............	A packaged functional unit consisting of one or more circuits made up of devices, which (in turn) may be part of an operating system or subsystem. A part of, or division of, the whole assembly or equipment. Examples of components are i-f amplifiers, counters, and power supplies.
9. Integrated circuit..........	The physical realization of a number of circuit elements inseparably associated on or within a continuous body to perform the function of a circuit.

Table 1-1. Definitions of Terms in Microelectronics (*Continued*)

Terminology	Definition
10. Thin-film integrated circuit	The physical realization of a number of electric elements entirely in the form of thin films deposited in a patterned relationship on a structural supporting material.
11. Semiconductor integrated circuit	The physical realization of a number of electric elements inseparably associated on or within a continuous body of semiconductor to perform the function of a circuit.
12. Magnetic integrated circuit	The physical realization of one or more magnetic elements inseparably associated to perform all or at least a major portion of its intended function.
13. Hybrid integrated circuit..	An arrangement consisting of one, or more, integrated circuits in combination with one, or more discrete devices. Alternatively, the combination of more than one type of integrated circuit into a single integrated component.
14. Subsystem...............	See Component.
15. Packaging...............	The process of physically locating, connecting, and protecting devices or components.
16. Packaging density........	The number of devices or equivalent devices per unit volume in a working system or subsystem.
17. Module..................	A unit in a packaging scheme displaying regularity and separable repetition. It may or may not be separable from other modules after initial assembly. Usually all major dimensions are in accordance with a prescribed series of dimensions.
18. Morphology, integrated...	The structural characterization of an electronic component in which the identity of the current or signal modifying areas, patterns, or volumes has become lost in the integration of electronic materials, in contrast to an assembly of devices performing the same function.
19. Morphology, translational.	The structural characterization of an electronic component in which the areas or patterns of resistive, conductive, dielectric, and active materials in or on the surface of the structure can be identified in a one-to-one correspondence with devices assembled to perform an equivalent function.

TABLE 1-1. DEFINITIONS OF TERMS IN MICROELECTRONICS (*Continued*)

Terminology	Definition
20. Substrate	The physical material upon which a circuit is fabricated. Used primarily for mechanical support, but may serve a useful thermal or electrical function.
21. Active substrate	A substrate for an integrated component in which parts of the substrate display transistance. Examples of active substrates are single crystals of semiconductor materials within which transistors and diodes are formed, and ferrite substrates within which flux is steered to perform logical, gating, or memory functions. It can also serve as a passive substrate.
22. Passive substrate	A substrate for an integrated component which may serve as physical support and thermal sink to a thin-film integrated circuit, but which exhibits no transistance. Examples of passive substrates are glass, ceramic, and similar materials.

In search for improvements in miniaturization of individual component parts or circuits, it is often valuable also to know their *volumetric efficiency*. This figure is expressed as the ratio of the volume of the electronically active part of a component or a circuit "stripped" from connection leads, mechanical support, encapsulation, etc., to the total volume of the same component or circuit. Figure 1-2 shows the volumetric efficiency of some common component parts.

2. Mean Time between Failures

This term has been well accepted in the electronics industry as an index of reliability of parts, electronic functions, equipment, and systems. The microelectronic activity should, accordingly, use this index for assessment of the reliability capability of its electronic "black boxes," whether these are called modules, circuit functions, functional blocks, or anything else. The use of such common language will permit quantitative assessments of this important performance feature of modern electronics.

3. Manufacturing Cost per Electronic Function

This term reflects the total cost of materials, parts, and all labor necessary to fabricate a ready-to-use circuit, irrespective of the class of microelectronics to which it belongs. Here again, the imperfection in

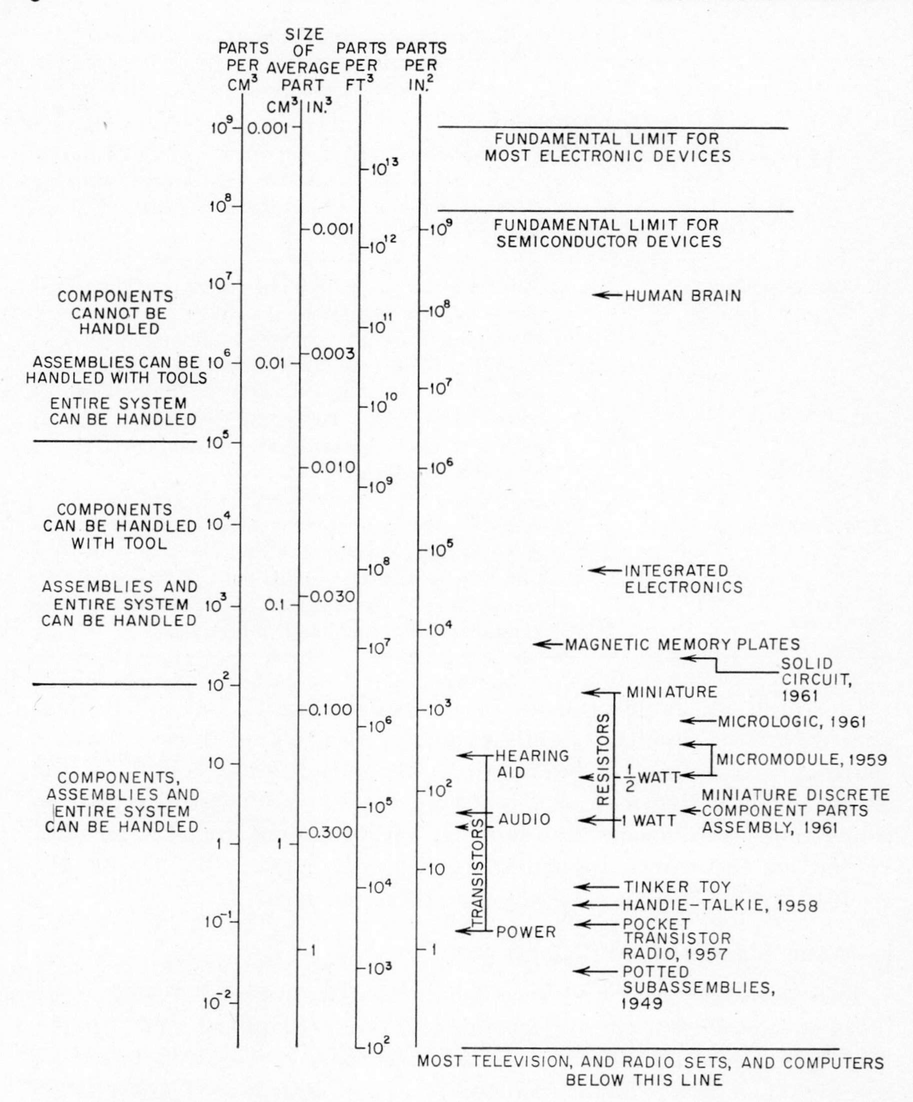

FIG. 1-1. Packaging density of electronic devices and circuits. (Ref. 8, Chap. 2.)

definition can cause misunderstanding since circuit complexities vary widely. In the discrete component parts, the circuit complexity may be defined in terms of a *ten part* for a circuit function; admittedly arbitrary as a base, the ten-part circuit function (or module) may be considered as reflecting the average of many of the common circuits in use today. In other concepts, perhaps some other base may be established. Once again, the usefulness of this costing index is measured by the tolerance of the user, who should recognize its general utility for comparison

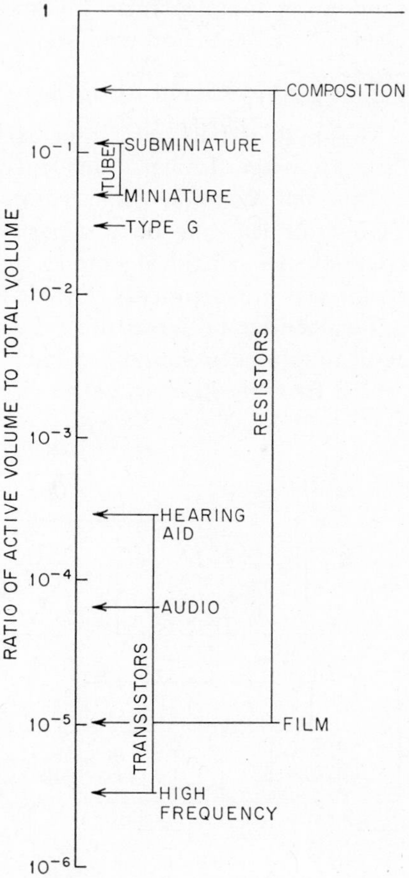

FIG. 1-2. Volumetric efficiency. Ratio of active to total volume for typical electronic devices. (Ref. 8, Chap. 2.)

purposes, and be guided by common sense in its application. Like the other units above, its careless or indiscriminate use can only yield distortions and unrealistic perspectives.

Cost, as used above, must be considered for all phases of an equipment's evolution. Though stress is usually placed on the final large-scale production cost, it is essential that experimental-model, developmental-model, and production-prototype-model costs be recognized. In many cases, particularly in peacetime military procurements and many industrial applications, a large-scale production usually is not required.

1-3. CLASSES OF MICROELECTRONICS

1. Discrete Component Parts Approach

Scaled-down separate component parts such as resistors, capacitors, inductors, diodes, transistors, and other separate electronic parts with random or uniform form factors are used to assemble microminiature electronic circuits and equipment.

2. Integrated Circuit Approach

Component parts are integrated into one single (monolithic) circuit. This approach has been developed along two major technologies:

Thin-film Circuits. The component parts are evaporated, electroplated, or the like on a separate substrate, which performs only the function of mechanical support. Usually the component parts are interconnected in the process of their fabrication.

Semiconductor Integrated Circuits. The semiconductor material is used to fabricate component parts within a single piece of semiconductor, which then becomes an entire circuit or part of a circuit.

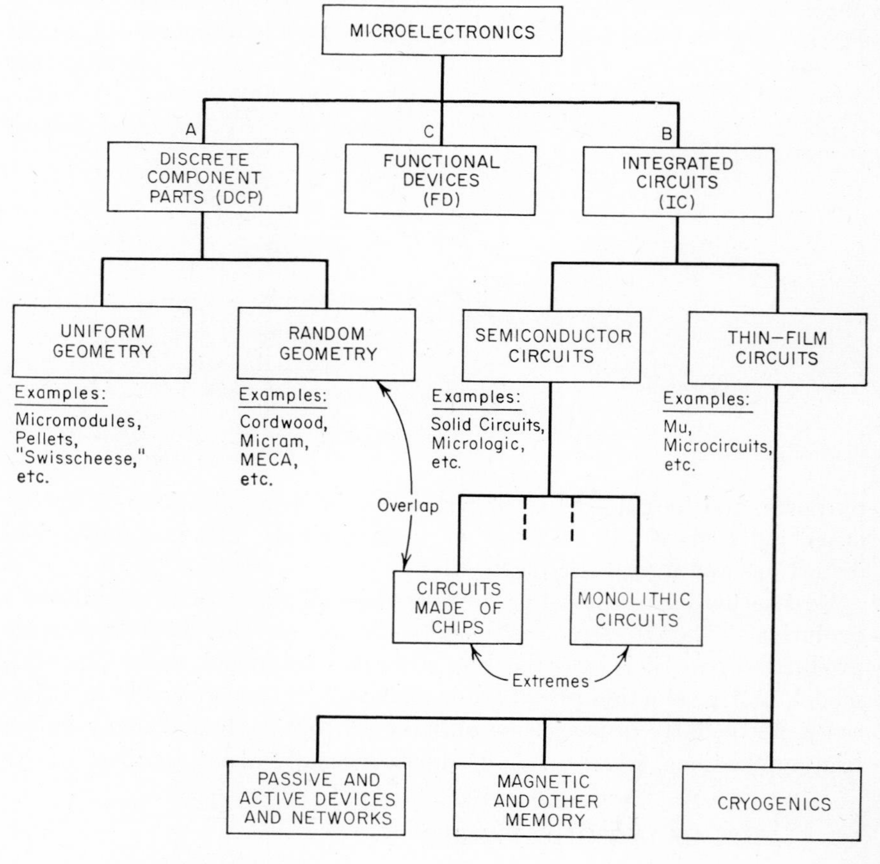

Hybrid circuits:

Combination of classes A and B, as well as combinations within class B

FIG. 1-3. The field of microelectronics.

3. Functional Devices

A monolithic piece of material is treated in such a way as to possess an electronic circuit function. Unlike previous categories, component parts in functional circuits cannot be distinguished from one another, and the structure itself cannot be divided without its stated electronic function being destroyed (although the indivisibility of the structure is also one of the characteristics of integrated circuits).

It should be noted that there can be many instances where the microelectronic circuit may combine more than one of these approaches in a single structure. For example, many thin-film circuits use individually attached active semiconductor devices, such as diodes and/or transistors. There are also circuits which combine the semiconductor integrated circuits with thin-film component parts. These and similar combinations of various approaches are commonly referred to as *hybrid approaches* or *hybrid circuitry*. (Some typical examples of hybrid circuitry are described in Chaps. 3 to 5.)

Figure 1-3 summarizes all the above-mentioned classes of microelectronics.

CHAPTER 2

BASIC CONSIDERATIONS IN MICROELECTRONICS

By J. Torkel Wallmark

2-1. INTRODUCTION

In this chapter a number of basic considerations are examined; these are general and apply to many or all of the approaches which are presently being pursued, such as integrated semiconductor circuits, cryotron memory circuits, and thin-film magnetic circuits. Because these are basic factors, they probably will be useful for some time, even though microelectronic methods and techniques change rapidly and soon outdate any particular microelectronic construction. The considerations fall into two general categories. The first category considers relations between various parameters which affect the design of present-day circuits, such as packaging density versus operating temperature or yield of integrated circuits versus shrinkage in fabrication. The second category considers the ultimate limitations to those parameters which will determine the possible design of such circuits in the future, such as the maximum packing density of semiconductor circuits and minimum size of magnetic memory cores. Technological and practical considerations, fabrication methods, interconnection design, electrical characteristics, etc., will be dealt with in separate chapters. From this point of view, the present chapter may be labeled the theory of microelectronics.

Section 2-1 deals with two of the basic reasons why small size is a desirable feature of microelectronic circuits, namely, cost of transport and speed of operation.

1. Value of Miniaturization per Se

Introduction. Of the advantages claimed for microelectronics, reduced size and weight, reduced cost, increased reliability, etc., the first mentioned is the most obvious. In this section, the value or reduced size and weight alone is estimated by considering size (or weight) a variable parameter which is independent of other parameters, such as reliability, signal-to-noise ratio, power, and life, which are assumed constant. Thus the title—miniaturization per se.

It is possible to evaluate quantitatively the advantage of miniaturization by analyzing the relative cost of transporting the extra weight. It is easy to do this for nonmilitary applications, as the necessary figures are well known and well defined. It is possible to do the same for military applications except that the transport "rates" are influenced by various logistic factors which are less well known. Therefore the analysis here will be made in terms of nonmilitary cases.

Method of Analysis. If a certain weight or volume is removed through miniaturization from the electronic equipment of an airplane, an automobile, etc., without sacrificing performance, the same weight or volume may be used for other useful purposes, say for revenue cargo. The time during which this extra capacity may be used is the average service life of such vehicles. If this is long, the value of future revenue at the time of building the electronic equipment must then be corrected for interest. Assume an interest rate of 5%. It will be assumed that the fill factor of the vehicle, i.e., the percentage of available capacity that is filled with revenue cargo, is unchanged when the capacity is increased.

As the fill factor F is available only for airplanes, the same figure will be used for motor vehicles. For scheduled United States air carriers in 1961, $F = 0.53$ (ton-mile load factor 50.5%, passenger load factor 55.4%).[1]

Aircraft Equipment. First consider an aircraft equipment application. If C_1 stands for the cost of air freight per unit weight disregarding the terminal costs, and D_1 is the average distance traveled per year by a commercial airplane, then the cost A_1 of carrying a unit weight in such an airplane during the entire year is

$$A_1 = C_1 D_1 F \qquad (2\text{-}1)$$

If, further, the median service life of such a plane is N_1 years, the total cost B_1 of building one unit of unnecessary weight into the electronic equipment is

$$B_1 = \frac{A_1}{G}[1 - (1 + G)^{-N_1}] \qquad (2\text{-}2)$$

Now, from available statistics,

$D_1 = 6.0 \times 10^5$ miles/plane-year (970×10^6 miles flown by 1,611 scheduled United States air carriers in 1961)[2]

$C_1 = 1.1 \times 10^{-4}$ dollar/lb-mile (1.1 cent per 100 miles and pound on distances exceeding 500 miles)[3]

$N_1 = 10.2$ years (51.9% of active civil aircraft recorded with the Federal Aviation Agency in use in 1961 were built before 1951)[2]

Substitution in Eq. (2-2) gives

$$B_1 = \$274 \text{ per lb, or } \$0.60 \text{ per g}$$

The figure would be somewhat higher if higher-quality freight such as passengers are considered.

Vehicular Equipment. A similar analysis may be carried out for vehicular electronic equipment using data for truck transport. Here:

D_2 = 59,590 miles/truck-year (Class I intercity motor carriers of property in 1960)[1]

C_2 = 3.15×10^{-5} dollar/lb-mile ($4.25 per 100 lb and 1,000 miles, including terminal cost, average for all articles; terminal cost eliminated by subtracting $1.43 per 100 lb and 100 miles, average for all articles, including terminal cost)[4]

N_2 = 6.7 years (average age of trucks in use in United States in 1956)[1]

Substitution in Eq. (2-2) gives

$$B_2 = \$5.6 \text{ per lb, or } \$0.012 \text{ per g}$$

Portable Equipment. As an example of human carrier, consider an instrument carrier for electronic prospecting. Here the figures are:

A_3 = $140 per lb-year [average weekly earnings in mining industries of production workers and nonsupervisory employees in 1961: $107.18; carrying capacity 40 lb (estimated)[1]]

N_3 = 5 years [service life of portable electronic equipment (estimated)]

Substitution in Eq. (2-2) gives

$$B_3 = \$605 \text{ per lb, or } \$1.33 \text{ per g}$$

Rocket Equipment. The real case for miniaturization per se is, of course, in satellite and rocket applications. Here the figures are quite large and show a wide spread, partly because of rapid development toward lower costs, partly because of large differences in cost for different missions. For 1960, the total budget of the National Aeronautics and Space Administration was 401×10^6 dollars.[1] The same year a total of 19,700 lb of payload was placed in orbit.[5] This gives a figure

$$B_4 = 2.0 \times 10^4 \text{ dollars/lb or } \$45 \text{ per g}$$

This figure is in good accord with figures given by NASA[5] on cost per pound payload for vehicle alone which run as shown in Table 2-1. The

TABLE 2-1. COST FOR VEHICLE ALONE IN DOLLARS PER POUND IN ORBIT

Vehicle	300-mile orbit	Deep-space probe
Juno II	$15,000	$100,000
Delta	5,000	50,000
Vega (future)	600	3,200

cost for vehicle alone is an underestimation, and to be comparable with the other categories, cost of ground equipment, administration, failures, etc., should be added.

The figure of \$45 per g may be approximately a factor of ten higher if the mission is a deep-space probe; a factor of approximately ten lower if the mission is a short-range nonorbiting trajectory.

Stationary Equipment. For an analysis of the case of stationary equipment, consider a piece of equipment such as a computer used in a typical factory. If E is the cost of building a unit volume of factory, M the cost of maintenance per unit volume per year, G the pertinent interest rate, H the service life of the computer, and I the service life of the building, then the cost B_5 of building an unnecessary volume into the electronic equipment is

$$B_5 = \frac{HE}{I} + \frac{M}{G}[1 - (1 + G)^{-H}] \quad (2\text{-}3)$$

Now, from available statistics,

E = \$1.2 per ft.^{3} (1.8×10^9 dollars spent in 1961 on construction of 150×10^6 ft^2 of industrial floor space; average floor height is assumed 10 ft)[1]

M = \$0.34 per ft^3-yr

H is assumed 10 years

I is assumed 50 years

G is, as before, 5%

B_5 = \$2.9 per ft^3, or 1.0×10^{-4} dollar/cm^3

For comparison with earlier figures a specific weight of 1 will be assumed, i.e., cubic centimeters and grams may be used interchangeably.

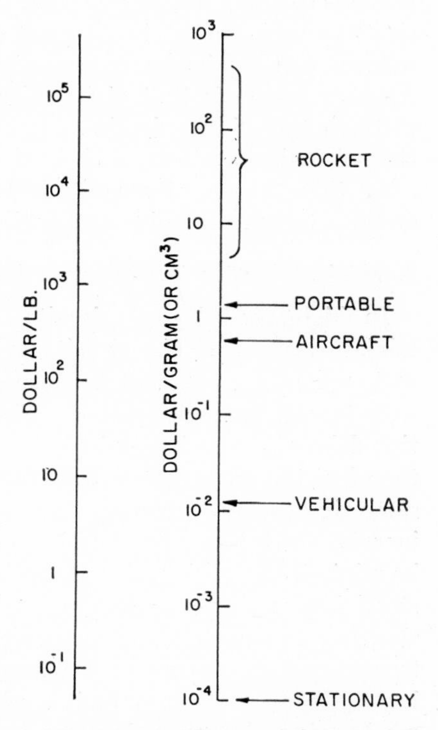

FIG. 2-1. Justifiable cost for removing through miniaturization one gram (one cubic centimeter) of unnecessary weight (volume) from various types of electronic equipment.

The different values of B have been summarized in Fig. 2-1. To put the figures in Fig. 2-1 on a more familiar basis, it may be pointed out that the weight of a typical small transistor (TO-18 enclosure) is about 0.5 g.

Conclusions. The value of miniaturization of electronic equipment, when all other factors, such as power consumption, reliability, and maintenance, are the same, may differ by as much as six orders of magnitude for different categories of equipment. Rocket applications not unexpectedly justify a cost of miniaturization which is 1 to 3 orders of magnitude higher than in aircraft or portable equipment. For stationary equipment, the gain of miniaturization is negligible.

Although military applications have not been analyzed, the corresponding figures are expected to be higher because of smaller payload, increased logistic support, etc.

In the final analysis, the power supply is an integral part of any mobile electronic equipment, and therefore should be considered in the same manner as above. This will be done in Sec. 2-6.

An incongruence seems to exist between the results of the analysis and projects to apply microelectronics to stationary computer systems, where size has been shown to be least significant. The explanation is that reduced cost and improved performance are the important goals, not size. Another reason is that digital circuits are simpler to make with these methods than linear circuits. In this connection it is interesting to note that biological systems, such as the brain, which have packing densities some three orders of magnitude higher than most advanced electronic systems today, use pulse modulation extensively.

2. Miniaturization in High-speed Circuits

In high-speed logical circuits, operating at 50 Mc/sec or higher, miniaturization of the circuits is desirable not only for advantages in cost, reliability, etc., but because *miniaturization becomes a necessity in order to reach* the high speed.[6,7] The reason for this is that signal-propagation delays become increasingly important, the higher the frequency, and therefore the propagation paths have to be kept to a minimum. At the same time, stray reactances have to be kept sufficiently low so as not to increase the delays, and undesired coupling between adjacent leads has to be minimized; this is easier, the smaller the system. This may be understood from a simple scaling argument. If the system is scaled down in size, but the electric and magnetic fields are kept constant, the upper-frequency limit increases. Consequently scaling down the size but keeping the upper-frequency limit constant allows leeway in reducing stray reactances and/or reducing coupling.

In miniaturizing for high speed, it is not sufficient to reduce the size of the components,* but the interconnections have to be made small at the same time, or the advantage of having small components will be wasted. The problem of making small interconnections is no less difficult than that of making small components, particularly if temporary disconnections are required.

In the design of a complex logic system, such as a computer, with conventional logic methods, it is desirable to reach as many gates as possible from any one gate in a time that is short compared to a cycle of

* Here and in Chap. 2 the word "component" is synonymous with "device."

operation. In the extreme case, it is desirable to reach all components from any one component. Then, as the signal-propagation velocity cannot exceed that of light, no component in a computer should be farther away from other components than the distance light travels in one cycle, i.e., 33 cm or about 1 foot in 10^{-9} sec. In reality, it has to be closer.

This may be expressed as

$$l = vt \tag{2-4}$$

where l = distance between ends of the system
v = signal-propagation velocity
t = longest allowed delay

For a system built on printed-circuit wafers which are stacked together, the signal-propagation path has to be approximately orthogonal in three planes. Then the longest path is that connecting two diametrically opposite corners, or

$$l = 3D \tag{2-5}$$

where D is the side of the cube containing the system.

The best insulating material at the present time for high-speed pulses, Teflon, has a dielectric constant of 2.1. Then the maximum signal-propagation speed is

$$v = \epsilon^{-1/2}c = 0.69c \tag{2-6}$$

where c is the velocity of light, 3.10^{10} cm/sec, and ϵ is the relative dielectric constant.

If the maximum allowed delay is one-tenth of a pulse length, if the pulse length is equal to the pulse interval, and the operating frequency is f, we have, from Eqs. (2-4) to (2-6),

$$3D = \frac{c}{20f\epsilon^{1/2}} \tag{2-7}$$

from which $$Df = 3.5 \times 10^8 \text{ cm/sec} = 1.4 \times 10^8 \text{ in./sec} \tag{2-8}$$

For $f = 10^9$ cps, $D = 0.35$ cm.

In Sec. 2-2 the minimum possible size of electronic devices is derived. If this minimum size is $d_{\min}$ cm on a side, the maximum number of components $N_{\max}$ in a system consisting of identical devices with 10% packing efficiency would be

$$N_{\max} = 4.3 \times 10^{24} f^{-3} d_{\min}^{-3} \tag{2-9}$$

This relation is shown in Fig. 2-2.

For tunnel diodes where $d_{\min} = 1{,}000\ \mu$, $N_{\max} = 4{,}300$ at $f = 10^8$ cps.

At the present time, however, a more typical value is $d = 5{,}000\ \mu$ (0.02 in.). Then either N decreases very markedly or else f decreases, as seen from Fig. 2-2.

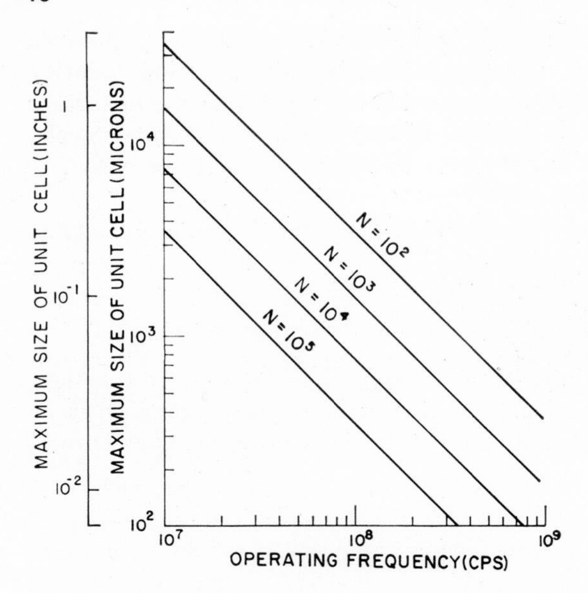

FIG. 2-2. The maximum size of the unit cell (component plus its surrounding space) of a logic circuit with N components versus operating frequency.

In this analysis, the additional delay caused by stray capacitances and stray inductances has not been considered, nor the delay in the devices. In a practical high-speed system these can usually not be neglected.

Another consequence of high speed which has to do with miniaturization, namely, the excess heat dissipation at high speed, will be treated in Sec. 2-3.

2-2. MINIMUM SIZE OF ELECTRONIC DEVICES

In evaluating and planning new advances in electronic miniaturization, it becomes of considerable interest to determine whether natural limits in one form or another will stop further progress toward ever smaller size and higher packing density or whether limits will be set entirely by economical considerations of diminishing return. It will be shown in this section that physical limits exist.

If the limiting size was very small, say of atomic dimensions, an analysis could not be very convincing, as the development of fabrication, handling, and measurement methods undoubtedly would require breakthroughs which could not be easily predicted, and which could completely change any assumption on which the analysis was based. If, on the other hand, the limiting size was close to the size of existing devices, the analysis would carry considerable weight, as the fabrication, handling, and measurement methods would be known.

This section will deal with the question of such limits, particularly those based on cosmic rays, heat dissipation, fabrication inexactness, and impurity fluctuations. It will be shown that for some existing devices, particularly semiconductor devices, these limitations have already been reached.[8]

1. Minimum Size of Passive Devices

It will be shown in this section that for passive devices, such as resistors, capacitors, and inductors, a minimum device size exists, set by basic laws of nature under certain reasonable conditions. This limit is within a factor of ten of devices being made in the laboratory today, and for a cubic form would correspond to a device 3 microns on a side. The conditions under which this minimum size is valid are believed pertinent to future microelectronic circuits in which a large number of such devices would have to be fabricated simultaneously. They would not be valid for a single device or a few selected devices. Of the many physical phenomena which may determine the minimum size of devices, it is believed that the definition of the edge in fabrication of the devices presents the most serious obstacle.

Edge Definition in Fabrication. In order to make a meaningful analysis, certain simplifying assumptions are necessary. These assumptions have been summarized in Table 2-2 and have been lettered A, B, C, . . . , I as they appear in the text. The first such assumption has to do with the physical appearance of resistors, capacitors, and inductors, which varies widely, depending upon the particular fabrication method used. For the purpose of this analysis, the device geometry, whether thin-film resistors, composition resistors, wirewound resistors, semiconductor resistors, wire coils perhaps with ferrite cores, electrolytic capacitors, etc., will temporarily be set aside. Instead, the simpler problem, namely, that of finding the *smallest possible cube*, (A) or square piece of thin film on a substrate, that can be cut out from, or formed in, the material will be investigated. Certainly the device could not be smaller than this cube, but instead must be larger, because of more

TABLE 2-2. SIMPLIFYING ASSUMPTIONS FOR MINIMUM-SIZE ANALYSIS OF PASSIVE COMPONENTS

A	Devices of cubic shape (arbitrary shape allowed subsequently).
B	All devices identical.
C	The system contains 10^5 devices.
D	No redundancy.
E	No negative feedback in circuits or fabrication.
F	No shrinkage caused by accidents in fabrication.
G	Ideal material, no surface effects.
H	Allowed tolerances $\pm 10\%$.
I	Equal and independent deviations in device parameters.

complex geometry. To form the element out of a solid piece of material, or from a thin film on a substrate, some kind of tool or process must be used. An example of such a tool or process is the photoresist method in combination with masking and etching, as used in semiconductor-device fabrication (see Chaps. 4 and 5), or for making masks for evaporation of metal or insulator patterns. Other examples are the laser and electron-beam machining tools recently developed. The real "tool" in these techniques is a beam of photons, or charged particles, which of course can be made much finer than a mechanical tool.

The energy of the particles has an optimum for highest resolution. Too low energy means that the focus cannot be minimum because of space-charge effects for charged particles, or diffraction for photons. Too high energy, on the other hand, is not desirable either, because of creation of secondary particles in the target material and therefore poor resolution. An optimum energy must exist between these two extremes. It is easy to see that this must be somewhere in the range used for electron-beam machining, i.e., some 10^2 to 10^6 electron volts. For this reason, the electron-beam tool may be the most ideal tool possible.

In using an electron beam, the edge of the beam defines the device side, and therefore must be closely controlled. The edge of the beam is not entirely abrupt. Rather, the intensity falls gradually over a short distance from a high value inside the edge of the beam to zero outside. When the beam has been focused to a minimum spot size, only these two sides of the intensity distribution remain. The minimum spot size is therefore a convenient measure of the edge falloff, or the edge definition.

The minimum spot size[9] of an electron beam of 50 kv, considering spherical aberration and diffraction, with the best-known magnetic objective, has a diameter of about 10 A. However, because of chromatic aberration and, more important, scattering of electrons in the target, 100 A is a more realistic value which has been approached in experimental tests.[10] The problems of maneuvering this beam, of registering with the target, of maintaining focus over a sufficient time, etc., will be neglected.

Analytical method. This idealized working tool will now be considered for the fabrication of a microelectronic circuit. For simplicity, it will be assumed that *all devices in the circuit are identical.* (*B*)

Certainly making the devices different would mean that some would exceed the minimum size.

It will further be assumed that the circuit is of reasonable complexity, say *comprising* 10^5 *components,* (*C*)

assembled in an essentially nonrepairable package. This assumption is based on the fact that very small elements can only be practical in large numbers. A much larger number than 10^5 is unrealistic at the present time. On the other hand, a much smaller number, say 10 to 10^2 components, as is typical of integrated circuits today, would not sufficiently realize the advantages offered by microelectronic methods, which increase with complexity. Furthermore, the problem of handling a circuit consisting of 10 to 10^2 minimum-size components, of connecting it, etc., makes such a small number unlikely. A verification by example is offered by the work in superconductive and magnetic memories where already the number of integrated components often exceeds 10^2. The choice of the number 10^5 is not critical, and even several orders of magnitude change up or down does not change the results very much.

A further assumption is *absence of redundancy,* (*D*)
a condition met by the overwhelming majority of circuits today. In future microelectronic circuits, this may not necessarily be true, but it will be assumed here to simplify the analysis. Use of redundancy would allow smaller components, as some of them could be faulty, but, on the other hand, would require more components, thereby tending to offset the advantage as far as packing density is concerned.

It will also be assumed that *no negative feedback is used,* (*E*)
either in the circuits or, in a wider sense, in the fabrication of devices. For negative feedback, the same arguments as for redundancy are valid. Negative feedback would relax the requirements on the devices, thereby allowing them to be smaller. However, this would have to be paid for in reduced gain, making more devices necessary to make up for this reduction, and thereby offsetting the advantage as far as packing density is concerned. Negative feedback in fabrication may be thought of as a process of measuring each device as it is being made, and adjusting the fabrication procedure during the fabrication. This is, of course, an essential feature of all sound fabrication praxis, although usually on a batch rather than a component level. The meaning of assumption *E* is that this would be true also for future microelectronic circuits; i.e., only the complete system of 10^5 components would be measured, not each individual component, and adjustments of the fabrication process would be made only between batches.

It will further be assumed that *shrinkage may be neglected.* (*F*)
The term shrinkage here refers to unusable devices because of accidents in the fabrication process. The justification for this seemingly unrealistic assumption is that shrinkage will be treated separately in Sec. 2-4.

The *material* used in fabrication *will be assumed ideal,* (*G*)
completely homogeneous, free of defects, and with bulk properties extending all the way to the surface. Neglecting the influence of the surface anticipates the outcome of the analysis, namely, a minimum device size that is not too small.

It is further assumed that all the devices in the system have to be fabricated so that their characteristics are within certain tolerances. This *tolerance will be assumed* $\pm 10\%$. (*H*)

Quantitative Analysis. As the characteristics of each device depends on a number of parameters, these parameters would have to be given tolerances also. For a resistance

$$R = \frac{d_1}{\rho n_1 \mu d_2 d_3} \tag{2-10}$$

where ρ = electron charge
n_1 = charge density in the material
μ = mobility of the charge carriers
d_1 = dimension of the resistor in the direction of the current
d_2, d_3 = the dimensions of the cross section of the resistor

If each of these five variable parameters exhibited deviations from its mean value, and if the *deviations were equal and independent,* (*I*) the tolerance allowed each parameter would only be approximately $\pm 2\%$, in view of assumption *H*.

For a capacitance,

$$C = \epsilon_0 \left(1 + \frac{\alpha n_2}{\epsilon_0}\right) \frac{d_4 d_5}{d_6} \tag{2-11}$$

where ϵ_0 = permittivity of free space
α = polarizability
n_2 = density of polarizable atoms in the dielectric
d_4, d_5 = area of the capacitor
d_6 = thickness of the dielectric

A similar expression for an inductance with a ferromagnetic core, neglecting the leakage flux, is

$$L = \mu_0 F(\beta, n_3) N^2 \frac{d_7 d_8}{d_9} \tag{2-12}$$

where μ_0 = permeability of free space
$F(\beta, n_3)$ = an effective permeability factor
β = magnetic polarizability
n_3 = density of magnetic dipoles
N = number of turns of the coil
d_7, d_8 = area of cross section of the coil
d_9 = length of the coil

In capacitors with air as a dielectric ($\epsilon = 1$) and in inductors without magnetic material ($\mu = 1$), the polarization may be neglected. However, for microminiature components, reasonable values of capacitance and inductance probably will require high-dielectric-constant and high-permeability materials.

Thus for each device there are a minimum of five variable parameters which have to be controlled to better than $\pm 2\%$

Now it is possible to compute the fraction S of elements for which the value of resistance (or capacitance or inductance) deviates by more than the tolerance ϵ from the mean value, when the inexactness of the edge is σ and the distribution is normal. This expression is[8]

$$S = 1 - \left[\frac{2}{\sqrt{2\pi}} \int_0^{y_1} \exp\left(-\frac{y^2}{2} \right) dy \right]^3 \tag{2-13}$$

$$y_1 = \frac{\epsilon d}{\sqrt{2}\sigma} \tag{2-14}$$

where d is the side of the cube, and σ is the standard deviation (100 A).

Results. Inserting practical values and requiring that $S \leqq 10^{-5}$ and $\epsilon \leqq 0.02$, it is found from Table 2-3 that the smallest possible device has $d \approx 3\ \mu$.

This result is also applicable to thin-film components, d being the side of the square, if the thickness of the film and the standard deviation in this thickness leave y unchanged. This means a standard deviation of 10 A for a film thickness of about 1,000 A, or a standard deviation of 1 A for a film thickness of about 200 A, values which are certainly reasonable.

The result, therefore, is that the minimum size of passive components based on the edge uncertainty in fabrication of the devices, is about 3 microns on a side. In practical devices, several factors would contribute to a larger-than-minimum size. One such factor is geometry. For anything more complex than a cube, which at least for inductances seems inevitable, the size would increase. Another factor is power dissipation and related temperature rise. In order to handle the dissipated heat, the devices may have to be larger. A third consideration is connections, protection, and support. A fourth consideration has to do with economy. It will certainly cost a premium to be close to the edge-uncertainty limit, whereas sufficiently removed from it, more relaxed and therefore cheaper fabrication processes may be used.

TABLE 2-3. FAILURE RATE S FOR VARIOUS VALUES OF DEVICE DIMENSIONS d AND MAXIMUM ALLOWABLE FRACTIONAL DEVIATION ϵ WHEN THE UNCERTAINTY OF THE EDGE IS 100 A

ϵ	d, microns			
	1	2	3	4
0.01	0.86	0.40	0.10	0.02
0.02	0.40	0.02	6×10^{-5}	$<10^{-8}$
0.033	0.05	7×10^{-6}	$<10^{-8}$	$<10^{-8}$
0.05	8×10^{-4}	$<10^{-8}$	$<10^{-8}$	$<10^{-8}$
0.10	$<10^{-8}$	$<10^{-8}$	$<10^{-8}$	$<10^{-8}$

2. Minimum Size of Magnetic, Superconductive, and Dielectric Devices

It will be shown in this article that a minimum size exists also for active devices, and magnetic, superconductive, and dielectric devices will be used as examples. These devices can, of course, also be used as passive devices. The most important group of active devices, semiconductor devices, will be treated in Art. 3.

It is well known that memory, and therefore probably also logic, is possible *on a molecular level,* as witnessed by the reproductive process in biological cells, where the necessary information for the forming of a new cell is stored in molecular sites in the genes. However, methods to use such a system practically, to write in, to read out, to do logic, etc., are not known at the present time.

In technical systems using magnetic, superconductive, or dielectric principles, it has been shown[11] that about 10^2 *participating particles,* spins, or the like, are a minimum requirement for each bit stored. For smaller devices, the information will be lost through either thermal agitation or quantum mechanical tunneling.

If a *system of some* 10^5 *devices* is required, assuming all devices to be good within reasonable tolerances, without feedback or redundancy, as would be required in an advanced microelectronic system, each device must have about 10^4 *participating particles.*[8] For fewer particles per device, the statistical distribution of particles will cause some devices with too few or too many particles to fall outside tolerances. This is in analogy with the case of semiconductor devices where at least 5×10^4 charge carriers per device are necessary (Art. 3).

If it is further required that it shall be *possible to fabricate* a system comprising some 10^5 components without selection, without negative feedback, and without redundancy, each device must be a *minimum of about* $(3\ \mu)^3$. Otherwise, because of the limitations of fabrication methods, some components will be defective and the system nonoperative. This is the most stringent limitation by far, as $(3\ \mu)^3$ corresponds to 10^{11} to 10^{12} atoms. This last limitation is the same as was analyzed in Art. 1, and the identical method of analysis may be used here.

Edge Definition. In magnetic, superconductive, and dielectric devices, just as in passive devices, tolerances have to be applied to device parameters for satisfactory performance. The tolerance requirement may be formulated as a condition that the energy to switch the devices be sufficiently uniform from device to device. For *a magnetic core,*[12] the energy E_1 necessary for switching is

$$E_1 = WB\left(\frac{S_w}{T} + H_c\right) \tag{2-15}$$

where W = volume of the core
B = saturation flux density, 1,000 gauss
S_w = so-called switching constant (0.2×10^{-6} oersted-sec, or 1.6×10^{-5} amp-sec/m for typical ferrites)
T = switching time
H_c = coercive field for the material (0.5 oersted, or 40 amp/m for typical ferrites)

Consequently, for magnetic cores, there are five parameters which have to be controlled, three physical dimensions, the density of polarizable spins in the ferromagnetic material, and the polarizability of the spins.

A similar expression is valid for *ferroelectric devices.* The parameters to be controlled here are three physical dimensions, the density of polarizable atoms, and the polarizability of the atoms.

For superconductive devices such as a *thin-film cryotron*[13] the energy E_2 necessary for switching is

$$E_2 = \frac{4 \times 10^{-9} UwH^2}{w_g} \tag{2-16}$$

where U = volume of the active region of the cryotron
w = width of the control
w_g = width of the gate
H = critical magnetic field

Consequently, for thin-film cryotrons, there are also at least five parameters that have to be controlled.

Consequently, for $\pm 10\%$ tolerance on the switching energy for these devices, the corresponding tolerance on dimensions of a cubic device is $\pm 2\%$ as for passive devices, and, therefore, the minimum size of magnetic, superconductive, and dielectric devices is the same as for passive devices, namely, 3 μ on a side.

3. Minimum Size of Semiconductor Devices

In this article, the minimum size of semiconductor devices will be analyzed. It will be shown that this minimum size is within a factor of 2 to 5 of the dimensions of the active region of devices now being made. Because existing devices are very close to the limit, and because this limit is larger (more restrictive) than for other devices, a somewhat more detailed investigation is warranted. It will be shown that in addition to the edge uncertainty, two more considerations, namely, variations in the doping distribution and cosmic-ray bombardment, are important in semiconductor devices.

For the analysis, a few more simplifying assumptions are necessary and are indicated as before as they appear in the text by italic type and

TABLE 2-4. ADDITIONAL SIMPLIFYING ASSUMPTIONS FOR MINIMUM-SIZE ANALYSIS OF SEMICONDUCTOR DEVICES

J	Semiconductor devices are resistors.
K	Material is silicon, *n* type.
L	Speed cannot be sacrificed by waiting for temporary failures to clear up.
M	A minimum of 1 month mean time to failure is assumed.

lettering *J* to *M*. A summary is provided in Table 2-4. For the basic assumptions and the method of analysis, see Art. 1.

Edge Definition. For the edge-definition consideration, the functioning of the particular semiconductor device will be temporarily set aside, and only the simpler problem of determining the size of the smallest possible *cubic resistor made of semiconductor material* (*J*) will be investigated. Later in this article any arbitrary semiconductor device will be synthesized from a number of such resistors. For resistors, the limiting smallest size is known from Art. 1 to be 3 μ on a side. This is indicated in Fig. 2-3 by the line marked "edge uncertainty."

Statistical Variations in Doping Distribution. It will be assumed, as in assumption *G* above, that the semiconductor material is uniformly doped. However, this assumption applies only on a macroscropic scale. On a microscopic scale, the doping concentration will fluctuate statistically around a mean value. When the size of the element cubes decreases, the total number of impurities in each cube will also decrease and, therefore, the statistical fluctuations will be proportionately larger.

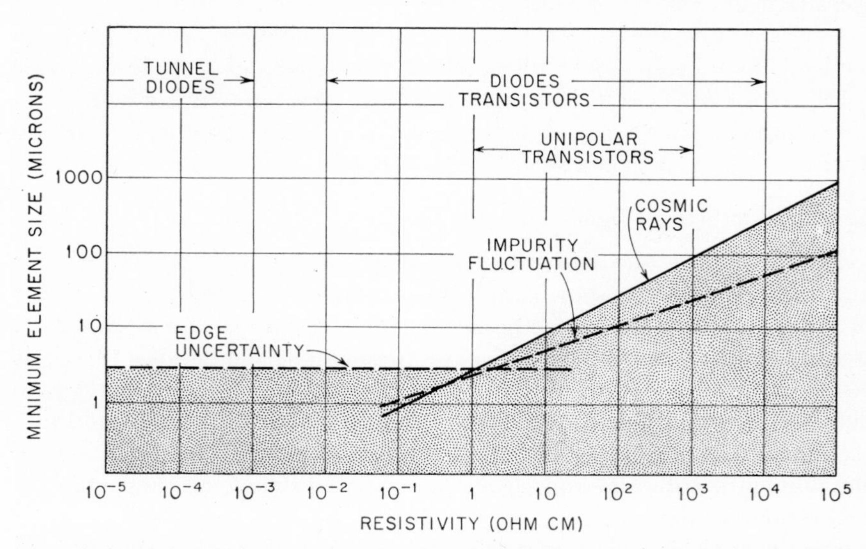

FIG. 2-3. Minimum device size versus resistivity of the material. Common ranges for some devices are indicated. Gray area is excluded.

TABLE 2-5. FAILURE RATE S_1 FOR VARIOUS TOTAL NUMBER OF IMPURITIES PER DEVICE N, ASSUMING A MAXIMUM ALLOWABLE FRACTIONAL DEVIATION ϵ OF N

ϵ	10^3	10^4	5×10^4	10^5
0.01	0.75	0.32	0.03	3×10^{-3}
0.02	0.53	0.05	1×10^{-5}	$<10^{-8}$
0.033	0.29	1×10^{-3}	$<10^{-8}$	$<10^{-8}$
0.05	0.12	6×10^{-7}	$<10^{-8}$	$<10^{-8}$
0.10	2×10^{-3}	$<10^{-8}$	$<10^{-8}$	$<10^{-8}$

For sufficiently small cubes, the probability of a deviation in impurity content above or below a certain limit can no longer be neglected.

If N is the mean number of impurities per cube, and ϵ is the maximum allowed fractional deviation from the mean, then the fraction of cubes S_1 in which ϵ is exceeded may be obtained from[8,14]

$$S_1 = 1 - \frac{2}{\sqrt{2\pi}} \int_0^{y_1} \exp\left(-\frac{y^2}{2}\right) dy \tag{2-17}$$

$$y_1 = \epsilon N^{1/2}$$

Table 2-5 gives some practical values.

The maximum allowed tolerance on doping density as found from Eq. (2-10) is $\pm 2\%$. With $\epsilon = 0.02$, this gives a minimum number of impurities from Table 2-5 of 5×10^4 for each element.

The condition that the number of doping agents N in a volume d^3 of a material doped to ρ ohm-cm exceeds 5×10^4 may be expressed as

$$N = \frac{d^3}{\rho q \mu} \geqq 5 \times 10^4 \tag{2-18}$$

where q = electronic charge (1.6×10^{-19} coul)
μ = carrier mobility (1350 cm²/volt-sec for n-type Si)
d = side of the cube, centimeters

In order to determine this equation, it will be assumed that the *material is silicon, n type.* (*K*)

p-type silicon would give a somewhat smaller size. Equation (2-18) is shown as a dashed line in Fig. 2-3.

Cosmic Radiation. The influence of cosmic radiation[15] on the minimum size of semiconductor devices will now be considered. It turns out that cosmic radiation is indeed the most severe limitation, particularly when the semiconductor material has high resistivity. Shielding against cosmic radiation is, of course, incompatible with miniaturization techniques, as the thickness of a lead shield reducing the radiation level 5 to 10 times would have to be of the order of 10 cm.

As cosmic rays are comparatively rare events, the average number of free carriers and displaced atoms in a volume of semiconductor is indeed very small, and would be negligible in a conventional-size component, such as an ordinary transistor or a diode. However, the fact that the disturbance in the trail of the cosmic ray is localized to essentially a line through otherwise undisturbed material becomes important when very small devices are considered. Then there is a definite probability that one element, or a line of elements, may be rendered inoperative through the passage of a cosmic ray, while all the other elements may still be operative and in spite of the average density of the disturbance over the entire volume being extremely small.

The influence of cosmic rays on semiconductor devices is threefold. First the ionization of the cosmic ray will create a temporary excess of hole-electron pairs, which will decay with the normal lifetime of minority carriers in the device. If this lifetime is of the order of, or larger than, the minimum operating time of the device, a false signal will result. In a nonredundant circuit, such false signals cannot be tolerated, even though the circuit returns to normal operation after the error.

Although a possible alternative, it will be assumed that *speed cannot be sacrificed* (*L*)
to cancel such error signals.

The second influence of cosmic radiation is atomic displacements in the lattice, which are known to introduce additional levels in the forbidden band of the semiconductor. In the case of silicon, these levels act as traps, temporarily removing majority carriers and gradually making both *n*-type and *p*-type silicon more intrinsic. Although these displacements show a certain amount of annealing already at room temperature, this is only partial, and the time involved is, of course, much larger than the operating time of the circuit. Therefore, annealing is of little help and may be neglected.

TABLE 2-6. DOSE RATES FROM COSMIC RAYS AT DIFFERENT ALTITUDES AND GEOGRAPHICAL LATITUDES

Altitude, feet	Dose rate at Equator, mrads/year*	Dose rate at latitude >40°, mrads/year
Sea level......	23	26
5,000	28	42
10,000	56	84
15,000	110	170

* 10^3 mrads = 1 rad, the unit of radiation dosage.

TABLE 2-7. TYPICAL DOSE RATES IN SATELLITES*

Location	Radiation	Dose rate, r/hr
Undisturbed interplanetary space	Cosmic rays	0.0006–0.0014
Center of inner Van Allen belt	Protons	24
Center of outer Van Allen belt	Soft X rays	200
Solar proton beam	Protons	$10–10^3$

* 1 r (roentgen) ≈ 1 rad.

The third influence of cosmic rays is nuclear events in which silicon nuclei are hit with sufficient energy to shatter them into fragments of various kinds. These fragments in their turn have large energies and ionize and displace atoms heavily. At the point of impact, a "star" of convergent tracks will be formed, raising the local density of ionization and displacement over that characterizing a single track.

The intensity of cosmic rays[16,17] is illustrated in Tables 2-6 and 2-7. At sea level the number of particles arriving is on the average 1 to 2 per cm^2 per minute.

In addition to cosmic rays the natural radioactive background[16] has to be considered, particularly for equipment on the ground. Therefore the figures for f for the sea level in Table 2-9 should be multiplied by approximately 3. The intensity of the natural radioactivity is illustrated in Table 2-8.

The number of excess (or deficit) free carriers n created in an element of silicon, d meters on a side, and the probability f that a burst should occur within a *mean time between failures of one month,* (M) are shown in Table 2-9.

For each category in Table 2-9, lightly ionizing particles, heavily ionizing particles, etc., two straight lines are obtained.

One is given by setting the probability equal to one that at least one element of 10^5 is traversed by a cosmic ray in one month, i.e., $f = 1$. This condition gives lines rising with increasing d in Fig. 2-4. To the left of these lines the element size is so small that less than one element will be hit. For $d = 1\ \mu$, for example, the probability of hit may be found by extension of the line to be 50, meaning that 50 elements of 10^5 will be hit in one month.

TABLE 2-8. DOSE RATES FROM NATURAL RADIOACTIVITY IN SOIL

Rock	*mrads/year*
Igneous	66
Shale	53
Sandstone	28
Limestone	11

TABLE 2-9. EXCESS (OR DEFICIT) OF CARRIERS n AND PROBABILITY OF ITS OCCURRENCE f AT VARIOUS ALTITUDES

Ionization	Altitude, km: 0 Sea level, n	0 Sea level, f	0 Natural radioactivity	10 Airplane height	100 Top of atmosphere
			f increases by a factor of		
Lightly ionizing particles....	$8 \times 10^7 d$	$2 \times 10^{13} d^2$	3	20	<1
Heavily ionizing particles...	$8 \times 10^8 d$	$4 \times 10^{10} d^2$	3	200	1,000
Atom displacements........	$2 \times 10^7 d$	$2 \times 10^{13} d^2$	3	20	2.5
Nuclear events............	$3 \times 10^9 d$	$7 \times 10^{12} d^3$	3	400	4,000

The second line is obtained by the condition that the excess, or deficit, carriers not exceed 2% (see discussion of assumption I) of the equilibrium carrier density, or

$$n \leq \frac{0.02 d^3}{\rho q \mu} \tag{2-19}$$

As shown in Sec. 2-3, ρ is 0.2 ohm-m (20 ohm-cm) for minimum size. Then Eq. (2-19) with results from Table 2-9 substituted gives four values

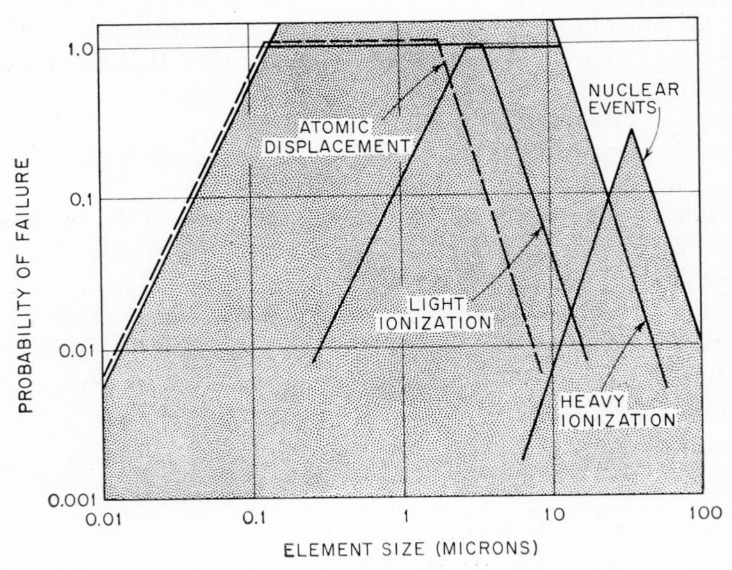

FIG. 2-4. Probability of failure through cosmic rays at ground level for 20 ohm-cm material versus device size.

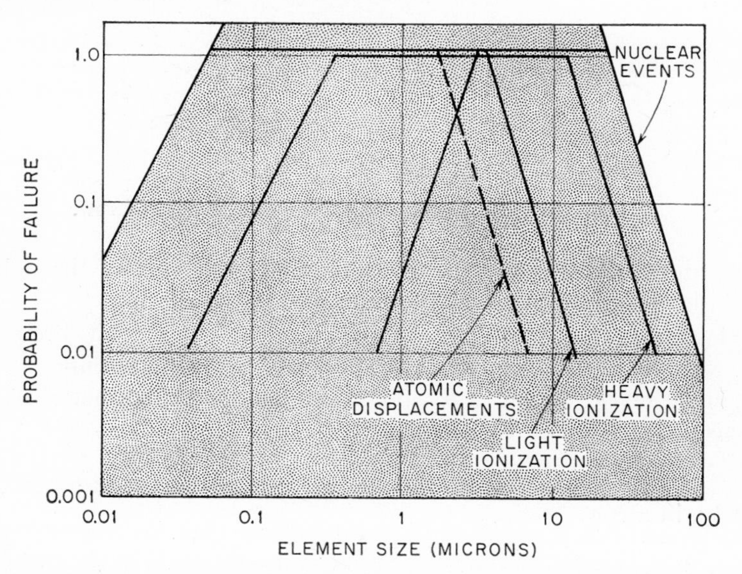

FIG. 2-5. Probability of failure through cosmic rays at airplane altitude (10 km, 33,000 ft) for 20-ohm-cm material versus device size.

for d. In addition, the slopes of the lines going through the points defined by these values are given by considerations of statistical fluctuations in carrier density. The lines are shown in Figs. 2-4 and 2-5. To the left of these lines, the influence of cosmic rays exceeds the tolerance limit, while to the right the influence may be neglected. Thus the regions between the lines, shown gray in Figs. 2-4 and 2-5, represent element sizes excluded by cosmic-ray failure.

For heavily ionizing rays, Eq. (2-19) gives

$$d^2 \geq 8.6 \times 10^{-10}\rho \tag{2-20}$$

Equation (2-20) is indicated in Fig. 2-3 and represents the most limiting condition at high resistivities.

At airplane altitudes, nuclear events have to be considered and give similarly

$$d^2 \geq 3.2 \times 10^{-9}\rho \tag{2-21}$$

Application of the Analysis to Practical Devices. Influence of Device Geometry. Now it is possible to relax the assumption of cubic geometry and allow any arbitrary shape, synthesized from a number of cubes. For example, a filamentary transistor may be thought of as a string of cubes, or a narrow-base junction transistor as a layer of cubes. Two extreme cases may be considered, namely, a cosmic-ray transit through the

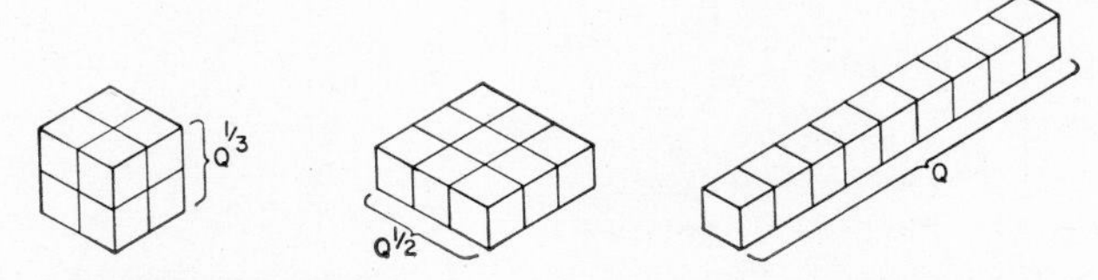

FIG. 2-6. Q elements arranged in cube, sheet, and row.

"short" dimension affecting only one cube, or a transit through the "long" dimension affecting a maximum number of cubes.

First, consider the worst case, namely, a cosmic-ray transit through the maximum number of elements. Assume Q elements in a cube, a sheet, or a row, as shown in Fig. 2-6. Then the maximum number of carriers in a sheet n_s, compared to that in a cube n_c, is

$$n_s = Q^{1/6} n_c \tag{2-22}$$

Similarly, in a row

$$n_r = Q^{2/3} n_c \tag{2-23}$$

The probability of transit in the long direction compared to a transit through the cube is proportional to the solid angle. Therefore, the probability of transit in the long direction of a sheet f_s, compared to the probability of transit in a cube f_c, is given by

$$f_s = Q^{-1/2} f_c \tag{2-24}$$

and similarly for a row,

$$f_r = Q^{-2} f_c \tag{2-25}$$

Some practical values are given in Table 2-10.

As geometries corresponding to $Q = 100$ to 10,000 are common in transistors, and $Q = 10$ to 100 in unipolar transistors, the minimum element size shown in Fig. 2-4 may have to be raised by a factor of 1 to 5, depending on the geometry of the devices. For more extreme geome-

TABLE 2-10. RELATIVE CHANGE IN EXCESS CARRIERS n AND PROBABILITY OF OCCURRENCE f FOR SHEET AND ROW GEOMETRY SYNTHESIZED BY Q CUBES

Q	Sheet		Row	
	n_s/n_c	f_s/f_c	n_r/n_c	f_r/f_c
10	1.5	0.3	4.6	0.01
100	2.2	0.1	22	0.0001
1,000	3.2	0.03		
10,000	4.6	0.01		

tries such as thin-film devices, the probability of transit may drop below the critical level. From Fig. 2-4 this may be seen to happen at a ratio $f/f_c = 0.001$.

Influence of Device Type. In considering various devices, a start will be made with the *unipolar transistor* (see Chap. 5) as a particularly simple case. In the unipolar transistor, the volume of the source, drain, and gate regions may be neglected compared to that of the channel region, as they can if desired be made smaller than the channel region. The channel region is obviously a semiconductor resistor with the added restriction that the resistor dimensions may be altered by the gate depletion layer. Therefore, the deductions for semiconductor resistors apply directly. The geometry of (the active region of) many unipolar transistors is a particularly unfavorable one corresponding to the row considered above. For typical medium-frequency units, such as shown schematically in Fig. 2-7, the factor Q (the number of cubes in the row) is approximately 20. This means an increase of the minimum device size of 7 times from $(10\ \mu)^3$ to $(70\ \mu)^3$.

The volume of the active region in Fig. 2-7 is $3 \times 10^5\ \mu^3$, or an element size of about $(70\ \mu)^3$, which is marginal. Higher-frequency units may violate the minimum-size requirement.

The next simplest device to consider is the *junction diode*. The volume of the highly doped side of the junction may be neglected, and only the side with the lowest doping considered, and only the part within one or a few diffusion lengths from the junction or, for a reverse-biased diode, the depletion layer. This then constitutes the *active region* of the diode. The geometry of the active region of diodes is usually somewhere between cubic and thin layer, dependent upon doping, area, signal amplitude, etc. The maximum allowed tolerance on diode characteristics, say, on current for a given voltage, is not easy to agree on, as many different circuits with different requirements exist. A figure somewhere between 10 and 100% is reasonable. The choice is not critical, since the difference between 10% and 100% corresponds to only a factor of 3 in element size. As intermediate values of tolerance combined with intermediate geometry (or 100% tolerance and a moderately thin layer) are equivalent to 10% tolerance and cubic geometry, the latter will be chosen for simplicity. As 10% tolerance in diode characteristics corresponds to 10% tolerance in carrier density, the curves in Fig. 2-3 apply.

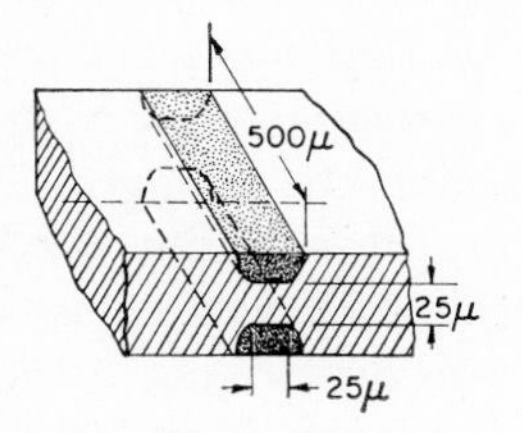

FIG. 2-7. Active region of medium-speed unipolar transistor.

In conventional (bipolar) *transistors*, the active region is the base region, and the volumes of the emitter and collector may be neglected. For transistors even more than for diodes, a maximum

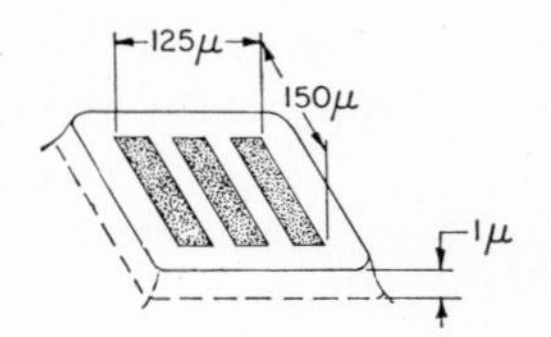

FIG. 2-8. Active region of mesa transistor.

allowed tolerance in carrier density in the base region is not easy to agree upon, because of the large variety of circuits with different requirements. However, a figure somewhere between 10 and 100% seems reasonable. The geometry of most transistors is much like the sheet considered above. Values of the factor Q (the number of cubes in the sheet) are 10^2 to 10^4 for most transistors, with the higher number for modern mesa transistors like the one shown in Fig. 2-8. The net result is that the influence of a large tolerance like 100% and the geometry largely cancel, and the values in Fig. 2-3 apply.

A further limiting phenomenon, heat generation, which is important, particularly at low resistivity, will be treated in Sec. 2-3.

2-3. HEAT CONSIDERATIONS

One of the most difficult problems in microelectronics is heat dissipation. Although present fabrication techniques are capable of building a system containing 10^6 devices in a cube one inch on a side, the operation of such a dense system would be impractical because of the following three important objections. First, if each device dissipates a few milliwatts, which is typical for present conventional circuits, it may not be possible to conduct the heat to the surface of the cube unless the temperature in the center of the cube rises to intolerable values, making the circuit inoperable, or at least less reliable.[20] Second, the heat exchanger necessary to transfer the heat to the surrounding air, or to space, has to be considerably larger than one cubic inch. Then the miniaturization of the system to a size below that of its heat exchanger faces diminishing return, except when a large surface for cooling is available, as in a satellite or an airplane. Third, the power dissipated ultimately has to come from a power supply, which has to be considerably larger than one cubic inch. Again, the miniaturization of the system below the size of its power supply faces diminishing return. The conclusion of this reasoning is that *miniaturization with regard to size must be accompanied by miniaturization with regard to power level*, and considerable attention must be given to cooling.

This section deals with the ultimate packing density of solid-state devices as a natural complement to Sec. 2-2. Then some of the work on heat removal internally through the package, or externally from the package, is reviewed. In Sec. 2-2, a basic assumption of fixed tolerances on devices was made. Here the influence of variable tolerances on heat dissipation is analyzed. Finally, a concluding section shows that reduced power level has to be paid for by reduced speed of operation, or by in effect faster devices operated at constant speed. The limitation imposed by the size of the power supply is dealt with in Sec. 2-6.

1. Maximum Packaging Density of Electronic Devices

An absolute upper limit to the packaging density of electronic devices is given by the minimum size derived in Sec. 2-2. It will be shown in this section that an even more stringent limit is set in some cases by heat dissipation. For this purpose, the same method of analysis as used in Sec. 2-2 will be continued, and particularly the assumptions in Tables 2-2 and 2-4 will be retained. One further assumption to be made is that the packaging density is reduced by one order of magnitude to allow for connections, insulation, support, etc. In other words, the *effective packaging density is* 10%. (*N*)

The network to be considered then is as shown in Fig. 2-9. If the minimum size of a component is a cube 3 μ on a side, then the minimum space that the component occupies is a cube about 6 μ on a side, corresponding to a maximum packaging density of about 4×10^9 devices/cm^3. This then represents an absolute upper limit for passive devices. Passive devices are not used by themselves but always in combination with active devices, and the heat dissipation in the passive devices depends entirely on the active devices used. Therefore, heat limitation on packaging density of passive devices will be deferred until the active devices are analyzed.

Magnetic, Superconductive, and Dielectric Devices. One obvious limitation to the maximum packaging density of magnetic, superconductive, and dielectric devices is given by the minimum size as for passive devices. This limit is indicated in Fig. 2-10 as a line marked "edge definition."

In contrast to semiconductor devices, magnetic, superconductive, and dielectric devices do not dissipate power in a quiescent state. For sufficiently low duty cycle, then, heat dissipation cannot be a limitation for these devices. However, power is consumed in switching, and for high frequencies and high-duty-cycle operation, heat dissipation may be a problem.

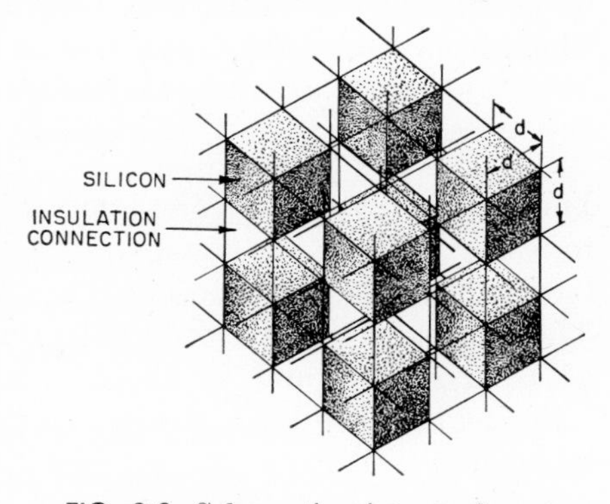

FIG. 2-9. Schematic picture of system to be considered, consisting of cubic elements surrounded on all sides by empty cubes signifying connections, insulation, support, etc.

Assume a number, *N*, of identical devices, each dissipating *E* joules

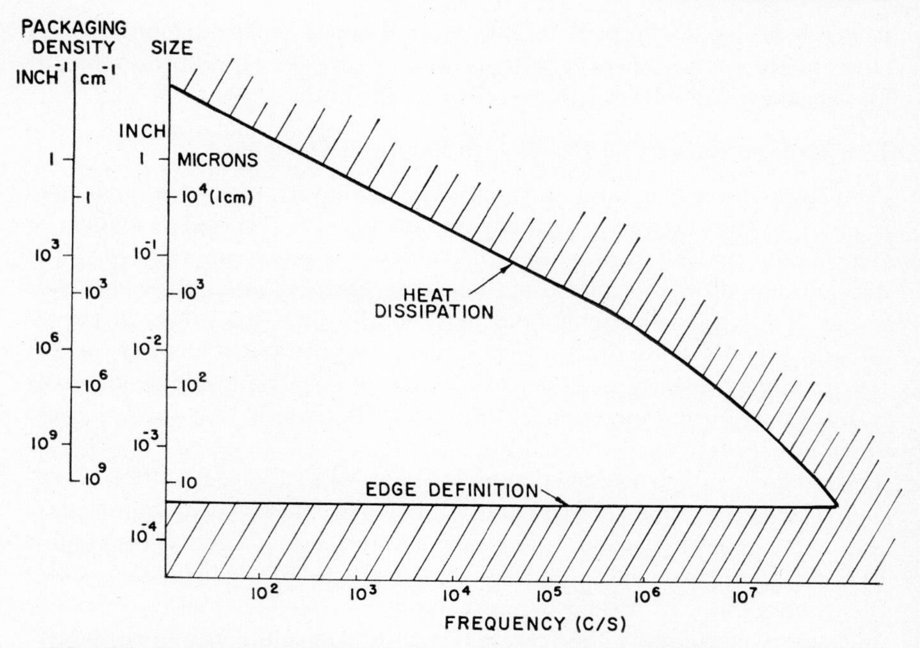

FIG. 2-10. Maximum packaging density and the corresponding minimum size of the space for each element versus the product of frequency of operation and duty cycle for magnetic devices. Shaded area is excluded.

for each switching cycle, being switched $f\beta$ times per second, where β is the duty cycle and f the operating frequency. Let these devices be assembled into a sphere with radius R, and with an effective heat conductivity, λ. Let the surface temperature of the sphere be θ_0, and the temperature in the center θ_i. Let each element occupy a volume d_0^3. Then, from Eq. (2-38),

$$d_0 = \left(\frac{3N}{4\pi}\right)^{2/3} \frac{f\beta E}{6\lambda(\theta_i - \theta_0)} \tag{2-26}$$

For magnetic materials, the temperature at the center cannot be allowed to exceed the Curie temperature for the material. For high-speed ferrites, the Curie temperature is relatively low, typically 80 to 100°C. The thermal conductivity of polycrystalline ferrites is fairly independent of composition, and is about 6 watts/(m)(C°) [1.5×10^{-2} cal/(sec)(cm)(deg)]. The heat dissipation for magnetic-memory cores is speed-dependent because of the necessary overdrive at high speed with resultant larger losses. The energy E to switch a core with rectangular hysteresis loop and which energy is converted to heat is obtained from Eq. (2-15). Assuming an ambient of 40°C, $\theta_i - \theta_0 = 40$°C. Assuming an effective heat conductivity of 3 watts/(m)(C°) and $N = 10^5$, Eq.

(2-26) gives

$$d_0 = 4.6(4 \times 10^{-7}f + 1)f\beta d^3 \tag{2-27}$$

Setting $d_0^3 = d^3$, assuming close packing, and $\beta = 1$ gives a line in Fig. 2-10 marked "heat dissipation." Magnetic devices with a size above this line cannot be assembled with the maximum packaging density, but must be spread out by increasing d_0 for constant d. Another alternative is to reduce the duty cycle β to less than one, thereby in effect allowing larger d. The requirement of low heat dissipation is, of course, the same as the requirement of low driving power—an even stronger incentive to small devices.

Superconductive Devices. For superconductive devices the temperature cannot be allowed to exceed the critical temperature of the material, for Pb 7.2, for Sn 3.7°K. Therefore, the temperature difference between the hottest and the coldest element cannot exceed about 1 °C. This means that the superconductive elements have to be in direct contact with the cooling medium: liquid helium. Therefore, although dimensions in two directions can be small, $\approx 3\ \mu$, dimensions in the third direction cannot.

Another even more serious limitation on packaging density of superconductive devices is the size of the cooling equipment. This cooling equipment is a serious handicap when small-size, low-weight, and low-cost applications are considered.

As an illustrative example, not necessarily representing the best that can be done, consider the data in Table 2-11 of a commercial liquid-helium closed-cycle refrigerator, designed for a cooling capacity of 250 mw at liquid-helium temperature. This capacity would be sufficient for a memory of some 10^5 cryotrons.

The power-supply figures have been calculated from data given in Sec. 2-6 for a lead-acid battery designed to operate for 10 hr.

Even assuming that these figures could be improved by perhaps a factor of 5 in as many years still leaves a considerable handicap when large packing density is required.

TABLE 2-11. APPROXIMATE VOLUME, WEIGHT, AND COST OF COOLING SYSTEM

	Power watts	Volume in.³	Weight, lb	Approximate cost
Cryotron system	0.25	10	0.5	$10,000
Refrigerator		8,400	135	70,000
Compressor		63,000	800	
Power supply	4000	20,000	2,000	10,000

Semiconductor Devices. The minimum voltage $V_{\min}$ that can be used with junction-type semiconductor devices in on-off operation is determined by

$$V_{\min} \gg \frac{kT}{q} \tag{2-28}$$

Assume

$$V_{\min} \approx \frac{10kT}{q}$$

where k = the Boltzmann constant ($k = 1.38 \times 10^{-23}$ watt-sec/°C)
T = temperature, °K
q = electronic charge ($q = 1.6 \times 10^{-19}$ coul).
For voltages smaller than $V_{\min}$, the device will be linear. At room temperature $V_{\min} \approx 0.25$ volt.

With a voltage V applied to each semiconductor cube of a material with resistivity ρ and with side d in a three-dimensional network as depicted in Fig. 2-9, the power density p with one resistor in each volume $10d^3$ is

$$p = \frac{V^2}{10d^2\rho} \tag{2-29}$$

From Eqs. (2-29) and (2-38) is obtained

$$R = d\left[\frac{60\lambda\rho(\theta_i - \theta_0)}{V^2}\right]^{1/2} \tag{2-30}$$

where θ_i is the temperature in the center of the sphere.

The total number of elements in the sphere N is

$$N = \frac{4\pi R^3}{30d^3} \tag{2-31}$$

Combination of Eqs. (2-30) and (2-31) gives

$$N = 195\left[\frac{\lambda\rho(\theta_i - \theta_0)}{V^2}\right]^{3/2} \tag{2-32}$$

Now insert practical values.

In the case of pure silicon, a temperature difference between the hottest and the coolest element $(\theta_i - \theta_0)$ may not exceed approximately 4°C, in view of assumption *I* and the known temperature dependence of the carrier mobility. However, the temperature variation of mobility decreases with doping, and is also smaller for germanium than for silicon. Therefore a temperature difference of 20°C may be possible and will be assumed here.

Assuming a construction with semiconductor elements mounted on printed circuits on ceramic wafers, the heat conductivity would be a compromise between that of air [0.025 watt/(m)(C°)], that of ceramic (alumina 2), that of semiconductor (Ge 58, Si 148), and that of metal [Ni 88, Au 300, Al 218 watts/(m)(C°)]. A fair estimate of the resulting heat conductivity may be 1 watt/(m)(C°). It is possible that this figure could be increased by special cooling. However, this would unavoidably add to the volume, offsetting at least partly the advantage gained. The minimum voltage for operation of junction-type devices is, as assumed earlier, 0.25 volt. Then Eq. (2-32) gives

$$N = 1.1 \times 10^6 \rho^{3/2} \tag{2-33}$$

With $N = 10^5$ we obtain $\rho = 0.2$ ohm-m (20 ohm-cm).

When ρ is reduced below 20 ohm-cm, the packing density also has to be reduced. This means that although the devices may be made smaller, the space they occupy must be enlarged. The volume occupied per device may be obtained from

$$d_0 = d_{0\,\min} \frac{0.2}{\rho} \tag{2-34}$$

where $d_{0\,\min}$ is the dimension at 0.2 ohm-m (20 ohm-cm). From Fig. 2-3, $d_{0\,\min} = 15\ \mu$. Equation (2-34) is indicated in Fig. 2-11. However, we

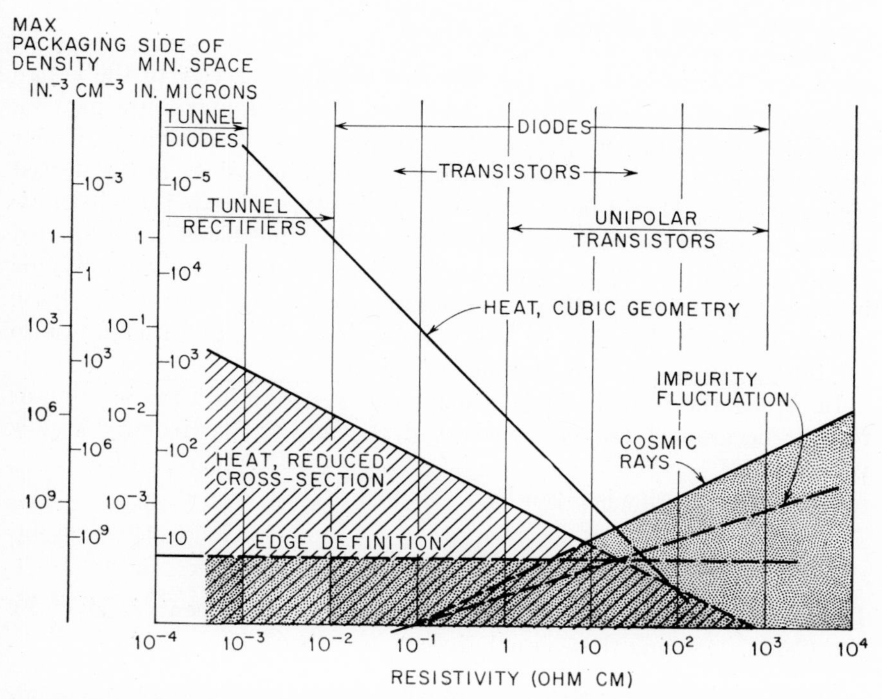

FIG. 2-11. Maximum packaging density and the corresponding minimum space for each element versus resistivity of the material for semiconductor devices. Gray area is excluded. When devices are close together, shaded area is also excluded.

also have to consider the influence of geometry in the following way. If the resistor is made with reduced cross section in the form of a stick or a sheet rather than in the form of a cube, the power density is reduced. Let us as a most favorable case consider a stick with cross section d_1^2, as before in a volume $10d^3$. Then the power density is

$$p = \frac{V^2 d_1^2}{10 \rho d^4} \tag{2-35}$$

Maximum power density is given by heat-removal considerations (see Art. 2). Then a reduction of d_1, by a factor k reduces d by a factor $k^{1/2}$. For semiconductor resistors with cross section reduced to $(3\ \mu)^2$, the minimum set by fabrication inexactness, we obtain the curve in Fig. 2-11 labeled "reduced cross section." This curve also satisfies the condition for stick geometry, i.e., $d \gg d_1$.

Sheet geometry is intermediate between cube and stick geometry.

The minimum element space is a factor of $10^{1/3}$, or about two times larger than the minimum semiconductor device size because each element occupies the volume $10d^3$. This reflects the assumption that wiring, insulation, support, etc., take up 90% of the element volume.

In applying these results to practical devices, a start will be made, as in the previous section, with *unipolar transistors*. These transistors are, in effect, variable resistors, so that the results apply directly to the active region of these devices. Figure 2-11 then gives the maximum packing density in terms of the minimum element space versus resistivity. Under assumptions A to N, unipolar transistors with dimensions in the dotted region of Fig. 2-11 will fail because of cosmic rays or inexactness in fabrication. In addition, unipolar transistors with dimensions in the shaded region will fail because of excess heat. Although the transistors may be made as small as $(3\ \mu)^3$, they must occupy a minimum element size lying outside the shaded region, so that the power density, and therefore also the temperature, does not become excessive.

In *diodes*, the heat dissipation may be lower than in a semiconductor resistor because of the nonlinearity of the former, particularly at low voltages.

The power density is given by

$$p = \frac{Vi}{10d} \tag{2-36}$$

The maximum power density for a given number of elements is given by Eq. (2-29). Setting these equal gives

$$i = \frac{V}{\rho d} \tag{2-37}$$

For typical values,

$$\rho = 10^{-3} \text{ ohm-m} \qquad (0.1 \text{ ohm-cm})$$
$$d = 10^{-4} \text{ m } (0.004'') \qquad i = 2.5 \times 10^{6} \text{ amp/m}^2 \ (250 \text{ amp/cm}^2)$$

However, a typical current density in a diode at 0.25 volt is only about 10^4 amp/m^2, representing a reduction in power density of about 100× and a corresponding reduction in minimum size of about 10×, or from $(1{,}000\ \mu)^3$ to $(100\ \mu)^3$. Also the diode area may be further reduced to the limit set by edge uncertainty, $(3\ \mu)^3$. This represents another reduction in element size by $10^{2/3}$ or about 5. The diode element size may therefore come close to the limit set by edge uncertainty, 3 μ.

The situation for the *tunnel diode* is somewhat analogous to that of the diode. Tunnel diodes require a doping corresponding to approximately 10^{-4} ohm-cm. Then a tunnel diode with

$$\rho = 10^{-6} \text{ ohm-m} \qquad (10^{-4} \text{ ohm-cm})$$
$$d = 5 \times 10^{-5} \text{ m} \qquad (0.002 \text{ in.})$$

would have a current density according to Eq. (2-37) of

$$i = 5 \times 10^{5} \text{ amp/cm}^2$$

However, a typical current density in germanium tunnel diodes with maximum peak-to-valley current ratio is about 5×10^3 amp/cm^2, representing a reduction in power of 100× and in size of 10×. In addition, the area of the tunnel diode may be reduced by nearly four orders of magnitude to the limit set by edge uncertainty, $(3\ \mu)^3$. Then from Eq. (2-35) the minimum size of the element may be reduced by another two orders of magnitude, or to just below $(100\ \mu)^3$. Tunnel diodes, then, although fabricated as small as $(3\ \mu)^3$, still require an element size of about $(100\ \mu)^3$ for proper cooling.

For (bipolar) *transistors* the results apply directly if the tolerance on base doping is the same as assumed earlier, ±10%. However, in many applications larger-tolerance base doping may be allowed, resulting in smaller size. Therefore the minimum size may be somewhere between $(3\ \mu)^3$ and $(10\ \mu)^3$.

In the same manner the results of the analysis may be applied to any arbitrary semiconductor devices.

Conclusions. The conclusion to this analysis is that already at the present time the active region of the smallest semiconductor devices is close to the fundamental limit.

At about 10 ohm-cm silicon in the active region an optimum exists, corresponding to elements about $(10\ \mu)^3$, and allowing a packing density of approximately 10^9 components/cm^3 (10^{10} components/in.^{3}) of most semiconductor devices such as transistors, unipolar transistors, diodes, and semiconductor resistors. When the resistivity is higher than 10 ohm-cm, the volume of the active region has to be larger, or else the devices fail from cosmic-ray bombardment, as shown in Sec. 2-2. When the resistivity is lower than 10 ohm-cm, the device can be made smaller, but the total volume occupied by the device must be larger than the minimum, or else the power density becomes excessive.

There are, however, ways in which these limitations may be circumvented. One of the most attractive alternatives has already been suggested by systems considerations, namely, redundancy (see Sec. 2-5).

Another possibility is the use of negative feedback in a general sense. By the use of negative feedback in circuits, the device tolerances may be wider. Also negative feedback could be applied in the fabrication of devices, as is now being done with tunnel diodes and some narrow-base transistors, where units are, in effect, measured during the fabrication and treated until they fall within tolerances. The price paid for closer tolerances is reduced performance, e.g., amplification or speed of operation or speed of fabrication. Another way in which negative feedback, in a general sense, could be applied is through the use of self-organizing systems which could, in principle, self-heal faulty circuits. Again the price to be paid is an increase in the number of devices and decreased speed.

2. Heat Removal

The heat generated by dissipation in the microelectronic elements, first, has to be transferred to the surface of the package and, second, has to be removed to the ambient. If the amount of heat is small, the removal need not consume extra volume or weight in the form of cooling arrangements or heat exchangers. However, in extreme microelectronic assemblies with high packaging density, heat is likely to be a limiting factor, in requiring not only cooling means *internally*, with accompanying volume and weight, but also special cooling means *externally*, in the form of a heat exchanger through which the heat may be removed by convection, conduction, and/or radiation.[18,19] In case of forced cooling, the weight and volume of cooling fans or circulating pumps, and perhaps of air inlet and outlet channels, have to be included in the packaging-density considerations. In many cases, the cost of sufficient cooling to bring the temperature of the components in a microelectronic assembly to the same value as for relatively spacious conventional assemblies cannot be accepted, a higher temperature with accompanying sacrifice in reliability cannot be tolerated, and, consequently, less extreme miniaturization must be resorted to.[20,21]

Exact calculations of temperature rise for certain cooling arrangements are difficult to make, and some experimental or semiexperimental procedure is usually necessary. However, some simple cases can be easily treated analytically and are useful as guidelines.[22–24]

Internal Heat Removal. From a heat-removal point of view, the devices in microelectronic circuits may be packaged in two different ways. The devices form either some sort of three-dimensional network, in its simplest form for analysis, a homogeneous sphere, or else a two-dimensional network such as a thin-film network on a substrate. As large systems are usually fabricated from several thin-film wafers packaged together, even thin-film devices often fall in the first category when more than a single circuit is considered.

The temperature θ at a radius r in a *homogeneous sphere* with radius R and a surface temperature θ_0 is given by

$$\theta = \theta_0 + \frac{p}{6\lambda}(R^2 - r^2) \tag{2-38}$$

where λ is the heat conductivity and p is the power developed per unit volume. Let p_1 be the power developed per element, each in a volume $10d^3$,

$$p_1 = p \times 10d^3 \tag{2-39}$$

Let D be the packing density of elements per cubic inch,

$$D = 1.64 \times 10^{-6}d^{-3} \tag{2-40}$$

Combining Eqs. (2-38) to (2-40) gives

$$p_1 D = 9.8 \times 10^{-5}\frac{\lambda(\theta_i - \theta_0)}{R^2} \tag{2-41}$$

where θ_i is the temperature in the center.

Equation (2-41) is shown in Fig. 2-12 for various values of the parameters. As can be seen, the general trend is a reduction in packing density with the same factor as the power per element is increased, or vice versa, leaving the product unchanged.

Let us now consider a homogeneous *thin film*, cooled from one side only, and far enough from other heat sources so that their influence may be neglected. Cooling from two sides may be analyzed simply by combining two such cases. The temperature θ at a distance x from the heat sink which has the temperature θ_0 is

$$\theta = \theta_0 + \frac{px^2}{2\lambda} \tag{2-42}$$

where as before λ is the heat conductivity, and p is the power developed per unit volume.

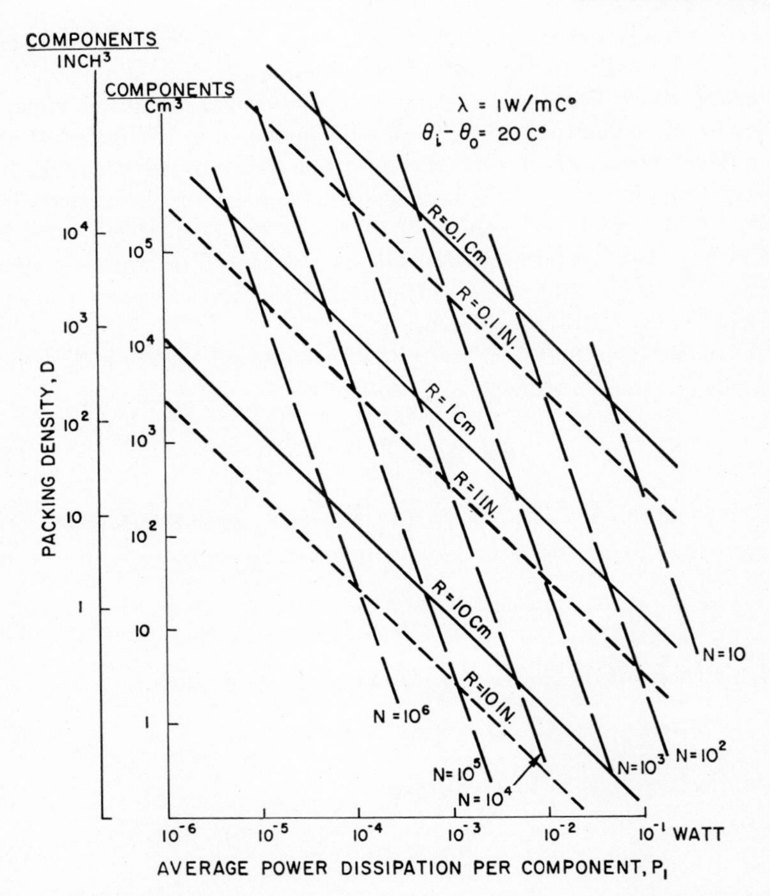

FIG. 2-12. Packing density versus average power dissipation per component for a homogeneous sphere with a uniform heat conductivity of 1 watt/(m)(°C), glass or ceramic, and temperature differential between center and surface of 20°C. Parameters are N, the number of components in the sphere, and R, the radius of the sphere. The diagram is approximately valid for a cube with side $2R$.

If p_1 is the power developed per element in a volume $10d^3$,

$$p_1 = p \times 10d^3 \tag{2-43}$$

and

$$x^2 = \frac{20\lambda(\theta_{\max} - \theta_0)d^3}{p_1} \tag{2-44}$$

If $\lambda = 10$ watts/(m)(°C) $\qquad \theta_{\max} - \theta_0 = 20°\text{C}$

$p_1 = 10^{-3}$ watt $\qquad d = 3\ \mu$

then $x = 10\ \mu$

Then with minimum-size elements no more than two layers may be placed on the heat sink. Keeping the thickness constant but increasing the area of the devices, so that they become flatter, increases x so that more layers are permitted.

When the power dissipation and heat conduction are not uniform throughout the microelectronic assembly, the analysis becomes more complex.[22] In a package like the *micromodule* (see Chap. 3) composed of insulator wafers sandwiched together with an encapsulant and joined by riser wires, as shown schematically in Fig. 2-13, the resulting heat conductivity may be obtained from

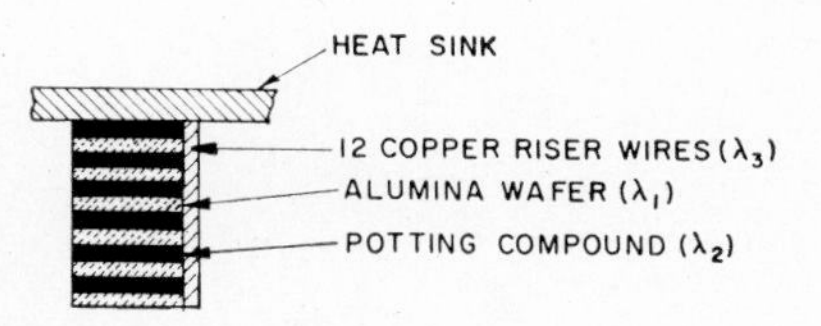

FIG. 2-13. Schematic picture of stack of wafers for derivation of effective heat conductivity.

$$\lambda = \frac{2\lambda_1\lambda_2}{\lambda_1 + \lambda_2} + \lambda_3 \frac{A_1}{A} \tag{2-45}$$

where λ_1 = heat conductivity of the insulator [alumina, $\lambda_1 = 2$ watts/(m)(°C)]
λ_2 = heat conductivity of the encapsulant [$\lambda_2 \approx 0.2$ watt/(m)(°C)], thickness assumed equal to that of insulator
λ_3 = heat conductivity of the riser wires [copper, $\lambda_3 = 395$ watts/(m)(°C)]
A_1 = cross-section area of the riser wires
A = cross-section area of the module

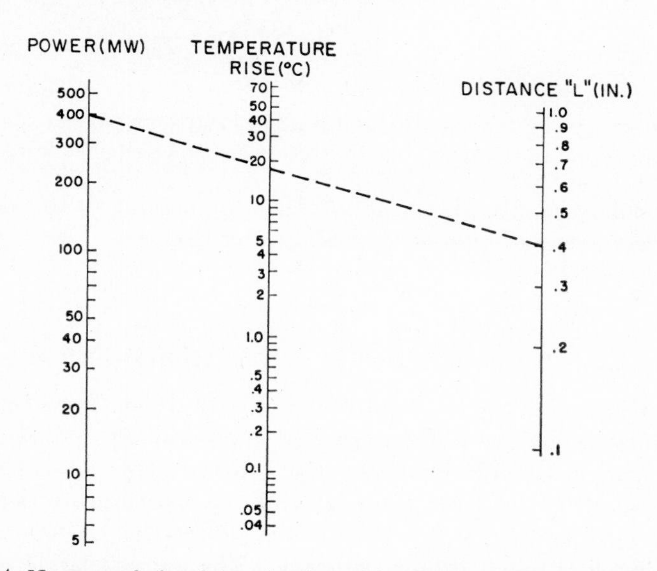

FIG. 2-14. Nomograph for determination of internal temperature elevation above temperature of heat sink in a stack of wafers, cooled by conduction only, when the chief heat-generating wafer generates P mw and is located L in. from the heat sink. Effective heat conductance is 2.9 watts/(m)(°C) (2.5 kcal/(m)(h)(°C).

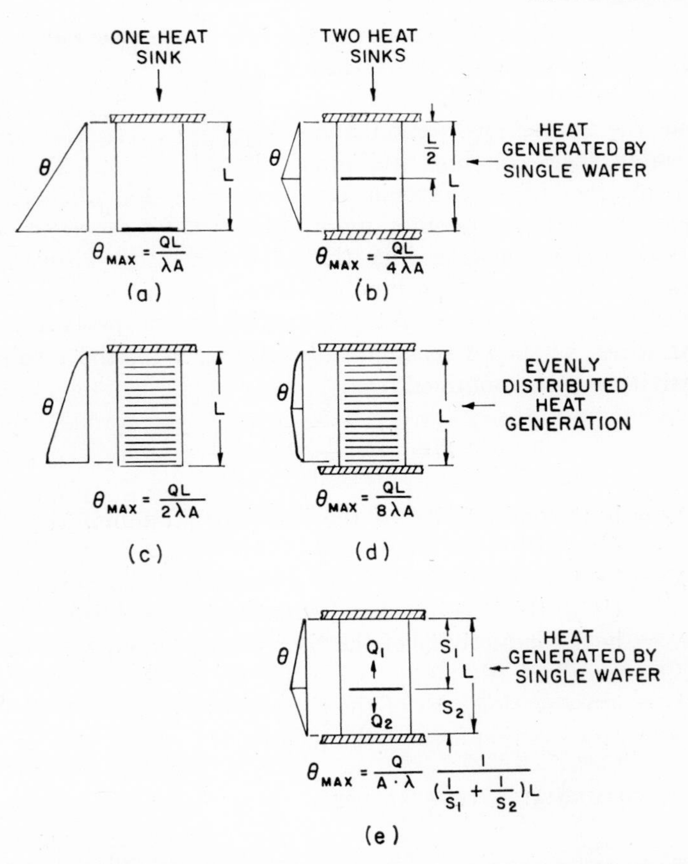

FIG. 2-15. Simple cases of temperature distribution and heat conductivity of stacked wafers with one and two heat sinks and with localized or distributed heat generation.

The thermal resistance along the insulator wafers out to the riser wires may be taken into account schematically through

$$\frac{1}{\lambda_{\text{eff}}} = \frac{1}{\lambda} + \frac{1}{\lambda_{\text{series}}} \tag{2-46}$$

Then $\lambda_{\text{eff}} = 2.9$ watts/(m)(°C) [2.5 kcal/(m)(h)(°C)]

This value is used in the nomograph of Fig. 2-14 which shows the internal temperature rise on one wafer assembled into a module and located L in. from the heat sink and generating P mw. If the wafer generates 400 mw and is located 0.4 in. from the heat sink, its temperature rise is 18°C.

Figure 2-15 shows five cases with one or two heat sinks, and with localized or distributed heat generation. The corresponding maximum temperature, $\theta_{\max}$, is also shown. In the case of evenly distributed heat generation, Eq. (2-42) applies and the temperature distribution is parabolic. For lowest possible θ_{max} anywhere in the package, the chief heat-generating wafers should be localized as close to the heat sink as possible.

Numerical Method. When the geometry is too complex for simple analytical solutions, or when the influence of variations in the parameters is to be determined, numerical methods on the basis of relaxation calculations may be resorted to.[25] For this purpose, a map of the circuit is prepared and covered by an analytical grid of, for example, 10-by-10 equidistant lines. The heat-flow equations are set up for each nodal point of this grid and solved under the conditions of temperature matching everywhere in the grid. Since, for sufficient accuracy, the grid must be sufficiently fine (i.e., must have many nodal points), and several successive approximations are necessary to determine the temperature at each point, much computation work is required. To arrive at a solution in a reasonable time, therefore, a computer is usually necessary.

Results for a microelectronic NOR circuit consisting of three resistors (the contributions from the diodes and transistors are neglected) deposited on a glass substrate are shown in Figs. 2-16 and 2-17. The ambient temperature is 60°C (140°F). In Fig. 2-16 is shown a single wafer cooled from all sides, and in Fig. 2-17 an infinite stack of wafers corresponding to cooling only from the edges of the wafers. Stacking many identical wafers together is shown to increase the hot-point temperature from 204°F

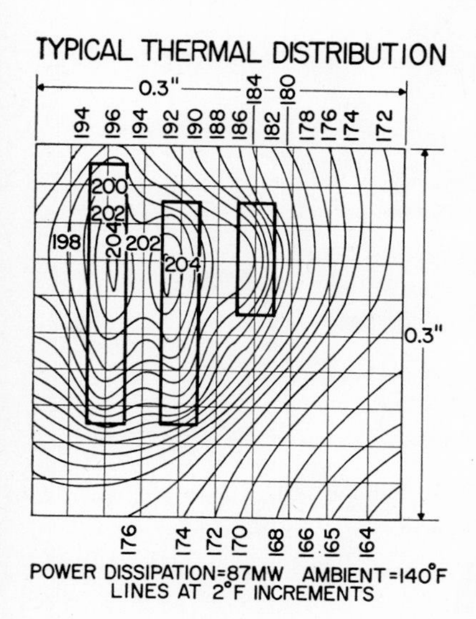

FIG. 2-16. Computed isotherms in single wafer containing a NOR circuit and cooled on all sides. Glass wafer, 0.3 by 0.3 by 0.02 in., power generation 87 mw.

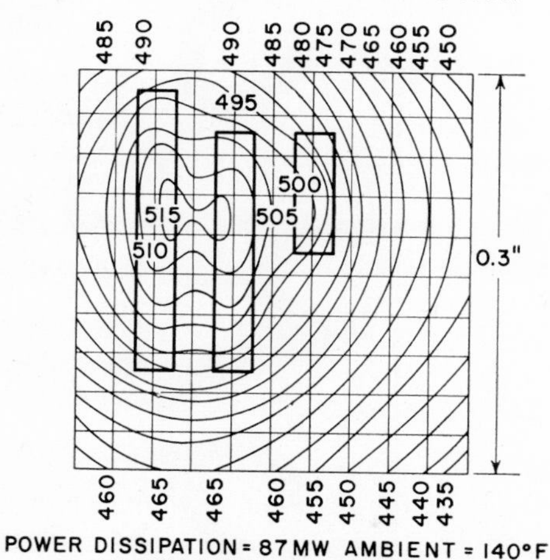

FIG. 2-17. Computed isotherms in wafer containing a NOR circuit. The wafer is in an infinite stack and is cooled only from the edges. Glass wafer 0.3 by 0.3 by 0.02 in., power generation 87 mw.

(96°C) to 515°F (268°C) at the same time that the temperature gradients are increased.

Experimental Method. Another method[26] for analyzing the heat pattern in microelectronic assemblies makes use of a heat-sensitive radiation detector to measure the actual temperature distribution on the wafer or on a model. The detector may be an InSb photocell with a long-wavelength cutoff near 6 μ. It is mounted at a fixed distance over the wafer and focused onto a small area of the wafer, which may be moved by a micrometer screw so that the entire wafer may be successively scanned. The photocell is calibrated by giving the wafer a known elevated temperature. To correct for different thermal emissivities of different surfaces, a calibration run may be used in which the wafer is held at a uniform elevated temperature. The result from the measurement takes the form of a temperature plot, which may be converted to an isothermal map as shown in Fig. 2-18 for a microelectronic circuit consisting of a number of deposited resistors.

The method is suited for plane assemblies but not without modification for three-dimensional problems. It has the advantages that the hot spots may be localized directly on an actual circuit and it is convenient and accurate in use.

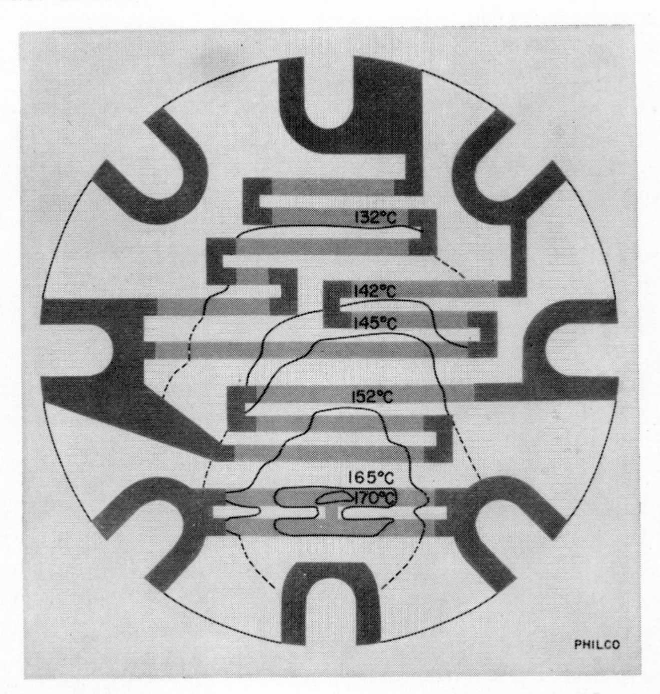

FIG. 2-18. Isotherms in wafer with evaporated resistor network, measured with radiation-detector probe.

Radiation. If return radiation is neglected, the power radiated from a surface at a temperature T_1 is

$$p = \epsilon \sigma T_1^4 \qquad \text{watts/m}^2 \qquad (2\text{-}47)$$

where σ is the Stefan-Boltzmann constant, 5.66×10^{-8} watt/(m²)(C°)⁴, and ϵ is thermal emissivity which is 1 for a perfect black surface, 0 for a perfect reflecting surface. Typical values of ϵ at room temperature are shown in Fig. 2-19. For a cube 1 cm on the side and

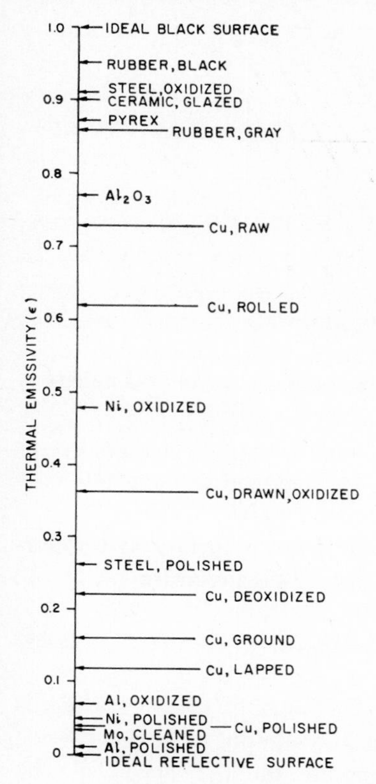

FIG. 2-19. Typical values of thermal emissivities for different surfaces and surface treatments.

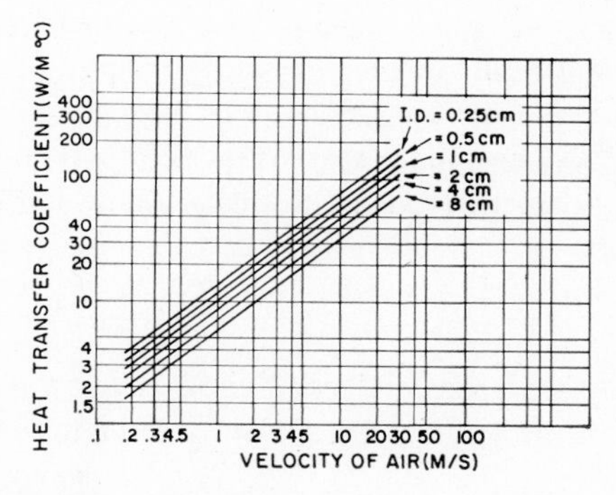

FIG. 2-20. Heat-transfer coefficient versus airspeed in tubular channels.

with the surface at room temperature (300°K), a thermal emissivity of 0.7, the total radiated power would be 0.2 watt. If each component dissipated 1 mw, this would only allow 200 components/cm³ (1,640 per in.³).

TABLE 2-12. HEAT-REMOVAL COEFFICIENT α FOR FREE CONVECTION IN AIR OF ROOM TEMPERATURE

Cooled surface	α
Horizontal surface upward	2.5
Horizontal surface downward	1.4
Vertical plane with height h:	
$h < 0.3$ m	$1.4h^{-1/4}$
$h > 0.3$ m	1.9
Horizontal cylinder, diameter d	$1.4h^{-1/4}$
Sphere, diameter d:	
$d > 0.15$ m	$1.3\left(1 + \frac{0.002}{d}\right)^{1/2} d^{-1/4}$

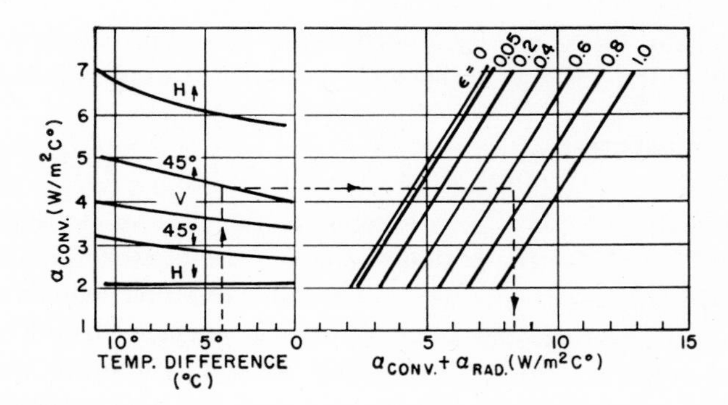

FIG. 2-21. Total heat-removal coefficient, including both convection and radiation, for plane surface in air of room temperature. For example, a surface with a temperature 4°C above the ambient, inclined 45°, and cooled by convection from the top side, with a thermal emissivity of 0.7, would have a total heat-removal coefficient of 8.3 watts/(m²)(C°).

Convection in Air. The heat carried away from a surface at temperature T_1 to the surrounding air of temperature T_2 is given by

$$p = \alpha(T_1 - T_2)^{5/4} \qquad \text{watts/m}^2 \tag{2-48}$$

The heat-removal coefficient α for various inclinations of the surface is given in Table 2-12. For forced convection, the amount of heat Q carried away depends on the speed of the airflow

$$Q = h(T_1 - T_2)$$

where h is the heat-transfer coefficient and may be obtained from Fig. 2-20 (strictly valid for tubular air channel).

Free Convection and Radiation. In a practical case, both convection and radiation have to be considered. The nomograph in Fig. 2-21 gives the total heat removed from plane surfaces by convection and radiation in air of room temperature.[22] An approximate figure is given by

$$p = 0.3(T_1 - T_2)[2 + (T_1 - T_2)^{1/4}] \qquad \text{watts/cm}^2 \tag{2-49}$$

This relation is shown in Fig. 2-22. The abscissa is the length of a micro-

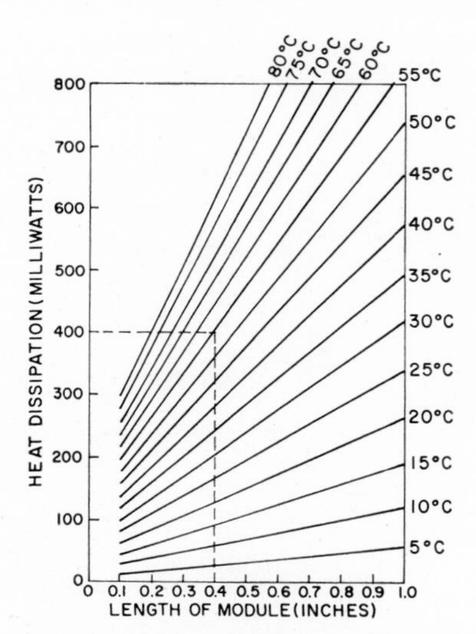

FIG. 2-22. Heat dissipation versus height of wafer stack of micromodule wafers. Parameter is surface temperature.

module package (cross section 0.35 × 0.35 in.). For a length of 0.4 in. and a dissipation of 400 mw, the surface temperature rises to 54°C above the ambient.

Review of Practical Alternatives. In most cases, the microelectronic assembly is not alone and able to convect or radiate freely, but is part of an equipment which introduces certain constraints on the heat removal. The manner in which this is done is shown schematically in Fig. 2-23, which shows the 15 possible alternatives.[18] The first row with the hot volume in the open corresponds to the cases treated above. The approximate heat-removal capacity is given for each case for comparison purposes and corresponds to a cube, one inch (2.5 cm) on a side, with a surface temperature 40°C above ambient temperature. The thermal emissivity is assumed 0.7. From the second row, it may be seen how enclosure in a cabinet or in an encapsulation reduces the heat-removal

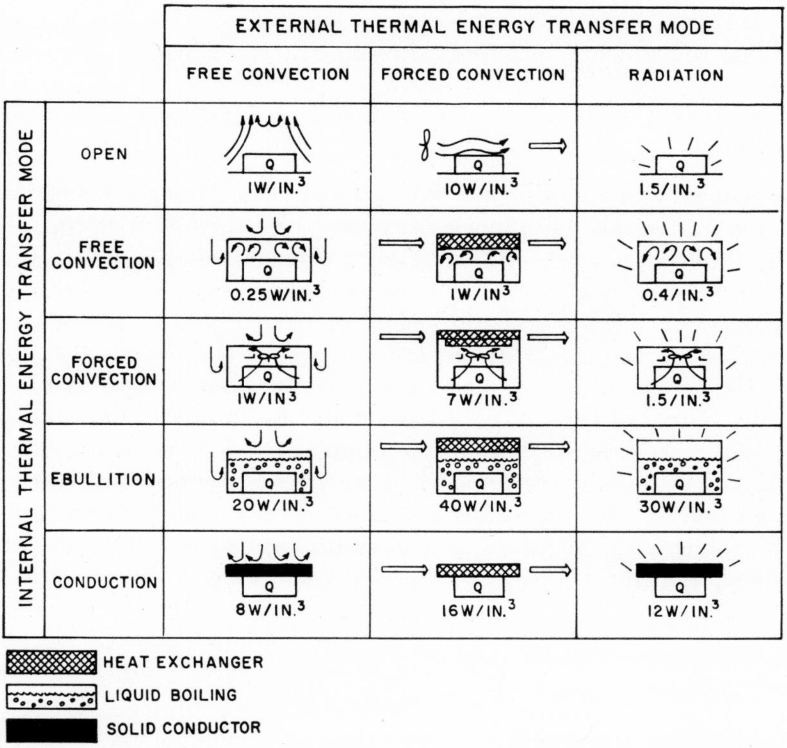

INTERNAL THERMAL ENERGY TRANSFER MODE	EXTERNAL THERMAL ENERGY TRANSFER MODE		
	FREE CONVECTION	FORCED CONVECTION	RADIATION
OPEN	1W/IN.^{3}	10W/IN.^{3}	1.5/IN.^{3}
FREE CONVECTION	0.25W/IN.^{3}	1W/IN^3	0.4/IN.^{3}
FORCED CONVECTION	1W/IN^3	7W/IN.^{3}	1.5/IN.^{3}
EBULLITION	20W/IN.^{3}	40W/IN.^{3}	30W/IN.^{3}
CONDUCTION	8W/IN.^{3}	16W/IN.^{3}	12W/IN.^{3}

HEAT EXCHANGER
LIQUID BOILING
SOLID CONDUCTOR

FIG. 2-23. Different alternatives for cooling a hot body Q. Horizontal rows represent different ways of heat transfer inside the equipment, vertical columns represent different outside cooling alternatives. The cooling-capacity figures are typical for a cubic hot body, one inch on a side, at a temperature 40°C above the outside ambient, with a thermal emissivity of 0.7.

capacity. From the difference between the first and the second column the value of a heat exchanger may be estimated.

3. Influence of Device and Circuit Tolerances

In determining the packing density from heat considerations, as was done in Sec. 2, various idealized conditions were assumed. In practice, deviations from these conditions are inevitable. This article will treat one such deviation, namely, the increased power dissipation caused by allowing the components of a circuit to have electrical characteristics which deviate by plus or minus a tolerance from nominal values.[20] It will be shown that allowing for engineering tolerances on device and circuit parameters has three consequences, all resulting from higher power and all detrimental to packing density, namely:

1. Increased power demand and therefore larger power supply, less space for other components
2. Increased power dissipation, necessitating more cooling, less space for other components; or improved cooling by spreading the components apart
3. Increased system complexity, meaning an increased number of components

The reduction in packing density is larger, the larger the tolerances. The reason for this may, in a general way, be compared with the second law of thermodynamics. Allowing wider component tolerances is equivalent to allowing a greater degree of randomness in selection of circuit components. From a statistical point of view, this may be thought of as equivalent to an increase in the entropy of the circuit. However, since the performance of the circuit cannot be allowed to deteriorate, its entropy is fixed. Then it is apparent that some form of increased energy expenditure, increased stress on the components, and/or increased number of components is the price to be paid for a more random selection, in a general way corresponding to the second law of thermodynamics.

In dealing with tolerances of devices three types of tolerances will be included, namely:

1. Initial deviations caused by the fabrication process
2. Temporary changes, caused by temperature, illumination, etc.
3. Permanent changes such as aging and mechanical damage

To illustrate the influence of tolerances in a more concrete way, two examples[20] will be chosen. As a first example, consider the simple circuit shown in Fig. 2-24, which has the base input of a transistor B, in series with a battery E and a current-limiting resistor R. In order for the transistor to turn on, it is necessary that the base current exceed a threshold value I_T. Now if the battery may vary by $\pm d_E$ (e.g., 10%), and if the resistor may vary by $\pm d_R$, and the base threshold current by $\pm d_B$, it is apparent that in order to satisfy the worst-case condition, R must be selected so that

$$R \leq \frac{E(1 - d_E)}{I_T(1 + d_R)(1 + d_B)} \qquad (2\text{-}50)$$

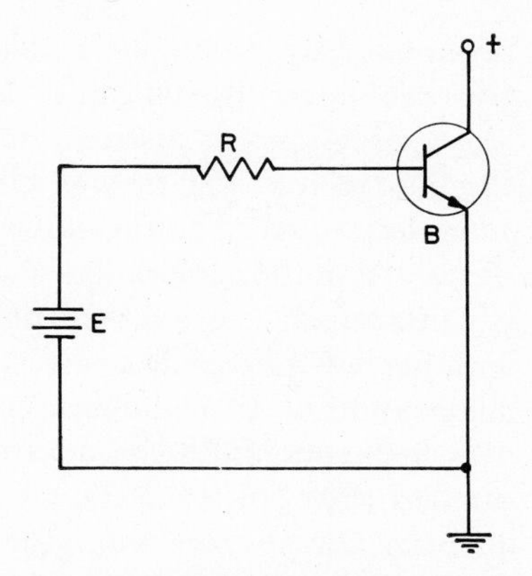

FIG. 2-24. Circuit containing the base input B of a transistor in series with a current-limiting resistance R and a battery E.

For least-power operation, R is selected so that its nominal value is just equal to the right-hand side of inequality (2-50). This ensures that the transistor will turn on for any expected battery voltage, and for any expected resistance. Then, with E, R, and I_T at nominal values, the steady-state power dissipation of the circuit p_{nom} is

$$p_{\text{nom}} = \frac{E^2}{R} = \frac{EI_T(1 + d_R)(1 + d_B)}{1 - d_E} \qquad (2\text{-}51)$$

The maximum power dissipation, when E is a maximum and R is a minimum, is

$$P_{\text{max}} = \frac{EI_T(1 + d_E)^2(1 + d_R)(1 + d_B)}{(1 - d_E)(1 - d_R)} \qquad (2\text{-}52)$$

Similarly, the minimum power dissipation is

$$P_{\text{min}} = EI_T(1 - d_E)(1 + d_B) \qquad (2\text{-}53)$$

Equations (2-51) to (2-53) are shown in Fig. 2-25 where normalized power, P/EI_T, is plotted versus tolerances under the assumption $d_B = d_R = d_E$.

The quantity of primary interest from the point of view of heat dissipation in closely packed circuits is the average power P_{av}, which is shown in Fig. 2-25. The average power has been obtained from

$$P_{\text{av}} = \frac{P_{\text{max}} + P_{\text{min}}}{2} \qquad (2\text{-}54)$$

using an error-function distribution for the tolerances. The fact that P_{av} is somewhat higher than P_{nom} is related to the fact that P_{max} shows a larger deviation from P_{nom} than does P_{min} for equal but opposite sign tolerances. As may be seen from Fig. 2-25, the design of the circuit to accommodate component tolerances of $\pm 10\%$ means an increase in average power dissipation of 40%. The wider the tolerances, the larger the average power dissipation. In addition, out of a number of similar circuits, a few will be hot spots, in that the tolerances happen to be cumulative, so that the power dissipation in those circuits approaches P_{max}. For 10% tolerances P_{max} represents an increase of 80%.

How much the average power dissipation increases depends on the number of physical quantities involved, their relationship and the magnitude of their tolerances. Although no study has been made of this for average circuits, it is clear that the number is larger than 3 and smaller than 10, 6 may be a reasonable figure. Then the average power increase for average miniature circuits is larger than indicated in Fig. 2-25, perhaps by a factor of 2.

In Fig. 2-26 is shown how the average power increases for the simple

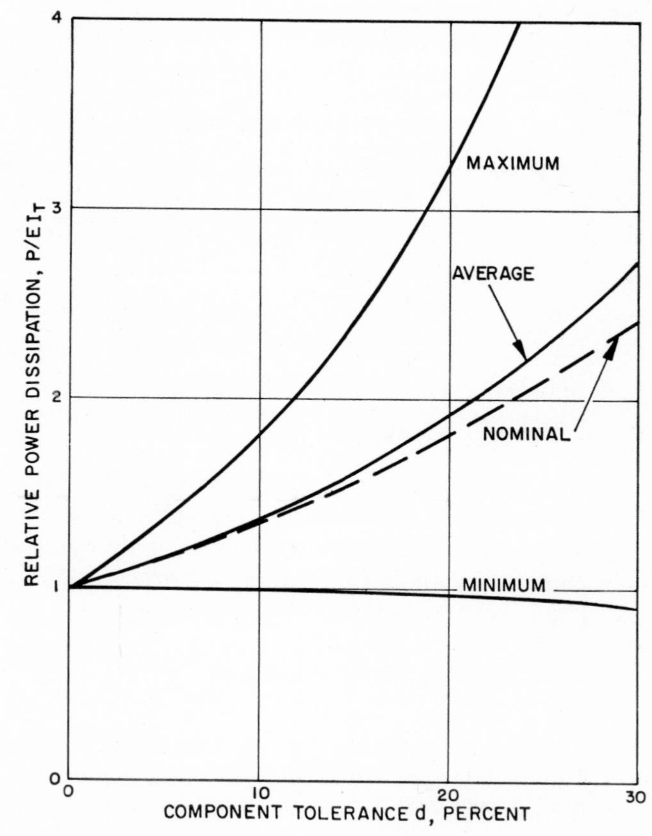

FIG. 2-25. Power dissipation in the circuit of Fig. 2-24 versus tolerances on device characteristics.

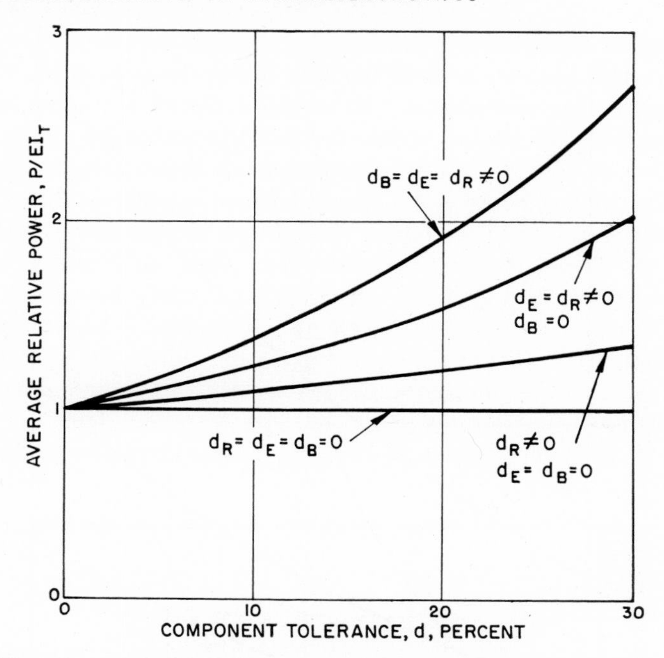

FIG. 2-26. Average power dissipation of the circuit of Fig. 2-24 when one, two, or three of the parameters are allowed tolerances.

example of Fig. 2-24 when one, two, or three of the components are allowed tolerances.

In the simple circuit of Fig. 2-24, an increase of component tolerances was not accompanied by an increase in the number of components. However, it is generally true that a deterioration through device tolerances of a voltage or current level, a pulse shape, or a waveform can

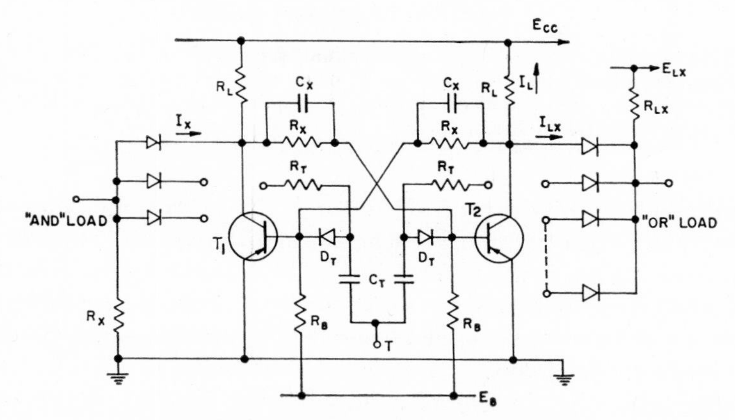

FIG. 2-27. Logic-driving flip-flop circuit.

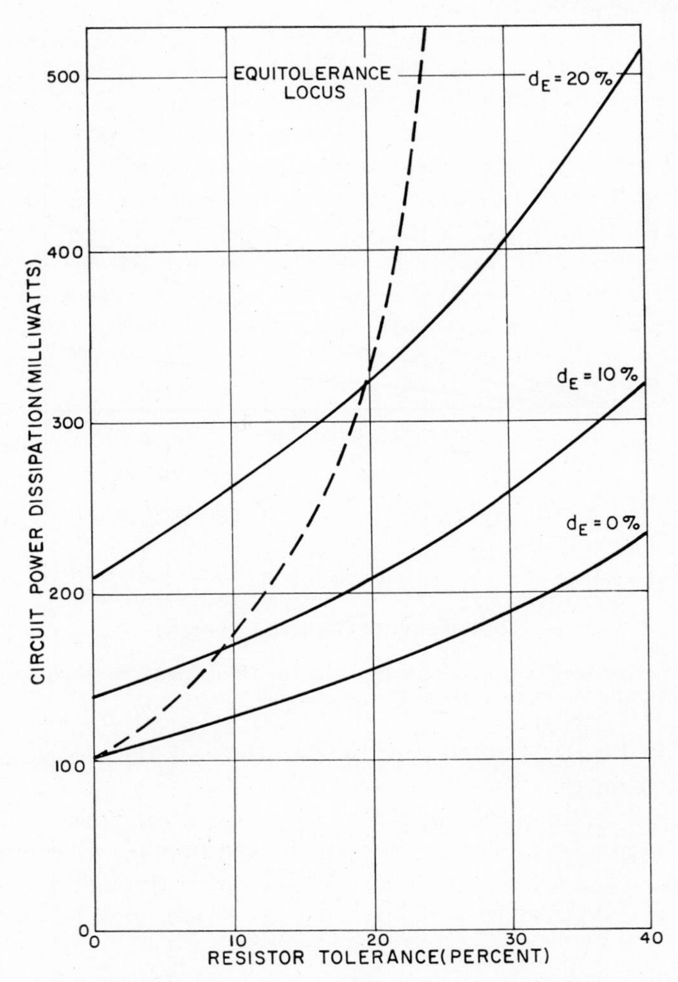

FIG. 2-28. Standby power dissipation (nominal) for circuit of Fig. 2-27 (d-c power dissipation, excluding power into the load), when resistors and supply voltages are allowed tolerances. The equitolerance curve applies when resistors and supply voltages have equal tolerances.

be counterbalanced by extra level restorers, by extra pulse clippers and shapers, or by extra negative feedback. In all cases, this requires extra components. Also, if the resistor in Fig. 2-25 could not be obtained at any power rating, but only for some moderate finite power dissipation, several would have to be used in parallel, to accommodate the higher power levels of the circuit.

Another example[20] is that of a logic-driving flip-flop, shown in Fig. 2-27. This circuit is typical of the kind used in shift registers or counters, where the stages are loaded by diode matrices or by various logic networks, in this case an AND and an OR circuit.

Figure 2-28 shows the standby circuit power dissipation (power dissipation of the circuit, but not including power into the load) versus resistor tolerances with power-supply tolerances as a parameter. Other parameters, such as transistor characteristics, capacitor values, ambient temperature, are assumed constant. The curves correspond to the nominal curves in the previous example. The average power would be somewhat higher. The dashed curve shows power dissipation versus tolerances if both resistors and power supply have equal tolerances. A still steeper curve would, of course, result if other parameters were allowed tolerances also. The circuit dissipates 100 mw for ideal resistors

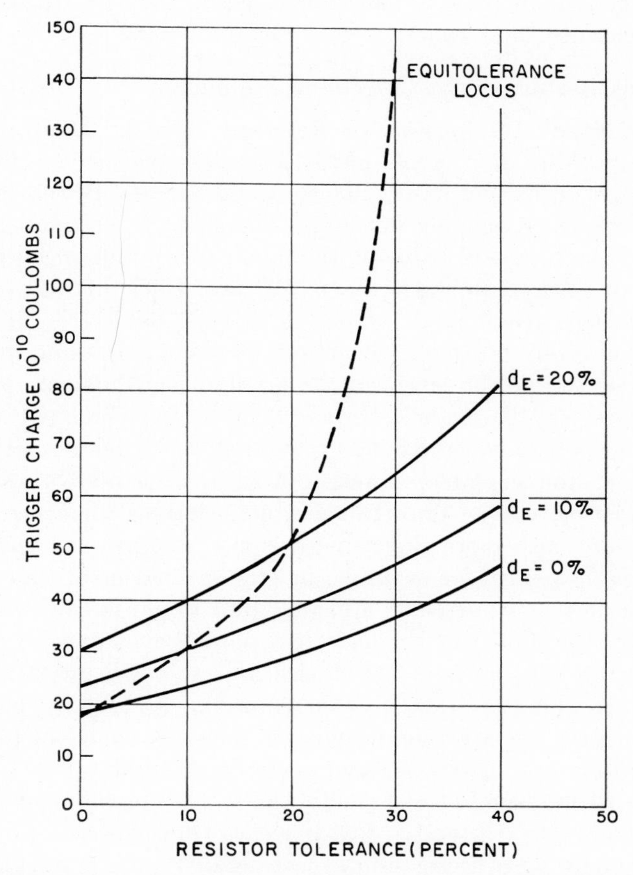

FIG. 2-29. The charge necessary to trigger the circuit of Fig. 2-27 versus resistor tolerances with supply-voltage tolerance as parameter. The equitolerance curve applies when resistors and supply voltages have equal tolerances.

and power supplies, but if tolerances of $\pm 12\%$ are allowed for both resistors and supply voltages, the circuit power dissipation doubles.

At the same time that the standby circuit power dissipation increases with increasing tolerances, the energy to trigger the flip-flop also increases. This is illustrated in Fig. 2-29, which shows trigger charge versus resistor tolerances, again with supply-voltage tolerance as parameter. Referring to the equitolerance locus, it is seen that for a tolerance level of 15%, the trigger charge doubles. Consequently, the tolerances cause an increase in both static and dynamic power dissipation of the circuit.

Although the flip-flop circuit is a rather typical one, the results are not typical, as we have allowed tolerances only on the resistor values and the supply voltages. If we allowed tolerances also on the transistors, the capacitors, the ambient temperature, the thermal resistance, and the diodes, a typical increase in power dissipation for 10% tolerance level might be close to three times.

4. Heat Dissipation versus Operating Frequency

Reduced Heat Dissipation at Reduced Speed. Heat dissipation limits the packing density of microelectronics equipment. The power supply which ultimately has to furnish the dissipated power, in its turn, limits the effective packing density. Therefore, it becomes important to consider the means of reducing the power dissipation in microcircuits. However, devices are designed to operate most efficiently at rated values of voltage and current, and the design of most devices is optimized as far as economically possible. Therefore, it is not surprising to find that reduction of the power level can be accomplished only at a sacrifice. This sacrifice may be made in the *cost of the device*, meaning that obtaining the same performance with less power requires a more exacting device construction, for example, transistors with scaled-down dimensions, smaller junction area in tunnel diodes, etc., making for a more difficult and, therefore, also more expensive design. The sacrifice may also be made in performance, for example, in *speed of operation*. Assuming an unchanged device, or, at most, a change that would not raise the cost of the device, operation can be made less power-consuming by, in effect, increasing the impedance of the device, but this will lower the highest frequency at which the device could be used. As will be shown, these two alternatives are equivalent, and it is therefore sufficient to treat the *relation between heat dissipation and operating frequency*.[20]

Change of the Number of Parallel Units. Let us first consider a simple circuit in which the active device, e.g., a transistor or a tunnel diode, may be thought of as a large number of parallel units in the same encapsulation. As is well known, a transistor or a tunnel diode may be divided into a number of such parallel units without affecting the operation, if the dissecting surfaces do not cross current paths. If first the device is cut into two parts, each part will, of course, deliver only half the power, but, on the other hand, will only dissipate half as much power as the whole unit. The impedance level will double. The upper-frequency limit of the device, if the encapsulation is not considered, is, of course, unchanged. Similarly, the device may be divided again and again within certain limits, until the power dissipation is sufficiently reduced. The limit as to how far the division may be carried is set by noise, by increased impedance level and its relation to noise, and by mechanical or physical reasons. The construction may be impossible or, at least, impractical mechanically, or the active part of the devices may be too close to the surface, with accompanying degradation of characteristics. However, before this limit is reached, another practical limitation may be set by the stray capacitance or series inductance of the encapsulation. Each time the device is divided, the device capacitance is cut in two. Sooner or later, this capacitance will be comparable to the capacitance of the encapsulation. When this happens, further division of the device will not leave the upper-frequency limit unchanged. Instead the upper-frequency limit will decrease with the same factor by which the device is divided, resulting in a diminishing return of further division.

The same argument may be made for inductance, although in reverse. Each time two devices are put in parallel, the inductance is cut in two. Sooner or later, this inductance will be smaller than the case inductance. Then further paralleling will not reduce the inductance, the L/R time constant increases, and the frequency response deteriorates. However, as paralleling units leads to higher, not lower, power, it is the division of units and, consequently, the capacitance consideration that is the pertinent one.

The input admittance of an arbitrary device may often be put in the general form

$$Z = R + jw(L + L_0) + \frac{1}{jw(C + C_0)} \tag{2-55}$$

where L_0, C_0 = inductance and capacitance characteristic of the encapsulation

L, C = inductance and capacitance characteristic of the device itself

R = a resistance characteristic of the device

For very small L, Eq. (2-55) reduces to an RC circuit equation with a time constant

$$\tau = R(C + C_0) \tag{2-56}$$

As a particularly simple example, consider a tunnel diode. In a tunnel diode, R is the absolute value of the negative resistance, C is the junction capacitance, C_0 is the case capacitance in parallel with the junction capacitance, L is the series inductance in the diode, L_0 is the series inductance in the diode leads. For a typical high-speed low-current ($I_p = 1$ ma) tunnel diode the values may be

$$R = 100 \text{ ohms}$$
$$C = 1.2 \text{ pf} \qquad C_0 = 0.7 \text{ pf}$$
$$L = 400 \text{ ph} \qquad L_0 = 50 \text{ ph}$$

Then
$$\tau' = 1.9 \times 10^{-10} \text{ sec}$$

For frequencies $\leq 5 \times 10^9$ cps, the inductances may be neglected. Here it is clear that dividing the unit into, for example, ten smaller units will result in lower I_p/C ratio and lower switching speed of the ten individual units compared to the original one,

$$\tau'' = 10R\left(\frac{C}{10} + C_0\right) = 8.2 \times 10^{-10} \text{ sec}$$

We may state as a conclusion to this argument that a reduction of a device, which is, in effect, a division, is useful for reducing power dissipation, but only to a limited extent, namely, until the encapsulation limits the frequency response. Any further such reduction of the size of the device reduces simultaneously the high-frequency limit. It is suggested from the example given that ultimately, when $C \ll C_0$, a reduction of power ten times requires a sacrifice in frequency ten times, or

$$f \propto p \tag{2-57}$$

where f is the maximum operating frequency, and p is the power dissipation per unit.

Change of Construction Other than Paralleling. Let us now consider changing an active device, such as a transistor or a tunnel diode, by other means than paralleling, so that the power dissipation is reduced. Such a change may be a reduction in the doping of the tunnel diode, lowering the peak current. However, for maximum speed, the doping has been brought to its physical upper limit in the construction of the diode, and any reduction in the doping will lower the I_p/C ratio and, consequently, the speed of the diode at the same time that the power dissipation is reduced. Similarly, because of the optimization in the construction of the device for high speed, other changes will work the same way.

Change of Operating Point. Another possibility for reducing the power dissipation is through change in the operating point of a device without any change in its construction. For example, by increased bias, the current in a unipolar transistor may be reduced. When this is done, the transconductance of the device is reduced while the input capacitance is nearly unchanged, and therefore, the g_m/C ratio is reduced and, consequently, also the speed of the device.

Another way of stating the same thesis is that, given a certain amount of power to use on a particular device type, increased frequency can be obtained only at reduced output power (and therefore increased losses). This may be expressed as [27]

$$P_{\text{out}} Z f_{\max}^2 = K^2 \tag{2-58}$$

where P_{out} = maximum output power
Z = impedance of device
$f_{\max}$ = maximum frequency of operation
$K = E_b v_{\max} = I_{\max}/C_{\text{barrier}} (= 3 \times 10^{11}$ volts/sec or 300 ma/pf)
E_b = avalanche breakdown voltage
$v_{\max}$ = thermal velocity ($5 - 10 \times 10^6$ cm/sec at room temperature)

Excess Heat Dissipation at High Speed. Although it is true that one of the means of obtaining reduced heat dissipation is a reduction of frequency, the inverse is also true, namely, that increased frequency response can be obtained at a sacrifice in excess heat dissipation.[20] This problem which is of interest in microelectronics, particularly when ultimate speed is required and heat is limiting the packing density, will be illustrated by the following two examples.

Consider first a logic stage, or an amplifier stage, using unipolar transistors, such as shown in Fig. 2-30. The upper-frequency limit of such

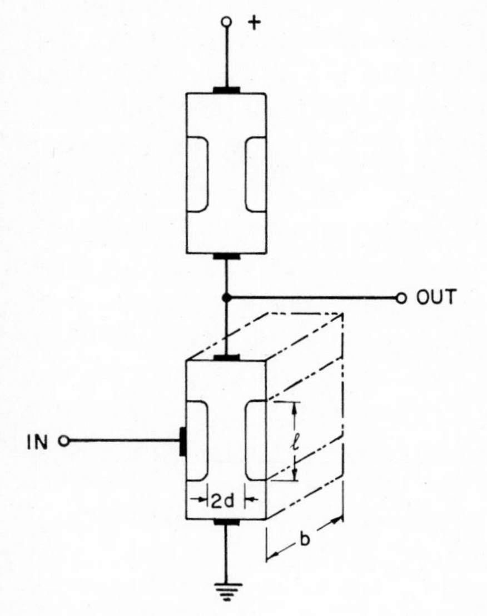

FIG. 2-30. Logic stage using unipolar transistors. Bottom unit is active, top unit passive.

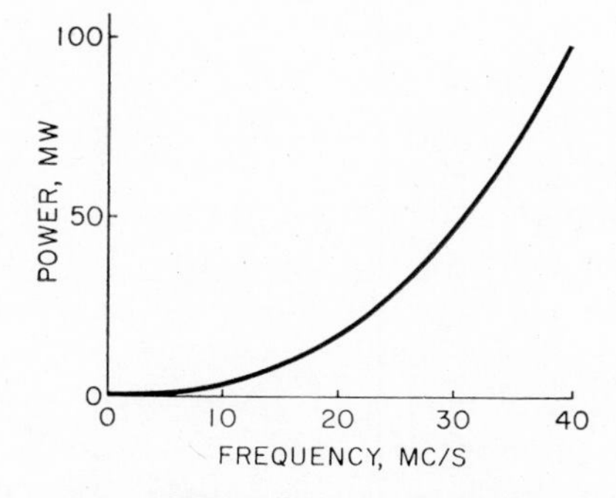

FIG. 2-31. Power dissipation versus frequency for the stage in Fig. 2-30, with silicon transistors with l = 0.010 in., b = 0.040 in., and ρ = 1 ohm-cm.

a stage, as is well known, is set by the RC time constant, where R is the channel resistance and C is the input gate capacitance.

$$f = \frac{1}{RC} \tag{2-59}$$

The d-c power dissipation of the stage P is set by the sum of two terms of the form

$$P = \frac{V^2}{R} \tag{2-60}$$

where V is the voltage applied from source to drain. Expressing R and C in dimensions shown in Fig. 2-30, we have

$$R = \rho \frac{l}{db} \tag{2-61}$$

$$C = \epsilon \frac{lb}{d} \tag{2-62}$$

where

$$d = (2\epsilon\rho\mu V)^{1/2} \tag{2-63}$$

$$f \propto \frac{2V}{l^2} \tag{2-64}$$

$$p \propto \frac{V^{5/2}}{l} \tag{2-65}$$

When a device has been designed for best possible frequency performance, further improvement can come only from increasing V. However, this causes the power to increase further, as plotted in Fig. 2-31, which shows power versus frequency with voltage as parameter for a practical case.

The second example of excessive heat dissipation at high speed is a flip-flop circuit for which the results are shown in Fig. 2-32. Here the

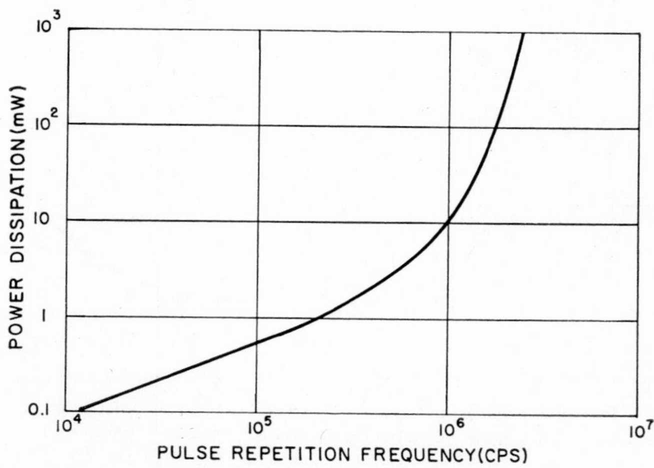

FIG. 2-32. D-c power dissipation versus pulse-repetition rate for a typical flip-flop stage.

d-c power dissipation is plotted versus pulse-repetition frequency. Below about 500 μw, the circuit is severely speed-limited because of drop-off of current-transfer ratio and therefore very low collector current. Between 500 μw and 50 mw, the speed is governed mainly by circuit *RC* values. Increasing the power above 50 mw, on the other hand, does not allow significant increase in speed because the transistors used are already operating at their upper-frequency limit, 2×10^6 cps.

2-4. SHRINKAGE IN FABRICATION OF INTEGRATED CIRCUITS

One of the most important considerations in the design of integrated circuits is component failures in fabrication.[28] The reason for this is that, as a rule, each of the components in an integrated circuit must function properly. Thus, the rejection of a few inferior or faulty components, which is possible when individual components are assembled, is seldom possible in integrated circuits, and the entire unit may have to be discarded.

In general, components may fail either (1) before actual use—in fabrication—or (2) in use. These two cases must be considered separately because the failure rates—the shrinkage—is completely different for the two. A very reliable device with low failure rate in use, say 0.01% failures in 1,000 hours, may be very difficult to fabricate so that, for example, 60% of the devices made have to be thrown away as shrinkage.

The analysis to follow is quite general and applies to all integrated structures, be they a few semiconductor components in the same enclosure, a plastic-encased subassembly, a ferrite memory sheet, or a superconductor array. The common feature is that several components have been combined into a subassembly, in which the individual components cannot be freely replaced.

For the purpose of comparison between individual units and integrated circuits, shrinkage may fall in two main categories. The first category is shrinkage caused by uncorrelated accidents. Here the chance that a unit is lost is independent of what happens to other units. Such accidents may be the breaking of a lead, dropping a unit on the floor, contaminants on a unit, etc. The second category is shrinkage caused by correlated accidents. Here the fate of one unit is the same as, or related to, that of other units. Such accidents may be the dropping of a batch on the floor, faulty raw material affecting many units, equipment breakdown, etc. Shrinkage falling in the second category affects individual units and integrated circuits equally in the limit when all units in a batch are useless, or when all units in a batch are good, and will be neglected as a favorable case giving no excess cost of integrated circuits. Shrinkage falling in the first category, however, represents the worst possible case with regard to integrated circuits, and will therefore be considered in detail. In reality, shrinkage is a compromise between these two extremes. The fewer the fabrication steps are for a particular circuit, the higher the correlation between failures. For most practical circuits, the number of fabrication steps is considerable, which tends to reduce the correlation.

1. Yield of Integrated Circuits

Consider now as an example an integrated circuit consisting of N identical components. Let the shrinkage in fabrication of individual

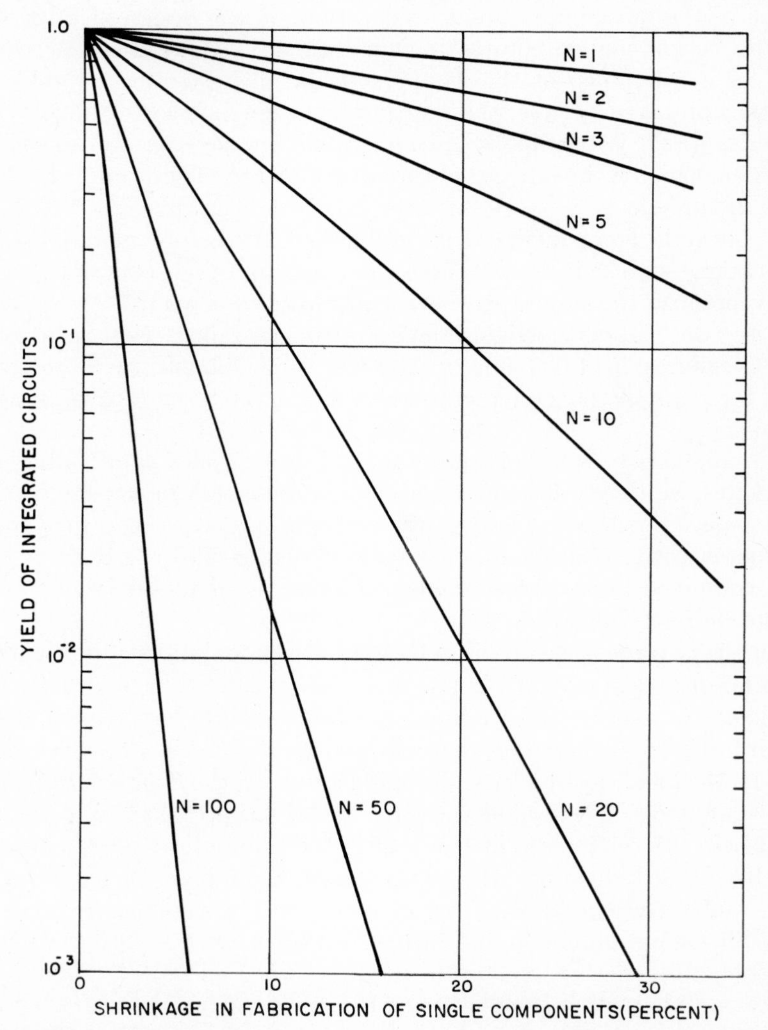

FIG. 2-33. Yield of integrated circuits with N components versus shrinkage in fabrication of components.

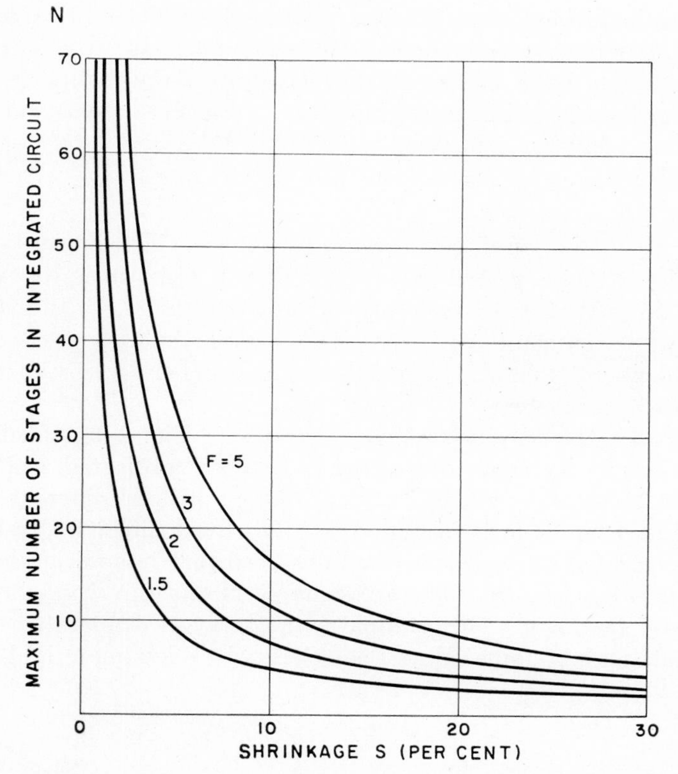

FIG. 2-34. The maximum number of components in an integrated circuit versus shrinkage in fabrication of components. The cost factor F is parameter.

components be S, where S is the ratio of the number of unusable components to the total number made. A typical figure might be $S = 0.17$, meaning that 17% of the fabricated components do not meet specifications. Then the yield Y_i of complete integrated circuits, assuming no correlation, is

$$Y_i = (1 - S)^N \tag{2-66}$$

For a comparison, the yield of circuits Y_s, using separate components which can then be replaced when found faulty, and assuming the same shrinkage in fabrication of individual components, is

$$Y_s = 1 - S \tag{2-67}$$

In Fig. 2-33, Eq. (2-66) has been plotted for some values of practical interest. As may be seen, the yield of integrated circuits for anything but a small number of components and a small shrinkage is extremely low.

Because of the low yield, integrated circuits tend to cost more in fabrication. To indicate how much more, a cost factor, F, may be used, defined as the ratio of the cost of an integrated circuit and the cost of an assembly of separate components. In a practical case, the increase in cost represented by F has to be weighed against the reduction in cost inherent in the use of integrated circuits. From Eqs. (2-66) and (2-67),

$$F = \frac{Y_s}{Y_i} = (1 - S)^{1-N} \tag{2-68}$$

For a particular application, such as satellite instrumentation, the advantages of an integrated circuit may allow a cost that is, for example, 1.5 to 5 times that of an assembly of individual units. The maximum number of components that may then be economically used in each integrated circuit is shown in Fig. 2-34. This number decreases extremely rapidly with shrinkage.

How Eq. (2-68) may be used is illustrated in Fig. 2-35, which shows four versions of the same 10-stage shift register, fabricated with a constant cost factor $F = 1.5$, but with shrinkage in four different ranges.

Correlated Faults in Fabrication of Integrated Circuits. In fabrication of integrated circuits, it is often observed that in addition to statistically distributed faults, there are some parts of a run where faults are particularly frequent. For example, in an array of evaporated circuits, more faults may be found along the edges, where less material has been

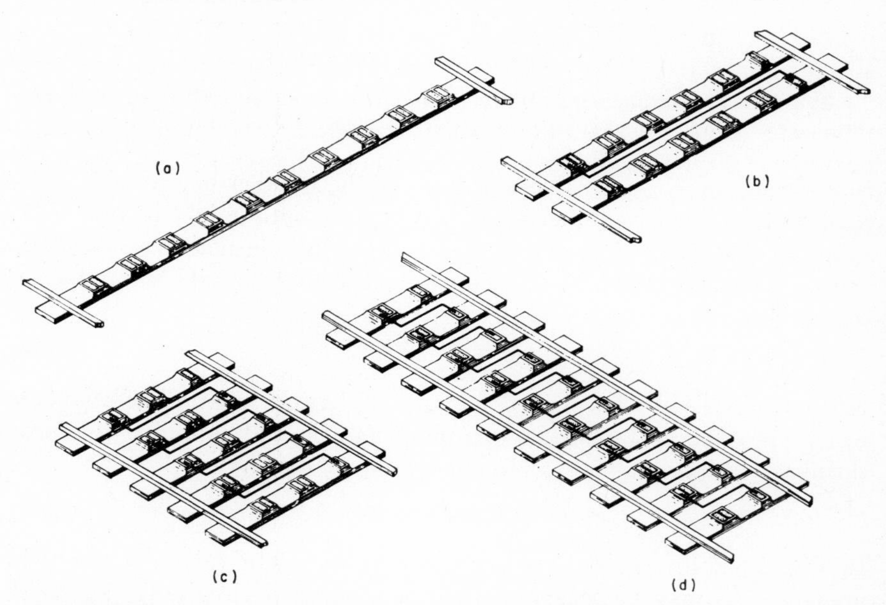

FIG. 2-35. Four versions of a 10-stage shift register fabricated with $F = 1.5$: (*a*) $S < 0.04$, $N = 10$; (*b*) $0.04 < S < 0.10$, $N = 5$; (*c*) $0.10 < S < 0.33$, $N = 2$; (*d*) $S > 0.33$, $N = 1$.

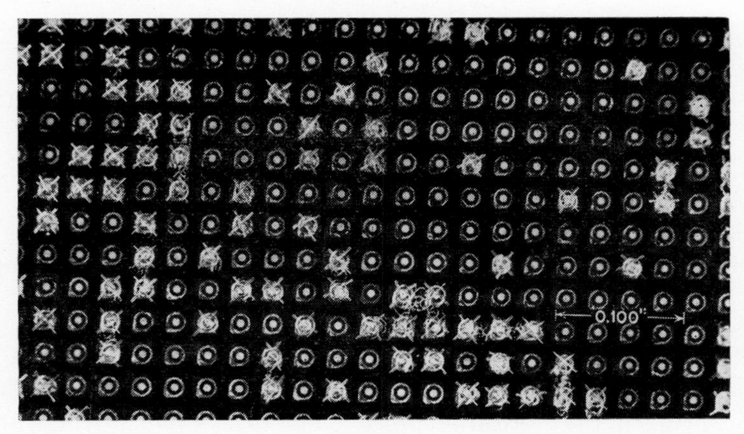

FIG. 2-36. Silicon wafer with transistors, some of which are faulty and therefore have been marked with scratched lines.

evaporated, than in the center. Faulty areas may also appear away from the edges, for example, in regions which have been imperfectly cleaned, and may be large or small, depending on circumstances. In the extreme case, when every unit is a failure in a certain area, this part of the wafer may usually be discarded at an early stage and need not concern us here.

An example of such correlated shrinkage (in distinction to uncorrelated or randomly distributed shrinkage) is given in Fig. 2-36, which shows a silicon wafer with a large number of transistors which have not yet been cut apart. These transistors have been measured individually, and the units with characteristics falling outside certain specifications have been marked by scratch lines across the transistor. As can be seen, one area of the wafer has close to 100% faulty units, while another area exhibits more or less random distribution of failures.

In such cases the yield Y is obtained from

$$Y = \sum_{i=1}^{a} \frac{A_i}{A_{\text{total}}} (1 - S_i)^N \tag{2-69}$$

where A_{total} = total number of units
A_i = number of units in an area with shrinkage S_i
N = number of components in the integrated circuit
a = number of areas

It is assumed that the total area may be divided into smaller areas, each with randomly distributed faults, but with different densities of such faults. For one area Eq. (2-69) reduces to Eq. (2-66). For two areas, one with A_1 components, A_1S_1 of which are faulty, and the other with

A_2 components, A_2S_2 of which are faulty, the yield is

$$Y = \frac{A_1}{A_1 + A_2}(1 - S_1)^N + \frac{A_2}{A_1 + A_2}(1 - S_2)^N \tag{2-70}$$

For $1 \approx S_2 \gg S_1$, the second term may be neglected, corresponding to rejection of the second part of the wafer.

Because of the importance of low shrinkage (high yield) in the fabrication of integrated circuits, some fundamental ways of increasing the yield will now be considered.

2. Yield Increase by Improved Fabrication

The most straightforward way of increasing yield is by continuous improvements in the fabrication process. In this section, it will be shown that a higher investment in the fabrication process is justified for integrated circuits compared to individual components. Let C, which is the cost per component of fabricating individual units and integrated circuits, consist of two parts so that

$$C = C_1 + C_2 \tag{2-71}$$

Here C_1 represents investment in the fabrication process, which is independent of the number of units made. This cost includes research and development, acquisition of machines, improvements in the process, automation, etc. C_2 represents processing cost, which is directly proportional to the number of units made. This cost includes material and labor. In an exact treatment, Eq. (2-71) should contain weight factors for the two parts, which have been neglected here.

According to Eq. (2-66)

$$C_2 = C_{20}(1 - S)^{-N} \tag{2-72}$$

where C_{20} is the cost per component with no shrinkage.

For C_1 let us make the simple assumption

$$C_1 = C_{10}(S^{-1} - 1) \tag{2-73}$$

where C_{10} is a proportionality constant which depends on over how many circuits the fixed cost can be distributed. According to Eq. (2-73), it is possible to reduce S arbitrarily by spending on C_1 as is found in practice.

Equations (2-71) to (2-73) have been plotted in Fig. 2-37 for two values of N, the number of integrated circuits. Equations (2-72) and (2-73) are shown as broken lines, Eq. (2-71) as solid lines.

The total cost per component for single circuits shows a minimum, which is the optimum point for fabrication of conventional devices. At this point, corresponding to a shrinkage of approximately 30% and a

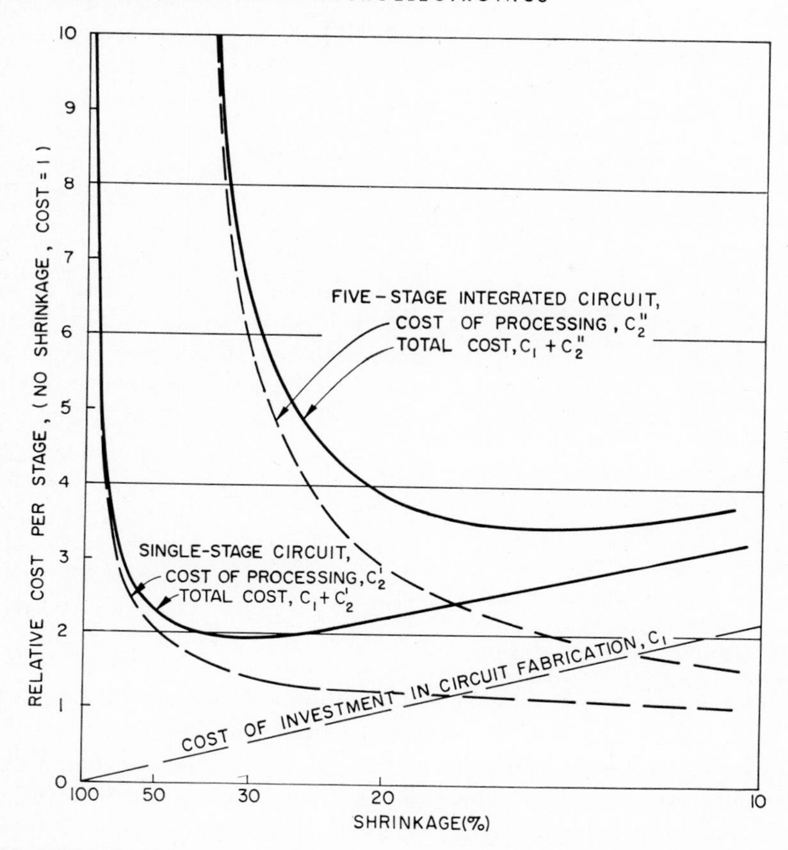

FIG. 2-37. Relative cost per component for integrated circuit versus shrinkage in fabrication of components.

fairly low investment, the cost of a five-component integrated circuit is very large. Increasing the investments in the fabrication method, however, sharply reduces the total cost for the integrated circuit until a minimum, representing the optimum point for fabrication of integrated circuits, is reached at a higher investment and lower shrinkage. Although this mathematically exact treatment represents an oversimplification of the practical situation, it is qualitatively correct.

3. Yield Increase by Redundancy

Passive Redundancy. A very attractive way of increasing yield is to use redundancy in one form or another, i.e., to supply extra components which ensure that a particular circuit function is performed, even if one or a few components are faulty. This is particularly pertinent in integrated circuits, where the cost of additional components may be

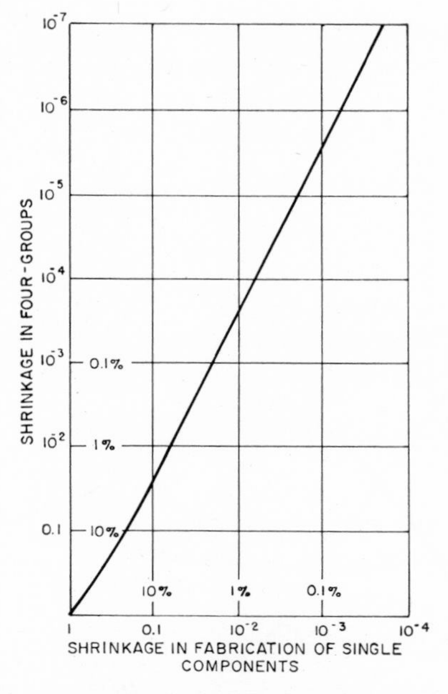

FIG. 2-38. Shrinkage of redundant four-groups versus shrinkage in fabrication of single components.

low. First, consider passive redundancy (see Sec. 2-5) of first order; i.e., each individual component is replaced by a four-group of components, either in a majority logic arrangement, or in a series-parallel arrangement. However, circuits are not redundant. In this case, the reduction of shrinkage with redundancy is moderate at high shrinkage, and becomes appreciable only when the shrinkage is low already to start with. This is illustrated in Fig. 2-38, which has been computed for the series-parallel four-group described in Sec. 2-5. The improvement in shrinkage is only a factor of 1.5 when S is 30%, but a factor of 40 when S is 1%. Against this gain must be weighed the cost of four times as many components.

Passive redundancy will be treated more fully in Sec. 2-5.

Active Redundancy. In this section, the case of reserve components, which may be connected instead of other components which have been found faulty, will be considered. As an illustration, consider an integrated circuit with N components. The yield of circuits with n good components, out of the total number N, is obtained from

$$Y_n = \binom{N}{n}(1 - S)^n S^{N-n} \tag{2-74}$$

where S as before is the shrinkage in fabrication of single components, and

$$\binom{N}{n} = \frac{N!}{(N - n)!n!}$$

Equation (2-74) is plotted in Fig. 2-39 for some values of practical interest and $N = 10$. If the fabrication was aiming at 10-component integrated circuits, the yield at a shrinkage of 20% would be only 11%. However, if instead the fabrication was aiming at 8-component circuits and if all the components could be measured and only good ones connected, the yield at 20% shrinkage would be

$$Y = Y_8 + Y_9 + Y_{10} = 31 + 27 + 11 = 69\%$$

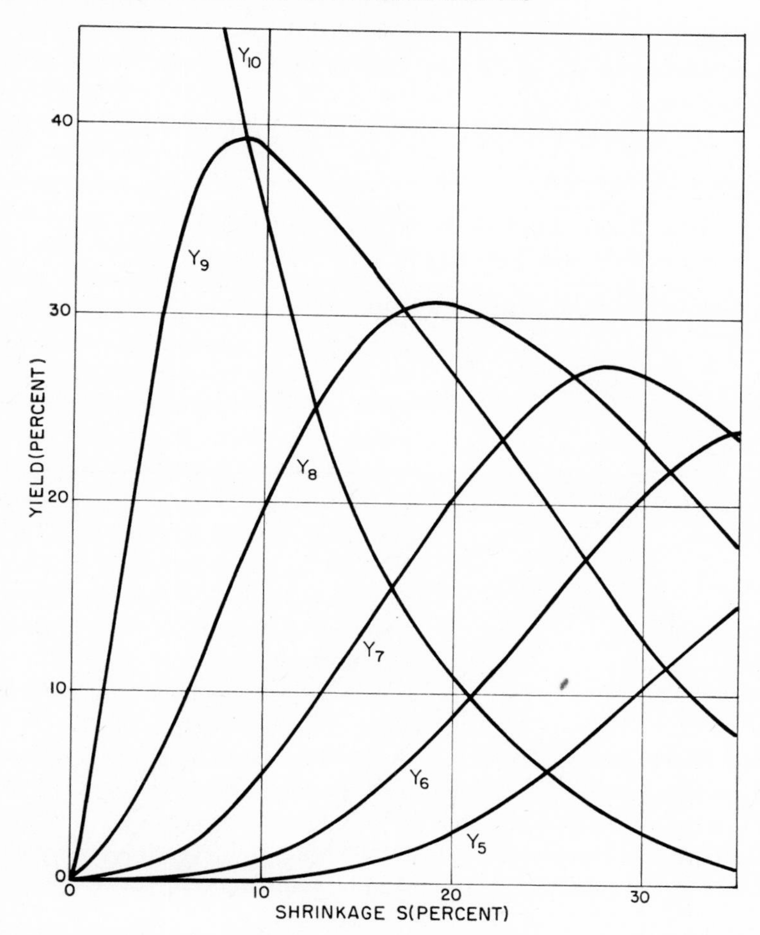

FIG. 2-39. Fractional yield of integrated circuits with 10, 9, 8, etc., good components out of 10 versus shrinkage in fabrication of components.

Therefore, reserve components are very useful as a means of increasing yield. The price to be paid for this increase in yield is an extra measurement of the components before connections are made, the cost of the redundant components, and the cost of connecting them.

Let us consider a practical example, namely a 10-stage integrated shift register such as shown in Fig. 2-35*a*. In this circuit, a faulty stage may be bypassed but at the cost of one extra stage (the two stages, one on either side of the faulty one, are paralleled, and together form one new stage). Then we may initially fabricate the circuit with 12, or perhaps 16, stages so that a maximum of one, or three, stages may be bypassed. Let us assume that this bypassing also is subjected to shrinkage. For simplicity, we will assume the same shrinkage as in the fabrication. Then, according to Eq. (2-74), the yield of usable circuits from a 12-stage lot is

$$Y = Y_{12} + Y_{11}(1 - S) = (1 - S)^{12}(1 + 12S) \tag{2-75}$$

and from a 16-stage lot

$$\begin{aligned} Y &= Y_{16} + Y_{15}(1 - S) + Y_{14}(1 - S)^2 + Y_{13}(1 - S)^3 \\ &= (1 - S)^{16}(1 + 16S + 120S^2 + 560S^3) \end{aligned} \tag{2-76}$$

From a nonredundant 10-stage lot the yield is

$$Y = (1 - S)^{10} \tag{2-77}$$

Relations (2-75) to (2-77) are shown in Fig. 2-40. The improvement by

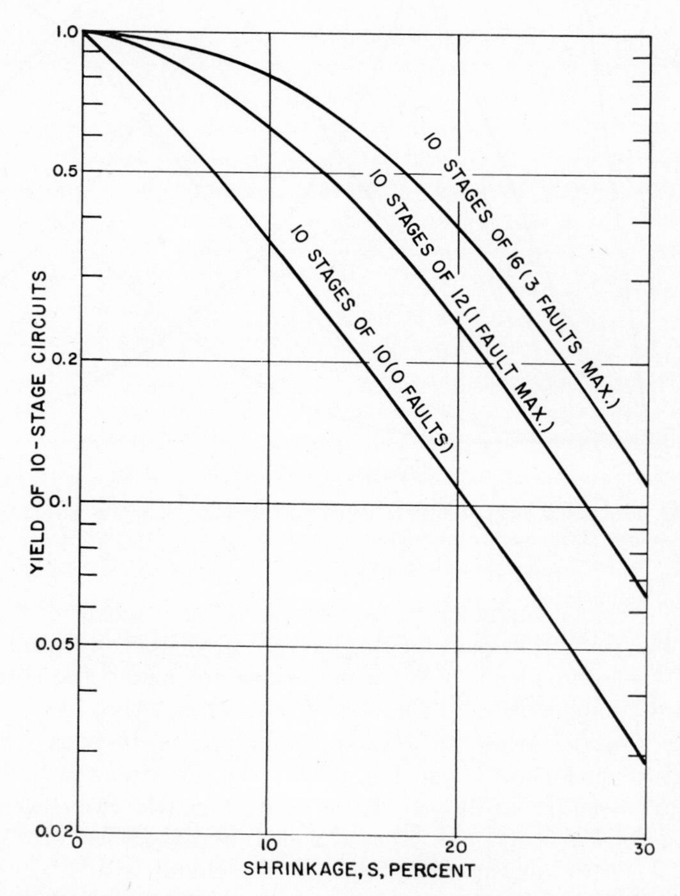

FIG. 2-40. Yield of good 10-stage integrated circuits versus shrinkage in fabrication of single stages, with 0, 2, or 6 redundant stages.

redundancy is about a factor of 2 and increases at larger shrinkage. Most of the benefit of active redundancy comes from correcting one fault. Adding many redundant stages for correction of many faults gives less and less return.

2-5. RELIABILITY CONSIDERATIONS

1. Reliability in Microelectronics

Along with reduced size and weight, reduced cost and reduced power consumption, one of the chief aims of microelectronics is increased reliability.[29] A reason for this is that two of the chief potential application areas of microelectronics are in space equipment, where reliability is of prime importance, and in large systems, where reliability is of importance because of the large number of components involved. Another reason is that the increased component densities which are possible by microelectronic methods lead to circuits and systems which are essentially nonrepairable, as analyzed in more detail in Sec. 2. Therefore, reliability is of crucial importance to the future of microelectronics.

Although most of the considerations of reliability for microelectronics are identical with those for conventional electronics,[30–34] some considerations specific to microelectronics should be indicated. The following six categories have been advanced to support claims for higher reliability in microelectronic circuits:

1. *Smaller size* of microelectronic circuits allows better protection, within constant weight and volume against hazards. The *strength-to-weight* ratio is increased.
2. *Fewer connections of dissimilar materials* have to be made in integrated circuits; therefore, there are less connection failures.
3. There is *lower cost* for microelectronic circuits, which may be traded for higher reliability by product improvement.
4. *Coupling between components is stronger* in microelectronic circuits, thermally, mechanically, chemically, and electrically, because of their close proximity to each other; therefore devices tend to exhibit identical changes in characteristics.
5. There is *less handling of components* in fabrication of integrated circuits; therefore, more uniform product.
6. Use of *redundancy is less costly* in integrated circuits than in conventional device assemblies. Redundancy will be treated more fully in Sec. 3 below.

It should be made clear that although these six categories represent good reasons why reliability of microelectronic circuits should be superior to conventional circuits, at the present time there is a long way to go to realize all these potential advantages.

Against these factors favorable to reliability in microelectronics, a number of factors detrimental to reliability should be weighed. The most important are:

1. Small components may be *more difficult to fabricate* than larger ones and, therefore, have more flaws. This will set an optimum size well above the minimum analyzed in Sec. 2-2. For sizes larger than this optimum, the objection may not be true.

2. *Small production volume* means less perfection in fabrication, and thus poor reliability, because sufficient engineering may not be possible within the given cost limits.

3. *Strong coupling* between elements, though favorable in some respects, is unfavorable in others, as explained below. A catastrophic failure would spread.

4. Small size means *small inertia,* thermal, mechanical and electrical, and therefore sensitivity to dynamic overload, either thermal, mechanical, or electrical.

In addition, two more factors which are less fundamental are:

5. *New constructions* may mean poor reliability initially until sufficient engineering has been done. This, of course, is a temporary drawback, but it is the main reason why microelectronics in its most advanced form has not yet found its way into space applications on any appreciable scale.

6. Excessive packing density may lead to *higher operating temperature* and, therefore, shortened life of components. However, higher operating temperature is not a necessary feature of microelectronics.

To evaluate more fully some of the considerations listed above, the following comments should be made.

Smaller Size. Many of the failures in electronic circuits are induced by high temperature, mechanical shock, chemical attack, etc. The smaller the circuit, the more effective the defense against these hazards can be made without increasing the over-all size. This defense may consist of making the structure more sturdy, or of providing improved encapsulation.

The weight of a solid body is proportional to the third power of its dimensions, whereas the strength is proportional to the second power. Therefore, reducing the size increases the strength-to-weight ratio.

Fewer Connections of Dissimilar Materials. Connections of dissimilar materials, in contrast to connections of identical materials, suffer from:

1. Thermal expansion differences which exert mechanical stress and may break the materials apart immediately, or after repeated temperature cycling, sometimes after considerable time.

2. Phase changes through reaction in the solid state, leading to a new stress situation and new and perhaps reduced strength.

3. Chemical reactions from the residues left in preparing the materials for the joint: fluxes, surface etches, etc.

4. Abuse in the joining operating, reducing the strength and electrical performance of the joint. Such abuse may be heating during soldering or welding; deformation during welding, crimping, or clamping; etc.

5. Many joining materials are chemically or physically less stable with time, temperature, or voltage; are less inert than germanium, silicon, tantalum, nichrome, etc. Solder melts, silver migrates on the surface, aluminum corrodes, etc.

Coupling between Components. As in microelectronic circuits, particularly in integrated circuits, the distance between components has been reduced to a minimum; the coupling between components—thermally, chemically, mechanically, and electrically—is very strong. By strong *thermal coupling* is meant that the temperature of one unit is closely related to the temperature of other units because of the good thermal contact between them. In similar manner, chemical coupling means that internal and external chemical factors are closely related for many units. Such chemical factors may be impurities on the surface, ambient gases in the encapsulation, moisture content, etc. Mechanical coupling means that an evaporated layer may peel off from unit to unit, stress may lead to a crack forming from unit to unit, etc. Finally, *electric coupling*, disregarding in this context the electric coupling involved in the circuit itself, may involve pickup of unwanted signals, defective grounding of shields, etc.

Coupling between components can be both an advantage and a disadvantage. It is quite clear that in the processing in *fabrication* of the units, extremely high correlation is desirable to ensure a uniform product. In *use*, strong coupling is desirable for moderate deviations from the normal. If, for example, the temperature of a component for some reason just exceeds the tolerance limit, so that the circuit ceases to function, the operation can often be restored if other components also change in an identical fashion. On the other hand, if the deviation from normal is large, say, for example, that a component melts from excessive heat, it is, of course, very undesirable to have such a strong thermal coupling that other components also melt, thereby enlarging a catastrophic kind of failure from one component to very many components. This is particularly undesirable in redundant circuits, since the entire advantage of redundancy would be lost, as will be described below.

The case for or against strong coupling between components must therefore be based on the type of failure that is most probable. If the most probable failure is a small deviation such as a moderate temperature rise, a moderate change in surface conditions, etc., then strong coupling is usually advantageous. If, on the other hand, the most probable failure is catastrophic—a large deviation from normal such as melting of a

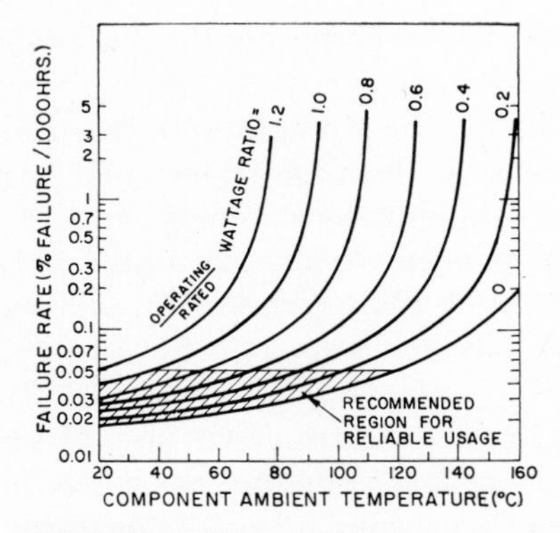

FIG. 2-41. Failure rate versus temperature for film resistor.[35]

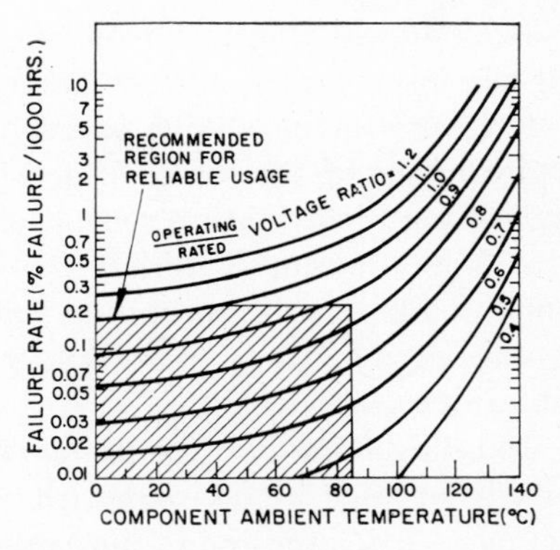

FIG. 2-42. Failure rate versus temperature for mica capacitors.[35]

component, cracking of a crystal, peeling off of a protective layer or contact—then the coupling between units should be small.

Small Production Volume. The substitution of microelectronic circuits for conventional circuits would not change the production volume. However, a particular kind of component, e.g., a 12-kilohm resistor, ½ watt, ±5% tolerance, could no longer be fabricated in one run of, say, 100,000 units. Instead, it would have to be spread out over a number of circuits made on different occasions, 6,000 of one, 8,500 of another, etc. In this sense, the production volume of the particular resistor would be small. That means that more tooling would be required, and since the cost is fixed, the tools cannot be so perfect as in the case of a single run.

Higher Operating Temperature. Excessive packing density may lead to higher operating temperature and, therefore, shortened life of components.[35] The failure rate versus temperature is shown in Fig. 2-41 for film resistors and in Fig. 2-42 for mica capacitors. If mica capacitors are used with an operating voltage which is 80% of the rated voltage at room temperature, the failure rate is approximately 0.06% in 1,000 hr. If the temperature is increased to 75°C, either the operating voltage must be reduced to 70% of rated voltage for the same failure rate, or else the failure rate will increase to about 0.13% in 1,000 hr. For this reason, it is clear that miniaturization must be implemented without increasing the component temperature.

2. Reliability of Nonrepairable Circuits

Discrete component assemblies offer the great advantage that replacement of faulty devices is simple and straightforward. From this point of view, microelectronic circuits are at a distinct disadvantage in that they are essentially nonrepairable. At present, this applies to relatively small aggregates of devices, typically 5 to 20. However, when in the future larger numbers of devices will become integrated into these nonrepairable packages, the reliability problem will be even more acute. This evaluation is analyzed below.

Let us assume a number of components N in an integrated circuit, consisting of components a, b, c, etc. Let the mean time between failure for these components as fabricated in the package be T_a, T_b, T_c, etc. The probability of failure S in a time T is, for each component,

$$S_a = \frac{T}{T_a} \qquad S_b = \frac{T}{T_b} \qquad S_c = \frac{T}{T_c} \qquad \text{etc.} \tag{2-78}$$

Similarly, the probability of survival P for each component is

$$P_a = 1 - S_a \qquad P_b = 1 - S_b \qquad \text{etc.} \tag{2-79}$$

If the failure rate γ is known instead of the mean time between failures, the probability of survival is

$$P_a = \exp(-T\gamma_a) \approx 1 - T\gamma_a \qquad \text{etc.} \tag{2-80}$$

(if $T\gamma_a \ll 1$). If the failure rate is given in per cent per 1,000 hr,

$$P_a = \exp(-T\gamma_a 10^{-5}) \qquad \text{etc.} \tag{2-81}$$

Then

$$S_a \approx T\gamma_a \qquad \text{etc.} \tag{2-82}$$

The probability of survival P_0 for the integrated circuit is then

$$P_0 = P_a P_b P_c \cdots \tag{2-83}$$

if all the failures are independent.

If the failure probability is small, the resulting failure rate may be obtained to a good approximation by simply adding the failure rates for the components

$$\gamma \approx \gamma_a + \gamma_b + \gamma_c + \cdots \tag{2-84}$$

As an example, consider the circuit in Fig. 2-43, which shows a conven-

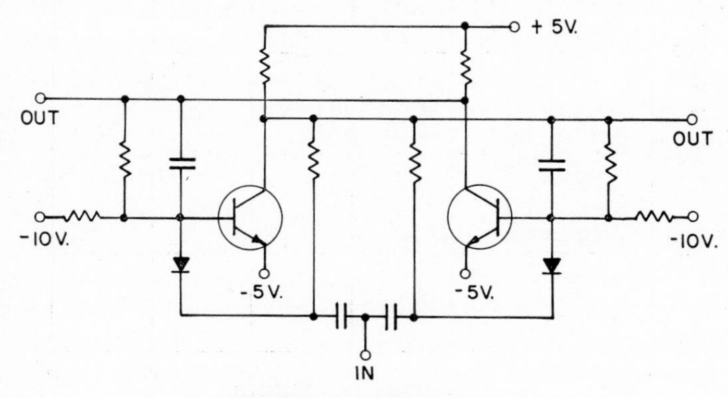

FIG. 2-43. Flip-flop circuit.

TABLE 2-13. TYPICAL FAILURE RATES OF FLIP-FLOP CIRCUIT DEVICES

Components	Number in circuit	Failure rate, % per 1,000 hr
Transistors, silicon	2	0.07
Diodes, silicon	2	0.02
Capacitors, paper	4	0.00125
Resistors, carbon composition	8	0.0043

tional flip-flop circuit consisting of two transistors, two diodes, four capacitors, and eight resistors. Typical failure rates[36] may be obtained from Table 2-13. Then the failure rate for the circuit may be obtained from Eq. (2-84) and is 0.22% per 1,000 hr. The mean time between failures is obtained from Eq. (2-78) and is 4.5×10^5 hr. The corresponding mean time between failures for more complex networks containing a number of such circuits is shown in Fig. 2-44. The probability

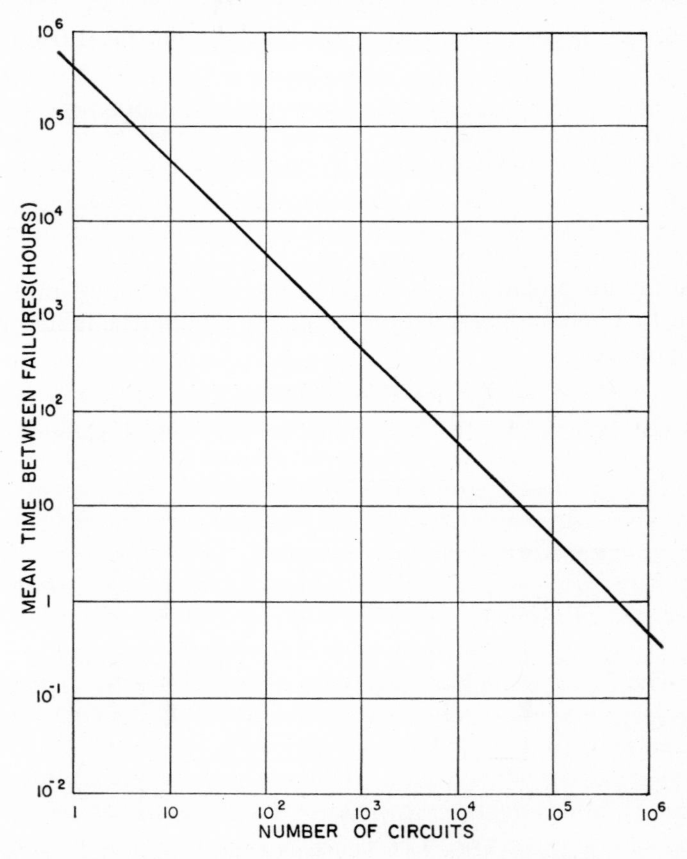

FIG. 2-44. Mean time between failures versus number of flip-flop circuits in system.

that the circuit will fail in one year (8,760 hr) may be obtained from Eq. (2-82) and is 0.019. For a package with 10 times more components, the survival probability after one year [Eq. (2-83)] is 0.82, and, consequently, the failure probability [Eq. (2-79)] is 0.18. In order to fabricate larger integrated circuits, it is, therefore, necessary to utilize methods that give appreciably lower failure rates for the components.

As the reliability of interconnections is an important objective of microelectronics, it is necessary to analyze them like components, measure their mean time between failure, etc. At present, most of the reliability data available are not directly applicable, since what is listed as component failures often includes connection failures—open emitter leads in transistors, broken resistor end-leads, etc.

3. Reliability Improvement by Redundancy

By the use of redundancy[37–40] in microelectronic circuits, two important advantages may be obtained.

First, the *reliability (life) of the circuits may be increased.* This is accomplished by the use of additional components, in such a way that if one or a few components fail, the electronic function of the circuit is taken over by other components. However, actual practical realizations of this principle are as yet few. On the other hand, it is believed that biological systems such as the brain make extensive use of redundancy, and that this is the reason for the reliability of the brain in spite of the unreliability of its components, a large number of which may fail (physical damage, surgical operation, etc.) without seriously impairing the normal functioning of the system.

By the use of redundancy, the *fabrication of microelectronic integrated circuits* is facilitated, in that all components do not have to be perfect. A certain fraction (usually less than 50%) may be faulty or absent at the start. This is of considerable importance, as the cost advantage expected of future microminiaturized systems fabricated with advanced methods of mass fabrication of interconnected components will increase with the number of components integrated. However, at the same time, the problem of shrinkage becomes more serious. In solving this problem, the first and least costly step is usually to improve the device and the fabrication method. However, sooner or later further improvement becomes too expensive, and redundancy is a better alternative.

Classification of Redundant Systems. There are three ways, in principle, in which redundancy may be used in electronic systems, namely, in *time*, in *power*, and in *hardware*. Redundancy in time in, for example, a computer means that, in effect, the computations are repeated several times, perhaps in different ways, and the results are compared for consistency. Another way is to allow more than the minimum number of bits (one) per digit, and therefore consume more time in the computations but with greater safety against errors. Redundancy in time works best against temporary failures.

Redundancy in power means that by using excessive power, the computer may be designed to run faster, thereby allowing more computations with no time sacrifice. This type of redundancy is not very effective except when a system is run below its speed capability to save power.

Redundancy in hardware, finally, is the most common form and, being most applicable to microelectronics, will be treated more fully below.

With respect to organization, redundancy may be classified into *active* and *passive* redundancy.[37] In active redundancy, the faulty circuit or component is disconnected, and a new circuit or component substituted. This requires several active processes: detection of a fault, localizing it, disconnecting the faulty part, and connecting a new part. This is most conveniently done when sufficient time and specialized equipment are available, namely, in the fabrication of the circuit. In passive redundancy a group of circuits or group of components are performing each circuit or component function, and when one or more fail, the rest in the group take over the operation automatically. This often requires a larger amount of redundancy than active redundancy, but functions instantly.

Redundancy may be applied to a system at different levels. Duplicating systems, so that one system may be disconnected when faulty and a spare system connected, represents redundancy applied at a system level. Duplicating components, so that when one component goes wrong, other components can take over the function, represents redundancy on a component level. Between these two extremes, redundancy may be applied on a circuit, subsystem, etc., level.

Applications of Active Redundancy. Examples of active redundancy are not uncommon. Spare components such as fuzes, indicator lamps, transistors, etc., are standard equipment among all users of electronic equipment. Many broadcast stations have reserve transmitters to be used in case the regular transmitter develops a fault, and many commercial airplanes carry reserve transceivers. In advanced microelectronics, this simple form of repair is not possible on a component or even a circuit level, as the microelectronic packages are in effect nonrepairable. However, on higher levels, this form of active redundancy may still be possible. More sophisticated forms of active redundancy may be used in microelectronic circuits in two cases. The first case is in fabrication as described in Sec. 2-4. The second is in repair when a faulty circuit may be disconnected and a spare circuit connected electronically. One such case involving cryoelectric circuits has been described.[41] In this case, the system was divided into a number of parts. Each part was duplicated in a few identical replicas, and a switching network was used to

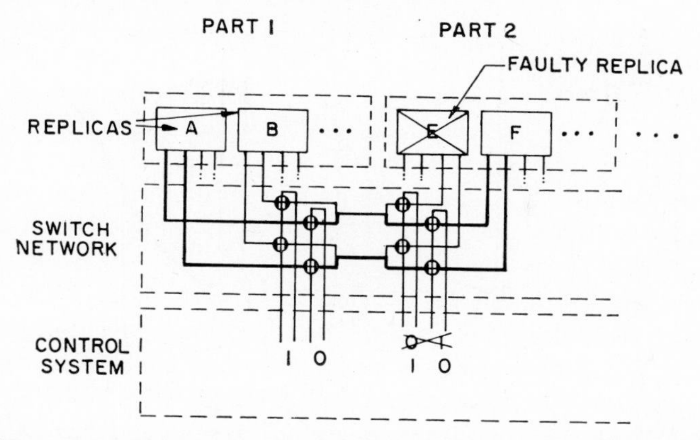

FIG. 2-45. Active redundancy in cryogenic computer. The computer has been divided into a number of parts, each duplicated in a number of replicas. When replica E developed a fault, it was disconnected by a switch network, and instead replica F was connected. The system for detecting the fault is not shown.

connect a replica to the other parts. When this replica developed a fault, the switching network disconnected it and connected another of the replicas. The principle is illustrated in Fig. 2-45, which shows how a tree-switching network may be used to connect A to F instead of to E when the latter has developed a fault. The control system for detecting the fault, localizing it, and steering the switching is indicated.

An exact comparison between active and passive redundancy schemes is not always possible because the reliability of detection, localizing and control mechanisms, often involving human intervention, is difficult to evaluate.

Applications of Passive Redundancy. *Theory.* The use of passive redundancy by the means of four-groups has been analyzed for the case of ideal relays.[42] By ideal relay is meant a relay in which failures occur only in the contacts while failures in the driving coil and in the insulation between input and output may be neglected. Figure 2-46 shows the general principle, namely, how one relay may be replaced by a redundant

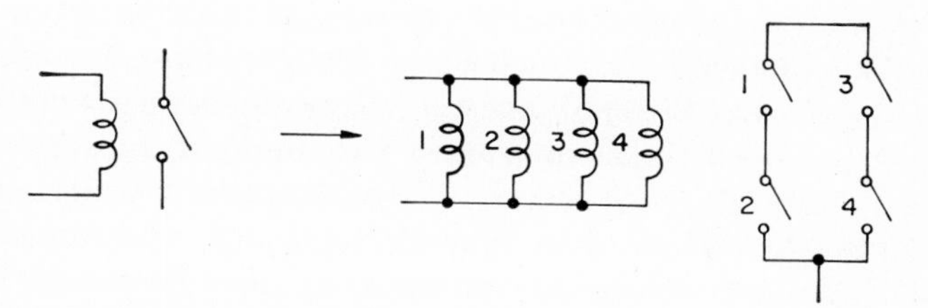

FIG. 2-46. Principle of passive redundancy according to Moore and Shannon.[42] One relay is replaced by four, connected in series-parallel.

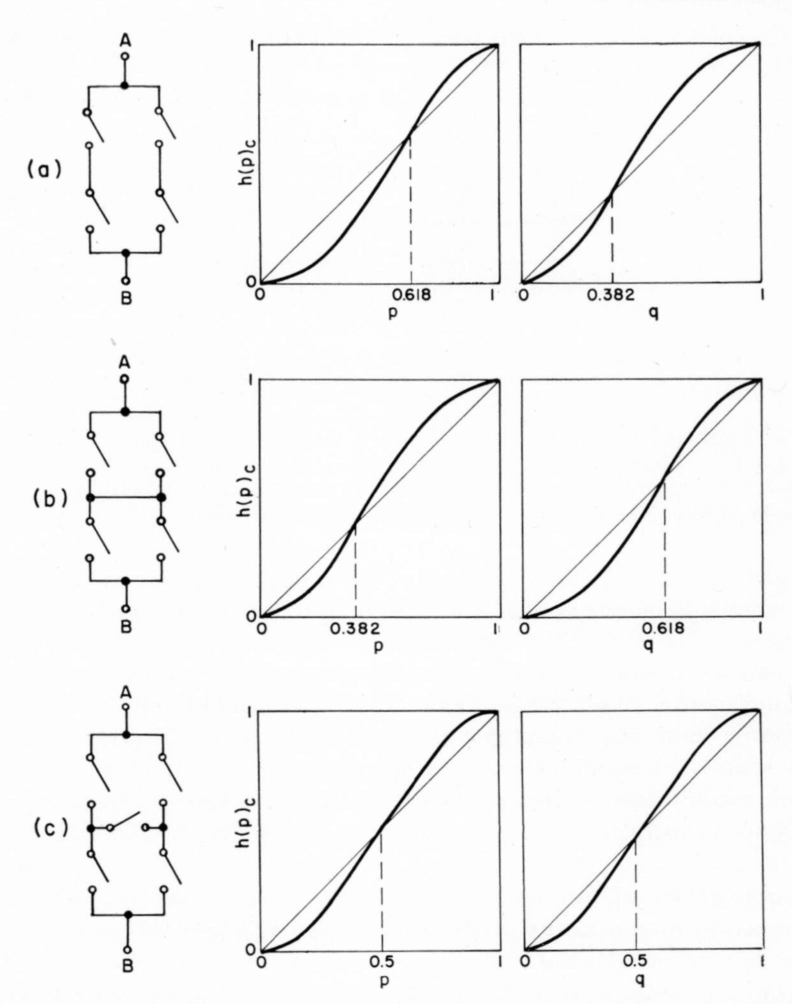

FIG. 2-47. Redundant relay nets according to Moore and Shannon.[42] Also shown is the probability of closure between A and B, when p is the probability that a single relay will close when it should, and q is the probability that a single relay will open when it should.

configuration of four relays in such a way that one, two, or, in some cases, even three relays may be faulty and the four-group yet performs perfectly.

Figure 2-47 shows three elementary redundant relay networks, each replacing a single relay. If p is the probability that a single contact will close when it should close, then the probability $h(p)$ that two contacts in series will close is p^2. The probability that two contacts in series are open is $(1 - p^2)$. Consequently, if two contacts in series are used to connect the path AB, and both are operated simultaneously, the redundancy improves the reliability for opening the path, but reduces the reliability for path closure. If four relays are used as shown in Fig. 2-46, the reliability improves both ways. Then the probability of the path AB being open is

$$h(p)_{\text{open}} = (1 - p^2)^2 \tag{2-85}$$

The probability of the path being closed is

$$h(p)_{\text{closed}} = 1 - (1 - p^2)^2 = 2p^2 - p^4 \tag{2-86}$$

This function is shown at the center in Fig. 2-47a. The diagonal represents the case of no redundancy. For $p > 0.618$, $h(p)_{\text{closed}}$ lies above the diagonal, meaning improved reliability for closure. If q is the probability that a single contact will open when it should, we have the equations

$$h(p)_{\text{closed}} = [1 - (1 - q)^2]^2 = 4q^2 - 4q^3 + q^4 \tag{2-87}$$

This function is shown at the right in Fig. 2-47a, and

$$h(p)_{\text{open}} = 1 - 4q^2 + 4q^3 - q^4 \tag{2-88}$$

The circuit shown in Fig. 2-47b is the dual of the one shown in Fig. 2-47a.

From Fig. 2-47a and b it may be seen that the circuit of Fig. 2-46a is best when the relays fail to open when they should (failures of type "short"), while the circuit in Fig. 2-47b is best when the relays fail to close when they should (failures of type "open").

The third circuit in Fig. 2-47 with five elements has the closure probability

$$h(p) = 2p^2 + 2p^3 - 5p^4 + 2p^5 \tag{2-89}$$

This represents a curve that crosses the diagonal at $p = 0.5$ and is symmetrical around that point. This circuit, therefore, increases the reliability of closing and opening the path AB equally, whereas the symmetrical curves in Fig. 2-47a and b increase reliability for one more than for the other.

These results may be generalized to apply to any complex redundant network between two points. Thus, if there are m contacts in a switching circuit between two points, and if n of them constitute a subset of closed contacts, the probability of path closure is

$$h(p) = \sum_{n=1}^{m} A_n \binom{m}{n} p^n (1 - p)^{m-n} \tag{2-90}$$

where A_n is the fraction of subsets that corresponds to a closed path and $\binom{m}{n}$

is the binomial coefficient defined by

$$\binom{m}{n} = \frac{m!}{n!(m-n)!} \tag{2-91}$$

For example, in the bridge circuit of Fig. 2-47*c* there are five contacts ($m = 5$). The probability that all five are closed is p^5. $A_5 = 1$. If any four close, the path AB will be closed. The probability of four of the five closing is $5p^4(1-p)$. There are five possible subsets. Consequently, $A_4 = 1$. The probability of three of the five contacts being closed is $10p^3(1-p)^2$, but only eight of the ten possible subsets result in a closed path between A and B; hence $A_3 = 0.8$. Similarly, the probability that two of the five contacts are closed is $10p^3(1-p)^2$, but only two of ten possible subsets result in a closed path; thus, $A_2 = 0.2$. If one contact closes, the path AB remains open. Consequently, from Eq. (2-90),

$$h(p) = p^5 + 5p^4(1-p) + 8p^3(1-p)^2 + 2p^2(1-p)^3 \tag{2-92}$$

which reduces to Eq. (2-89).

Practice. In practical applications of passive redundancy, several considerations must be made. The *cost* of redundancy is fairly high,

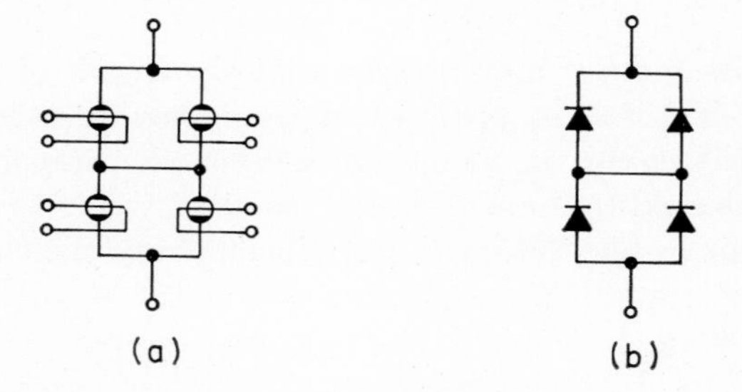

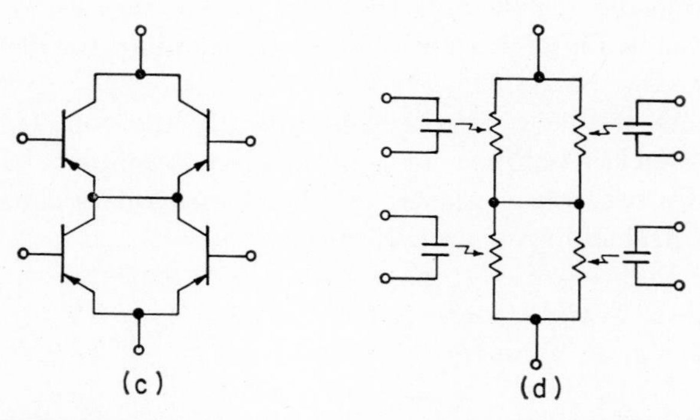

FIG. 2-48. Redundant four-groups using (*a*) cryotrons, (*b*) diodes, (*c*) transistors, (*d*) optoelectronic components.

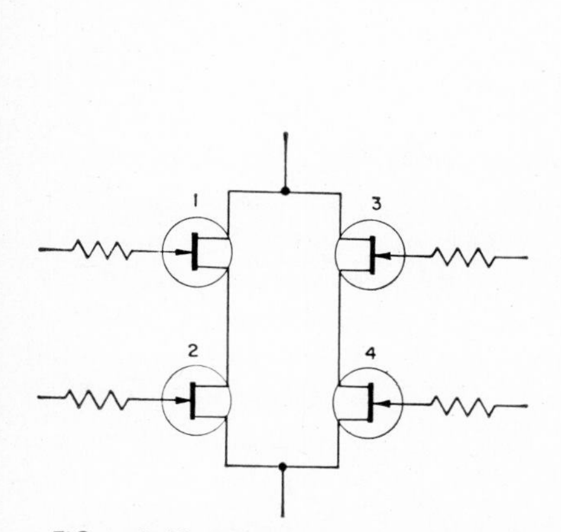

FIG. 2-49. Redundant four-group using unipolar field-effect transistors with resistors protecting against short input-output.

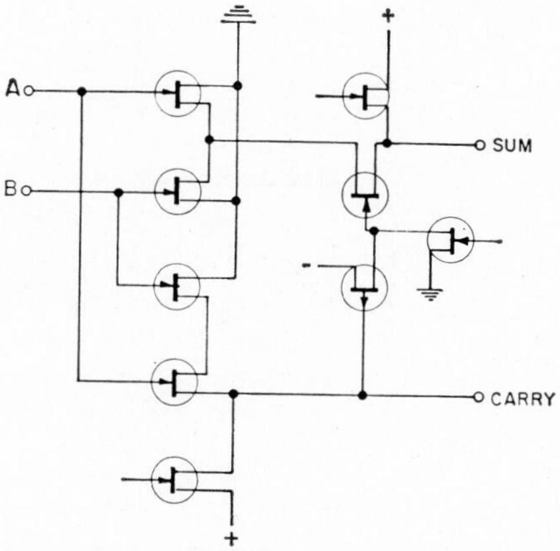

FIG. 2-50. Full-adder circuit using direct-coupled unipolar transistors.

requiring about four times as many components, two times as much wiring, two times the power supply, etc. *Power* goes up proportional to the number of components, both d-c power and power for driving. Many components *differ considerably from ideal relays* and require additional features. Figure 2-48 shows four different four-groups, using cryotrons, diodes, transistors, and optoelectronic components. Only the optoelectronic components are similar to the relays, in that the insulation between input and output is not likely to develop faults. The diodes, of course, do not have a separate input. For unipolar transistors, the insulation can be protected by additional redundancy, as described below. Further, the *circuit design* must tolerate the four-group characteristics differing when zero, one, two, or three elements are faulty. If one branch is open, the impedance will increase, and if two elements in parallel are short-circuited the impedance will decrease.

To illustrate some of these points, let us consider a specific example, namely, unipolar field-effect transistor logic circuits. Unipolar transistors, like conventional bipolar transistors, differ from ideal relays in that the insulation between input and output cannot be considered fault-free. Indeed, in some cases it may well represent the most common source of faults. To protect against this possibility, a four-group takes the form shown in Fig. 2-49. Here each input gate electrode has a series resistance which is small compared to the insulation resistance normally present between input and output, and yet large enough to prevent appreciable current from passing if a short circuit should develop. A typical value may be 10^5 ohms.

Figure 2-50 shows a circuit using direct-coupled unipolar transistors, in

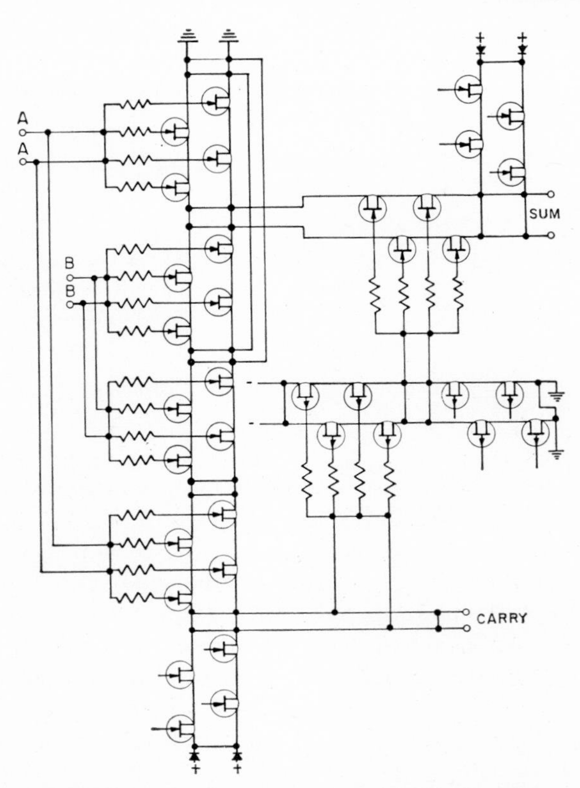

FIG. 2-51. Redundant version of the circuit in Fig. 2-50.

this case a full adder. Figure 2-51 shows the same circuit with redundancy applied. Note that the redundancy extends to the wiring and the power supplies. Fourfold redundancy in the components and the power supplies to protect against open and short circuits corresponds to twofold redundancy in the wiring to protect against open circuits (short circuits do not appear as the wire itself constitutes a short).

To compute the gain in reliability, the following assumptions will be made:

1. Failures completely random and uncorrelated
2. Short failure and open failure, equally probable
3. Input failure and output failure, equally probable
4. Failures in resistors and failures in transistors, equally probable

A transistor short circuit is defined as any change causing too large a current to flow, including aging, heating, etc. A transistor open circuit is defined as any change causing too small a current to flow. The median time to failure T_0 (probability of survival = 0.5) for the nonredundant system is

$$T_0 = \frac{\ln 2}{Nf_0} \tag{2-93}$$

where N is the number of components in the nonredundant system, and f_0 is the failure rate, fraction failing per hr.

As an example, for $N = 10^5$, $f_0 = 10^{-7}$, we obtain $T_0 = 69.5$ hr. The median time to failure T for the redundant system is

$$T = \frac{0.357}{fN^{1/2}} \tag{2-94}$$

At this time, the number of failing components n_T is

$$n_T = 2.86N^{1/2} \tag{2-95}$$

In a nonredundant system, this number would, of course, be one.

A figure of merit B is obtained by the ratio of the median times to failure for the redundant and the nonredundant system, and is

$$B = 0.515N^{1/2} \tag{2-96}$$

Table 2-14 and Fig. 2-52 summarize the results for some typical values of N.

An interesting result of importance in microelectronics is that the percentage of faulty units in a redundant system, n_T/N, decreases with increasing N. This means that even with redundancy, there is a limit

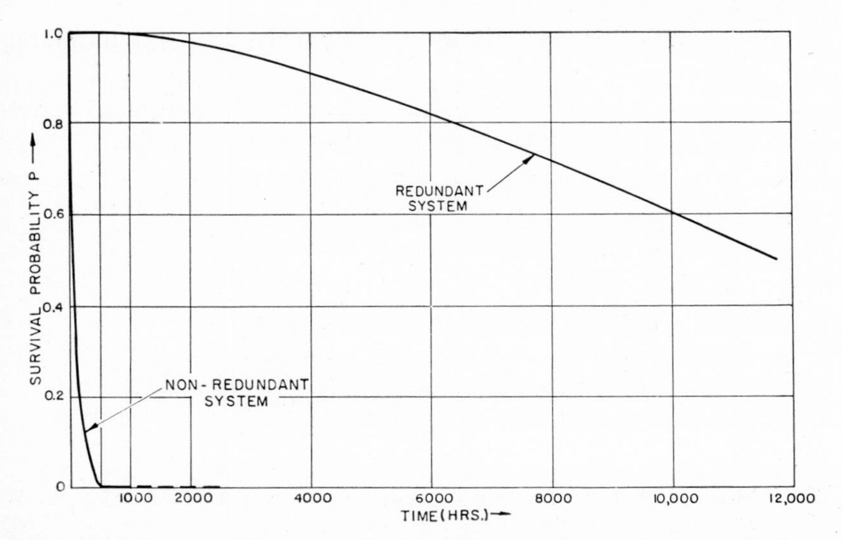

FIG. 2-52. Survival probability of nonredundant and redundant system with $N = 10^5$, $f = 10^{-7}$ per hr.

TABLE 2-14. NUMBER OF FAULTY DEVICES n_T AND IMPROVEMENT IN RELIABILITY IN REDUNDANT SYSTEM B

N	10^3	10^4	10^5	10^6	10^7
n_T	90	286	904	2,859	9,040
B	16	52	163	515	1,630

to the number of elements that can economically be integrated into a nonrepairable package when the shrinkage in fabrication of components is fixed.

The improvement in reliability may be used in two ways. Either all components of a redundant system are operating initially, and the life expectancy of the system is correspondingly long, or many components are defective initially, and yet the life expectancy of the system is at least equal to that of a nonredundant system. In practice, a compromise between these two extremes appears most promising.

Passive Redundancy Using Majority Logic. The application of redundancy to switching systems by a method that is independent of the type of components used has been described by von Neumann.[43] The principle is illustrated in Fig. 2-53. Here the input to a majority logic circuit M is not from one but from three or more identical circuits L. The majority logic circuit is so constructed that its output is determined by the majority of the inputs. Therefore, a minority of the circuits L may be allowed to fail.

The probability that the majority circuit will give a correct output may

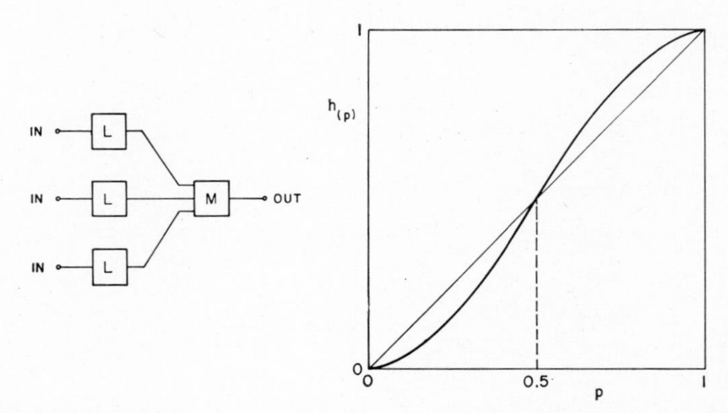

FIG. 2-53. Principle of simple majority logic according to von Neumann.[43] Also shown is the probability of correct output $h(p)$, when p is the probability that each circuit L gives a correct output.

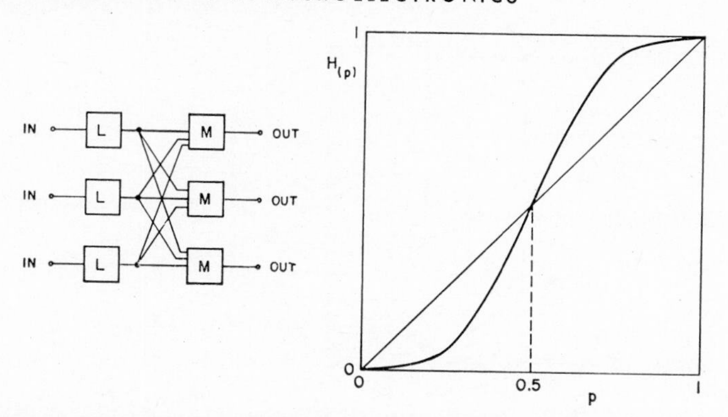

FIG. 2-54. Majority logic with redundancy extended to majority circuit. Also shown is the probability of correct output $H(p)$, when p is the probability that each circuit L gives a correct output.

be expressed mathematically in an equation similar to Eq. (2-90),

$$h(p) = \sum_{n=1}^{m} \binom{m}{n} p^n (1-p)^{m-n} \tag{2-97}$$

where m is the number of inputs to the majority circuit, n of which carry a correct signal, and p is the probability that each circuit L gives a correct output.

For $m = 3$,

$$h(p) = p^3 + 3p^2(1-p) \tag{2-98}$$

The first term on the right-hand side represents the probability that all three inputs are correct. The second term represents the probability that two of the three inputs are correct. This function is shown to the right in Fig. 2-53.

If the majority circuit is not perfect but is also subject to failure it may also be protected by redundancy, as shown in Fig. 2-54. Here any one of the circuits L and also any one of the majority circuits M may fail, and still the probability that at least two outputs are activated correctly, $H(p)$, is

$$H(p) = h^3(p) + 3h^2(p)[1 - h(p)] \tag{2-99}$$

where $h(p)$ is obtained from Eq. (2-98), and is the probability that one majority circuit be activated correctly. This function is shown to the right in Fig. 2-54.

A further development of majority logic is represented by *triangular recursive networks.*[44,45] In these networks the majority action is obtained without the use of a special majority circuit.

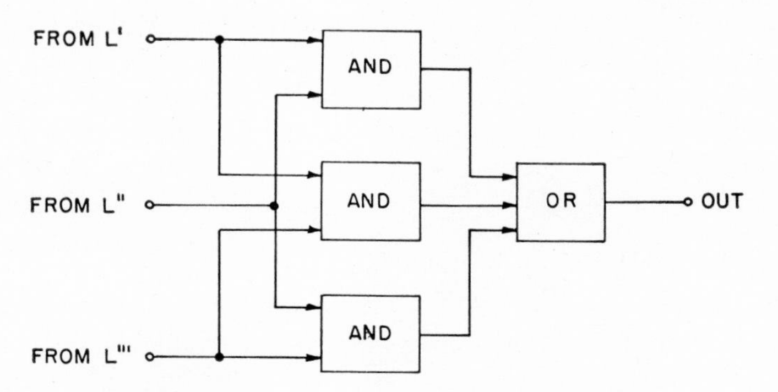

FIG. 2-55. Practical realization of majority circuit.

Practical Application of Majority Logic. In a practical application, the majority circuit marked M in Fig. 2-53 may take the form shown in Fig. 2-55. Here the inputs from the three identical circuits L are paired together in three AND circuits, the outputs of which go to a three-input OR circuit. If all circuits L give a correct output, all three inputs to the OR circuit will be activated, and an output signal will result. If one of the circuits L will not give an output, there will still be one AND gate with an output signal to the OR circuit and, therefore, an output signal from the entire circuit. If two or three circuits L fail, there will, however, not be a correct output signal.

A simple practical example of such a majority circuit is shown in Fig. 2-56. Each AND circuit is realized by two diodes and a resistor, and the OR circuit by three diodes and a resistor.

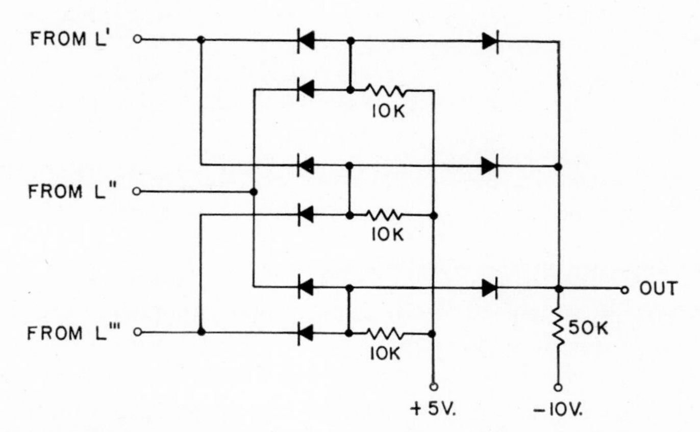

FIG. 2-56. Simple practical example of majority circuit of Fig. 2-55.

The following conclusions concerning redundancy in microelectronic systems may be drawn:

1. For *random failures* in components, wiring or batteries, large improvements in reliability may be obtained by *passive or active* redundancy, although at a cost.

2. For nonrandom failures, the preferred method is *active redundancy.*

3. The *improvement* in reliability is *larger, the larger the network* (proportional to $N^{1/2}$).

4. The improvement in reliability may be used for both *improved life* and *easier fabrication* (many components may be faulty).

5. An arbitrarily large system may be built with arbitrarily large shrinkage in fabrication of components (but usually less than 50%), provided that subassemblies of a *size determined by the failure percentage are pretested, and imperfect assemblies are rejected.*

2-6. POWER-SUPPLY CONSIDERATIONS IN MICROELECTRONICS

In stationary equipment, power may be obtained simply from a wall outlet, and no thought need be given to the volume and weight of the power supply. In mobile equipment, the power supply often has to be included in the equipment, and, therefore, the volume and weight of the supply become prime considerations. The reason for this is that the supply may be as large as, or larger than, the electronic equipment, and it may then be questioned what is to be gained from reducing the size and weight of the electronic equipment if the power supply is the chief factor which determines the over-all weight and volume. In most cases, there is the usual law of diminishing return after the equipment has been reduced until its size matches that of the power supply. This is true even in the cases when power may be tapped from large power packs intended primarily for locomotion, as in airplanes or in rockets. In these cases, an additional (even if small) capacity of the power supply is required to handle the electronic equipment. However, the larger the power supply, the more efficient it can be made in terms of watthours per pound.

In the following two sections data on power-supply limitation to miniaturization, and the value of power reduction per se, will be examined.

1. Power-supply Limitation to Miniaturization

A summary of weight efficiency, life, and cost of typical power supplies is shown in Table 2-15.[46–49]

The point at which the power supply limits the over-all packaging density may be found by equalizing the size of the system with the size of its power supply. This will be done on a component basis, assuming

TABLE 2-15. DATA ON TYPICAL POWER SUPPLIES

Power supply	Maximum capacity, whr/lb	Life, hours	Cost, whr/dollar
Chemical batteries	10–80	$1–10^3$	
Primary: LeClanché (zinc–manganese oxide)	30	$1–10^3$	10–100
Zinc–mercuric oxide	53	$1–10^3$	0.3–3
Secondary: Lead-acid	16–20	$1–10^3$	5–50
Cadmium–nickel oxide	10–12	$1–10^3$	0.2–2
Zinc–silver oxide	50–55	$1–10^3$	0.2–2
Fuel cells (future)	250–500	$1–10^3$	
Nuclear batteries (future)	1000–3000	$10^3–10^5$	
Mechanical (flywheel)	2.5*	10^{-1}	1
(spring)	0.03*	Large	10^{-3}
Solar cell	3000–6000	$>10^4$	10–50
Capacitor	0.1	10	10^{-1}
Chemical battery + solar cell	1500–3000	10^4	1–50
Chemical battery + motor generator	10–100	10^4	50–500

* Not including generator.

for simplicity that all components in the system are identical. Then

$$W = \frac{10pt}{C_1} \tag{2-100}$$

where p = power dissipated per component (average power if components are different)
t = operating time of the power supply
C_1 = capacity of the power supply, whr/lb
W = weight of a component (average weight if the components are different), assuming 10% packing efficiency

Using figures from Table 2-15, the curves in Fig. 2-57 have been computed for some typical cases. The curves may be interpreted in the following way. Given an average dissipation per component (plotted along the abscissa), the curves show the critical packing density (plotted along the ordinate in terms of the side of the cube containing one component). For components smaller than this critical size, the power supply is larger than the system; for components larger than this critical size, the system is larger than the power supply. Figure 2-57 shows how much operating power should be reduced when packaging density is increased, to obtain minimum over-all size of the equipment.

Typical order-of-magnitude figures for present systems with moderate miniaturization, such as pocket radios or airplane equipment, are

Packaging density: 1 component/cm^3
Power dissipation: 1 mw/component

It was found in Sec. 2-2 that the minimum possible linear size of a component is about 10 μ, representing a linear reduction in size compared to the figure given above of about 10^3 times. From Sec. 2-3 it is known that a reduction in power in most cases requires a sacrifice in speed. Therefore, it may be concluded that:

1. At the present time maximum over-all packaging density for self-contained equipment is generally limited by the size of its power supply.
2. Much orders-of-magnitude (10^6 to 10^9) improvement in the weight efficiency of power supplies is called for to make the power-supply size match the smallest possible size of the system at constant power level.
3. Reduction of power requirements of devices in most cases requires a sacrifice in speed. Therefore, electronic equipment with extreme over-all packaging density (both system and power supply small) is difficult to design for high speed.

The last conclusion is particularly interesting, in view of the capabilities of biological systems. The average size of a neuron in the brain is about 10 μ—comparable to the minimum size of solid-state components. The speed of neurons, however, is relatively slow, about 1 msec, whereas

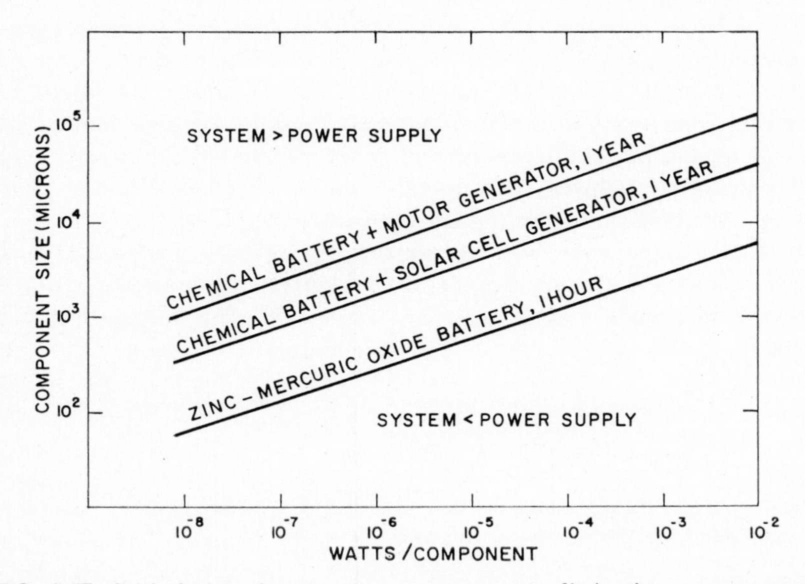

FIG. 2-57. Critical size of components versus power dissipation per component. Above the curves the size of the power supply is smaller than the system. Below the curves the size of the power supply is larger than the system.

electronic devices are capable of a speed of about 1 nanosec, a difference of a factor 10^6. However, this speed of electronic devices is obtained at a power level of about 1 mw or more. In order that the power supply not be larger than the system, a reduction in power requirements of about 10^6 would be necessary, resulting in a speed reduction of about 10^6 times, unless especially designed devices are used with more closely spaced electrodes and, therefore, requiring additional care in construction. In the end, therefore, biological systems and electronic systems are not very different from each other when packaging density and speed are considered.

2. The Value of Power Reduction per Se

In this section, the value of reduction in power requirements will be assessed quantitatively. For this purpose, the *power level of the equipment will be considered as an independent parameter*, variable at will, while all other parameters such as reliability, signal-to-noise ratio, frequency limit, etc., are left unchanged.

There are three direct and independent benefits to be derived from a reduction of the power dissipation, each with two aspects, as shown in Table 2-16. These benefits may be evaluated as cost savings. In most cases, size of the electronic system has no simple relation to cost, and may even be independent within reasonable limits. This aspect will, therefore, be left out. However, reduced size means a saving in transport or housing cost, as analyzed in Sec. 2-1.

Size of power supplies is rather directly related to cost for each type of supply, both physical size of complete supply and electrical size—capacity or stored energy. Therefore, assuming a linear relation, a reduction of the power consumption represents a proportionate saving of supply cost. Let this saving be C_1 dollars for each watt eliminated. Similarly, each watt eliminated reduces the transport or housing cost for the power supply. Let this saving be C_2 dollars/watt, where $C_2 \propto C_1$.

Heat exchangers may, in the majority of cases, be made part of the vehicle, e.g., the frame of a rocket, or the frame of an airplane, and so involve small extra cost or weight. Therefore, their influence will be neglected.

TABLE 2-16. ADVANTAGES OF REDUCED POWER LEVEL IN MICROELECTRONIC EQUIPMENT

Lower power makes possible:	*Cost saving*
1. *Smaller system,* if the packing density is limited by heat dissipation	
Reduced transport or housing cost for system	See 2-1
2. *Smaller power supply,* with reduced cost	C_1
Reduced transport or housing cost for power supply	C_2
3. *Smaller heat exchanger* (cooling equipment), with reduced cost	Neglect
Reduced transport or housing cost for heat exchanger	Neglect

Now the cost saving for one watt reduction of power consumption in the construction of a particular system will be calculated, using typical figures.

To compute the *cost of materials* for battery, generator, and fuel, let:

A = capacity of battery, in whr per watt power required
B = fractional discharge depth
D = relative cost of battery, dollars/whr (from Table 2-15)
l_1 = life of battery, years
L = life of equipment, years (from Sec. 2-1)
E = cost of generator per watt delivered, dollars/watt
l_2 = life of generator, years
F = cost of fuel per watt for life of equipment

Then, reducing the power requirements of a particular piece of equipment by one watt results in a material cost saving, C_1, dollars/watt:

$$C_1 = \frac{ADL}{Bl_1} + \frac{EL}{l_2} + F \qquad \text{(2-101)}$$

To compute the *cost of transport,* let G be the transport cost, dollars/lb

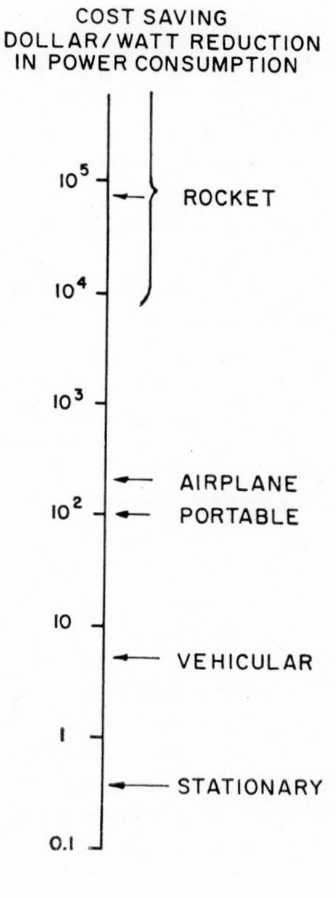

FIG. 2-58. Cost saving in dollar for each watt of power consumption eliminated in the design of typical microelectronic equipment.

(from Sec. 2-1), and H the weight of battery and generator per watt delivered, lb/watt.

Then the saving in transport cost for each watt saved amounts to

$$C_2 = GH \tag{2-102}$$

The total cost C is

$$C = C_1 + C_2 \tag{2-103}$$

For *stationary equipment* a somewhat different approach must be used.

Let I be the cost of power per kwhr, and n the number of hours of use of the system each day. Then

$$C = 0.375ILn \tag{2-104}$$

Results of such calculations with typical figures inserted are shown in Fig. 2-58.

REFERENCES

[1] U.S. Bureau of the Census, Statistical Abstracts of the United States: 1959 and 1962, Washington, D.C.

[2] B. S. Lee (ed.), "Aviation Facts and Figures," American Aviation Publications, Inc., Washington, D.C., 1959 and 1962.

[3] *Elec./Electron. Procurement,* April, 1961.

[4] Mail Order Freight Rates.

[5] NASA Authorization for Fiscal Year 1960 and 1962, Hearings before the NASA Authorization Subcommittee of the Committee on Aeronautical and Space Sciences, United States Senate.

[6] D. R. Crosby, Some Considerations in Kilomegacycle Computer Design, AIEE Winter General Meeting, New York, Jan. 28–Feb. 2, 1962, Special Publication S 136.

[7] J. M. Early, Speed, Power and Component Density in Multielement High-speed Logic Systems, International Solid-state Circuit Conference, Philadelphia, Pa., Feb. 10–12, 1960, Digest of Technical Papers, p. 78.

[8] J. T. Wallmark and S. M. Marcus, Minimum Size and Maximum Packing Density of Non-redundant Semiconductor Devices, *Proc. IRE,* vol. 50, p. 286, 1962.

[9] V. K. Zworykin, G. A. Morton, E. G. Ramberg, J. Hillier, and A. W. Vance, "Electron Optics and the Electron Microscope," John Wiley & Sons, Inc., New York, 1945.

[10] G. Mollenstedt and R. Speidel, Electron Optical Micrctool with Electron Microscopy Control (in German), *Physik Bl.,* vol. 16, p. 192, 1960.

[11] J. A. Swanson, Physical versus Logical Coupling in Memory System, *IBM J. Res. Develop.,* vol. 4, p. 305, 1960.

[12] J. A. Rajchman, Computer Memories; A Survey of the State-of-the-Art, *Proc. IRE,* vol. 49, p. 104, 1961.

[13] C. R. Smallman, A. E. Slade, and M. L. Cohen, Thin Film Cryotrons, *Proc. IRE,* vol. 48, p. 1562, 1960.

[14] E. Parzen, "Modern Probability Theory and Its Applications," John Wiley & Sons, New York, 1960.

[15] B. Peters, Cosmic Ray Physics, chapter in E. U. Condon (ed.), "Handbook of Physics," McGraw-Hill Book Company, Inc., New York, 1958.

[16] A. M. Brues (ed.), "Low-level Irradiation," American Association for the Advancement of Science, Washington, D.C., 1959.

[17] H. E. Newell and J. E. Naugle, Radiation Environment in Space, *Science*, vol. 132, p. 1465, Nov. 18, 1960.

[18] A. L. Johnson, How to Select an Adequate Cooling System, *Electronics*, vol. 34, p. 54, Oct. 20, 1961.

[19] A. E. Rosenberg and T. C. Taylor, A Thermal Design Approach for Solid-state Encapsulated High-density Computer Circuits, *IRE Trans. Military Electron.*, vol. MIL-5, p. 216, July, 1961.

[20] J. J. Suran, "Circuit Considerations Relating to Microelectronics," *Proc. IRE*, vol. 49, p. 420, 1961.

[21] J. D. Meindl, Power Dissipation in Microelectronic Transmission Circuits, *IRE Trans. Military Electron.*, vol. MIL-5, p. 209, July, 1961.

[22] G. Rezek and P. K. Taylor, Thermal Design Solutions for Micro-modular Equipment, *IRE Trans. Prod. Eng. Prod.*, vol. PEP-5, p. 71, June, 1961.

[23] D. M. Cawthon, A Technique for Transient Thermal Prediction in Microminiaturized Circuits, *IRE Trans. Prod. Eng. Prod.*, vol. PEP-6, p. 15, July, 1962.

[24] J. R. Baum, Thermal Design Considerations in Thin-film Microelectronics, *Electro-Technol.*, vol. 68, p. 92, July, 1961.

[25] H. C. Kammerer, Thermal Design for Microminiaturized Circuitry, *Trans. ASME, Ser. B, J. Eng. Ind.*, vol. 84, p. 1, 1962.

[26] M. Walker, J. Roschen, and E. Schlegel, Determination of Temperature Profiles in Microcircuits, Electron Devices Meeting, Washington, D.C., Oct. 25–27, 1962.

[27] J. M. Early, Speed in Semiconductor Devices, IRE International Convention, New York, Mar. 26–29, 1962.

[28] J. T. Wallmark, Design Considerations for Integrated Electronic Devices, *Proc. IRE*, vol. 48, p. 293, 1960.

[29] J. J. Suran, Progress and Pitfalls in Microelectronics, *Electronics*, vol. 35, p. 45, Oct. 19, 1962.

[30] H. S. Balaban, A Selected Bibliography on Reliability, *IRE Trans. Reliability Quality Control*, vol. RQC-11, July, 1962, p. 86 (260 references).

[31] Institution Report, Reliability of Electronic Equipment, *J. Brit. Inst. Radio Engrs.*, vol. 23, p. 287, 1962 (192 references).

[32] List of Literature on Reliability Problems (translated from Russian), *Automation and Remote Control*, vol. 23, no. 9, p. 1270, September, 1962.

[33] J. E. Shwop and H. J. Sullivan, "Semiconductor Reliability," Engineering Publishers, Elizabeth, N.J., 1961.

[34] W. H. Von Alven, "Semiconductor Reliability," vol. 2, Engineering Publishers, Elizabeth, N.J., 1962.

[35] "Reliability Stress Analysis for Electronic Equipment," Navships 900-193, Bureau of Ships.

[36] D. R. Earles, Reliability Growth Predictions during the Initial Design Analysis, *Proc. 7th Natl. Symp. Reliability and Quality Control*, Philadelphia, Pa., Jan. 9–11, 1961, p. 380.

[37] J. J. Suran, Use of Passive Redundancy in Electronic Systems, *IRE Trans. Military Electron.*, vol. MIL-5, p. 202, July, 1961.

[38] A. A. Sorensen, Improving Analog-circuit Reliability with Redundancy Techniques, *Electro-Technol.*, vol. 68, p. 66, October, 1961.

[39] A. A. Sorensen, Digital-circuit Reliability through Redundancy, *Electro-Technol.*, vol. 68, p. 118, July, 1961.

[40] R. H. Wilcox and W. C. Mann (eds.): "Redundancy Techniques for Computing Systems," Spartan Books, Washington, D.C., 1962 (23 articles, extensive bibliography)

[41] J. H. Griesmer, R. E. Miller, and J. P. Roth, The Design of Digital Circuits to Eliminate Catastrophic Failures, chapter in Ref. 40.

[42] E. F. Moore and C. E. Shannon, Reliable Circuits Using Less Reliable Relays, *J. Franklin Inst.*, vol. 262, pp. 191, 281, 1956.

[43] J. von Neumann, Probabalistic Logics and the Synthesis of Reliable Organism from Unreliable Components, in C. E. Shannon and J. McCarthy (eds.), "Automata Studies," Princeton University Press, Princeton, N.J., 1956.

[44] S. Amarel and J. A. Brzozowski, Theoretical Considerations on Reliability Properties of Recursive Triangular Switching Networks, chapter in Ref. 40.

[45] S. Levy, The Reliability of Recursive Triangular Switching Networks Built of Rectifier Gates, chapter in Ref. 40.

[46] C. K. Morehouse, R. Glicksman, and G. S. Lozier, Batteries, *Proc. IRE*, vol. 46, p. 1462, 1958.

[47] H. J. Sketch, Generation of Power in Satellites, *J. IEE London*, vol. 8, p. 148, 1962.

[48] H. A. Zahl and H. K. Ziegler, Power Sources for Satellites and Space Vehicles, *Solar Energy*, vol. 4, p. 32, January, 1960.

[49] *IRE Trans. Military Electron.*, vol. MIL-6, January, 1962, special issue on Direct Energy Conversion (17 up-to-date articles on power supplies).

CHAPTER 3

DISCRETE COMPONENT PARTS CONCEPT

By Stanislaus F. Danko, Robert A. Gerhold, and Vincent J. Kublin

3-1. INTRODUCTION

In the course of implementation of part-by-part serial assembly of electronic equipment and systems, seemingly endless improvements have been made in the parts themselves; the techniques of parts assembly (interconnection) have been substantially simplified, and even partial mechanization of the assembly operations has been achieved. It is common today to point to dip-soldered printed wiring as representing the state of the practicing arts of assembling electronic circuits from discrete parts, involving automatic, semiautomatic, and manual process steps at the different stages of the assembly operation (Fig. 3-1).[1] It is from this reference point of the current practicing arts and the many criteria of quality and performance that characterize today's applications of electronics that discussion in this chapter must start. This chapter underscores the various considerations, techniques, and characteristics of the discrete-parts approaches which collectively support the proposition that the first generation of microelectronics leans heavily on the classical part-by-part serial assembly concept. The significant progress in the elemental electronic parts themselves in terms of size, reliability, and cost is objectively considered as well as the several different techniques that come within the context of the discrete-parts concept. Finally, the value of the discrete-parts approaches is measured with the yardsticks of reliability, size, cost, producibility, timetable, and skills involved.

1. Philosophy

In the period 1956 to 1959 in the United States, in the first flush of microelectronics interest, it appeared that size reduction was the prime attribute that could open vast new potentials for the electronics of the future. Such emphasis on size in the literature of the day led to many enthusiastic projections about "pinhead" electronics, based on ultimate capabilities of certain technologies, without regard (at the moment) to

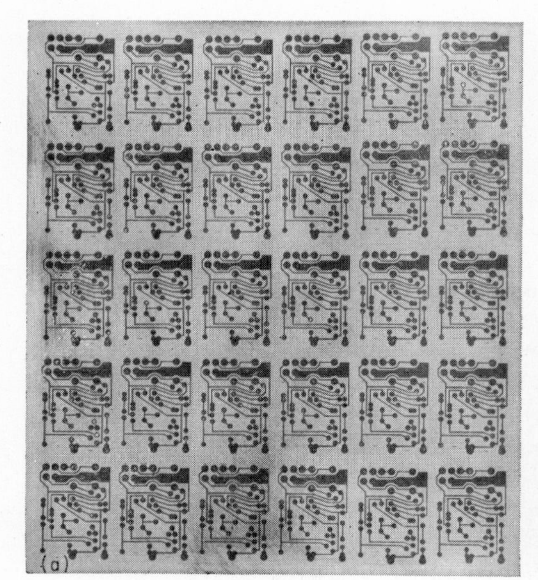

FIG. 3-1. A conventional dip-soldered printing-wiring assembly. Process steps include: (*a*) Fabrication of the conductor pattern in multiple arrays, usually by etching of copper-foil-clad plastic laminates: The individual patterns are cut out and perforated appropriately to receive leaded parts. (*b*) Axial or radial lead parts (or terminated assemblies of parts) are inserted into proper perforations. (*c*) The pattern side of the board is brought into contact with a molten solder bath to solder all terminations simultaneously. Many of these process steps are mechanized. Skills needed in assembly are low, production capabilities are high, costs are moderate, and a vast inventory of parts from multiple sources is available.

some of the practical engineering limitations. Today, however, the designers would not accept any of these microelectronic techniques for their particular projects without unequivocally demanding to know: (1) the reliability capabilities of the suggested construction, (2) the cost comparison with conventional miniaturization techniques currently being used, (3) the availability of the sources which could provide the new microelectronic technologies or circuits. Therefore, *size by itself is not the ultimate goal of microelectronics;* size is the expendable requirement in seeking the optimum compromise between size reduction and the economic, production, logistic, and operating requirements of industrial and military electronics. This thesis of size expendability is the root of the discrete-component-parts systems which recognize their moderate size-reduction capabilities in microelectronics as the trade-off for more immediate production, economic, and reliability advantages. Obviously such size trade-off must have a practical limit, beyond which the advantages of the microelectronic approach over the conventional approaches become marginal.

Before any analysis of the discrete-component-parts approaches, one additional point should be made with regard to markets for the products of the electronics industry, whether conventional or microelectronic. On the one hand, there is the mass production market involving the entertainment and industrial class of equipments (TV, radio, test equipment, record players, tape recorders, etc.); there is, on the other hand, the more limited production market of military, commercial, aviation, medical, etc., electronics, where the emphasis is on smaller production runs of selected designs. These production considerations raise important requirements for microelectronics, namely, *machine producibility* of the microelectronic circuit for large production runs, and economical (possibly machine or manual) producibility for limited production. It is in the hard, real terms of production flexibility and cost that new assembly concepts must also be assessed. These points are important to all civilian, industrial, and military electronic equipment users and explain their interest in the discrete-parts concept as a first step to microelectronics, with the logical and inevitable major contribution of thin-film concepts and integrated semiconductor circuits to succeeding generations of equipment and systems.

2. Motivation

The serial method of assembly of discrete electronic parts derives its principal attraction from the following attributes:

1. Availability of reliable electronic parts
2. Complete circuit capability
3. Relative simplicity of assembly
4. Ability to breadboard
5. Relative ease of production-line changes
6. Relatively simple mechanization and automation of fabrication processes

Since each of these points is an element for decision in selection of an acceptable microelectronic technique, each point warrants a brief discussion.

Availability of reliable electronic parts from multiple sources is a prime attribute because it removes the burden of technological know-how and quality manufacture of the basic elements from the assembler of the microelectronic circuit. The commercial sources of such manufactured parts generally have mechanized or automatic facilities, permitting mass production of uniform quality parts at moderate prices. Competition and strict consumer demands provide the incentive for progressive technological improvement of the parts by these manufacturing sources, yielding the highest-quality performance that the arts and sciences can produce. The lines of contemporary conventional Minuteman parts* reflect this progressive product refinement and availability of discrete parts with histories of proved reliability and dependability. Scaling down of such parts to dimensions, ratings, and even special shapes compatible with microelectronic construction is a major engineering effort, requiring the mass market motivation (large consumer demand) as a catalytic agent.

Complete circuit capability from direct current through intermediate frequency and video to radio frequency, including both linear and digital circuits, can be effectively realized using the discrete-parts concept. This broad range of circuit applications provides a strong attraction to this concept. By drawing upon the specialized skills of a large competitive electronic-parts industry for his microelectronic parts, the equipment producer can concentrate his engineering resources on equipment and system design, rather than spread this effort in developing and improving the entire range of parts and materials technologies.

Relative simplicity of assembly of discrete parts is based on the premise that the assembler of discrete parts has at his disposal the established arts of printed wiring and soldering, or the more recent disciplined techniques of welding to effect his intercomponent connections at low cost. This is not to minimize the difficulties inherent in the applications of these techniques on a microelectronic scale, but to suggest that just as parts can and have been adapted dimensionally to the smaller sizes of microelectronics, so can the printed wiring, the solder joint, and the welded joint be modified to be compatible with microelectronics.

Ability to breadboard is an inherent feature of the discrete-parts concept, and reflects the capability of the circuit designer to test out and modify (if necessary) his work product by actual do-it-yourself construction, using parts identical with those that will be used in the final packaged circuit. Certain qualifications pertain to translation of nonpackaged breadboards into highly compact microassemblies; these include such new influences as increased stray capacitances in the compact package, shorter leakage paths, intercomponent couplings, parasitics, and increased dissipations per unit surface in the finished microassemblies. However, these problems are common to all the microcircuit systems, varying only in degree among them. Within the discrete-parts systems, these problems are least difficult and are amenable to solution (in the general case) by recognizing and using *size* as a trade-off parameter against these difficulties (in other words, size is the expendable requirement).

* A major Air Force program with the Autonetics Division of North American Aviation Corp. has established sources of supply of families of high-reliability parts for the Minuteman ICB missile. These parts are generally regarded as reliability reference points in assessing other component-parts systems.

Relative ease of production-line changes is also inherent in the discrete-parts system and, in the ordinary circumstance, merely requires the substitution of one discrete-part value for another, possibly a change in the wiring matrix (printed wiring or welding connections) or, in the extreme, even substitution of an entirely different package design. Such flexibility in design and production is an important economic consideration in selection of a microelectronics technique.

Mechanization and automation of the microelectronic fabrication process on the production line is an almost inevitable last step from practical and economical points of view. It is almost axiomatic that a mass production complex, striving for uniformity and economy in the final product, should use the minimum of fallible human skills on the production line. To this end, the several categories of systems of discrete-parts assembly to be discussed differ in their adaptability for machine processing; those systems that impose certain disciplines relating to uniform geometry of the parts permit much simplified machine handling of such parts. From both the military and the commercial points of view of the electronic industry, this intimate relationship of small size and machine producibility is very important, and must be considered as a significant requirement in the evolution of a microelectronic fabrication system.

These six points reflect the realities of designing, testing, producing, and using microelectronics based on the discrete-parts concept. However, it is difficult to determine the true value of any microelectronic technique even on these points unless one uses some quantitative measure for evaluation, some common and well-known yardstick for reference purposes. The conventional printed-wiring assembly of current universal usage in military and commercial production meets this yardstick requirement. Such printed-wiring construction (Fig. 3-2) has:

1. Demonstrated a high level of reliability in critical commercial and military applications.
2. Permitted straightforward assessment of size by direct measurement.
3. Had sufficient use to permit accurate cost estimating.

It is essentially this level of electronic capability that present microelectronic efforts are dedicated to surpass. At this time (1963), industry's microelectronic efforts easily produce a significant size reduction when measured against this yardstick, but with serious penalties in terms of cost and in deficiency of reliability data. Present emphasis is clearly aimed at reducing this cost, and enhancing the reliability beyond that achieved by electronics to date. Size will be traded to achieve these two objectives.

For each specific application (civilian entertainment; industrial, commercial, or military usage), the superior assembly concept that will

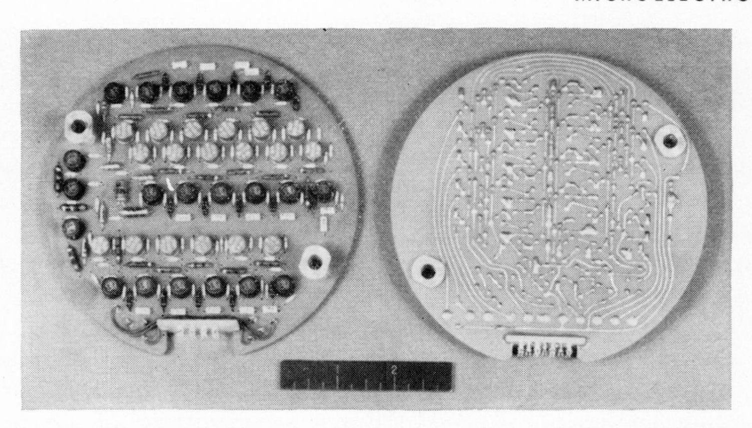

FIG. 3-2. A converter deck from Explorer VII satellite. The three digital-to-analog converters on this deck (prior to encapsulation) use conventional parts laid out on a 0.1-in. grid. The 225 components on the 5½-in. deck reflect a packaging density of about 40,000 parts/ft^3, and an estimated mean time to failure of 20,000 hr. (*Army Ballistic Missile Agency.*)

be selected will be found by weighing objectively the above factors (size, reliability, and cost) to arrive at an accurate quantitative figure of *value*. Such value analysis at the equipment or subassembly level most often indicates that *hybrid approaches* yield optimum return, thus dictating availability and usage of flexible assembly and production techniques. An evaluation of the discrete-parts approaches with this value yardstick in the succeeding pages provides perspective with regard to capabilities relative to the present practicing art, in terms of reliability, cost, and size. Other microelectronic approaches can likewise be assessed against these criteria.

3-2. PARAMETERS AND TECHNOLOGIES

Miniature discrete component parts are discussed for the following assembly systems:

1. *Random-shaped parts* packaged in three dimensions (e.g., cordwood)
2. *Random-shaped parts* packaged in two dimensions (e.g., Arma, DOFL 2-D, etc.)
3. *Uniform-shaped parts* packaged in three dimensions (e.g., micromodule, pellet system, etc.)

In systems 1 and 2, the random-shaped parts are mostly scaled-down versions of component parts used in conventional printed-wiring assemblies (Fig. 3-2). The distinguishing features of systems 1 and 2 are not in the parts themselves, but in the methods of assembling (integrating) these miniature parts to increase the volumetric efficiency of packaging. System 3 is unique in that it requires new families of parts having common and uniform geometries.

To provide a base for the discussion of these three assembly approaches, a panoramic view of the parts available for these systems seems appropriate. Since systems 1 and 2 utilize the same basic elements, a survey of the currently available parts for microelectronic practice of these systems may be quite revealing. Figure 3-3 is such a broadbanded display of available miniature parts in the random-shape category; also charted are the bands of parts characteristic of two of the disciplined geometry systems. The only characteristic for microelectronic usage that is featured here (for references purposes) is the *packaging capability* of the various classes of parts; Table 3-1 correlates this capability with actual physical sizes of some typical component parts.

Figure 3-3 is also useful to point out the weak areas (with regard to size compatibility) of such essential parts as relays, switches, variable reactances, and r-f connectors, though fortunately these parts are minority elements in the parts population of typical circuits. These disparities in sizes will be treated individually in subsequent discussions.

Each of the available families of miniature parts characterized in Fig. 3-3 is reviewed briefly on the following pages to provide reference base points for assessing their capabilities and limitations for microelectronic circuit designs. The tedious detail relating to numerical dimensions and multivendor sources of these parts is not included in the tables accompanying these reviews, since such information is readily available from

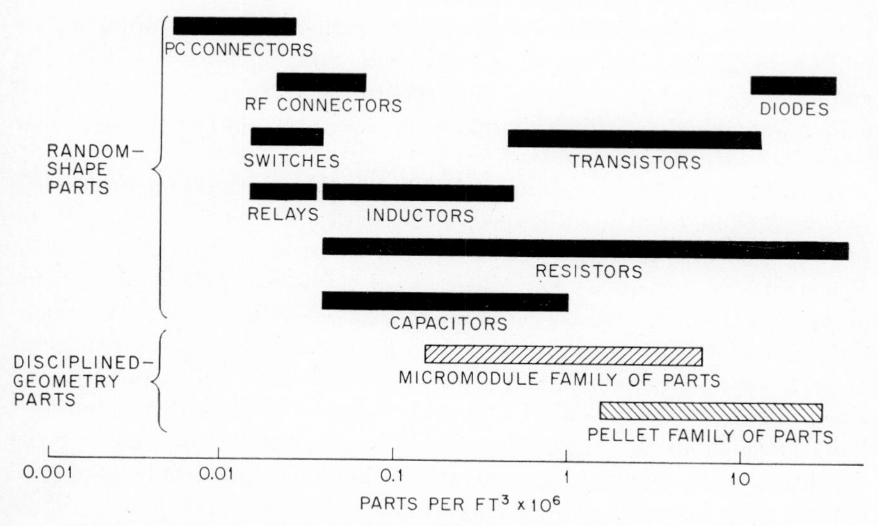

FIG. 3-3. Approximate packaging capabilities of miniature discrete parts.

TABLE 3-1. BODY PROFILES OF TYPICAL DISCRETE PARTS REPRESENTING DIFFERENT PART PACKAGING CAPABILITIES

Packaging capability, parts/ft^3 x 10^6	Typical body profiles, actual size	Typical part approximating packaging capability
10*		RN-55 (MIL-R-10509D)
1*		RC-07 (MIL-R-11D)
0.1		Ceramic capacitor, 100 volt, 0.047 μf
0.01		General-purpose variable resistor

*Parts with packaging capabilities in these decades (up to about 1.5 x 10^6 parts/ft^3) find ready usage in the cordwood system where the feature requirement is a controlled length of the part. The dimensions of the RC-07 shown above represent a particularly popular size for passive parts for the cordwood system.

the commercial catalogs of United States parts vendors. Rather, the tables are intended to emphasize the physical sizes of typical available elements in each component-part category and their ranges of pertinent electrical characteristics.

1. Component Parts

Fixed Resistors. The fixed resistor, one of the most frequently used parts in circuit assemblies, warrants first review of some of the features pertinent to microelectronics usage. Transistor circuit applications seldom require resistances of greater than 100,000 ohms, so that the gamut of resistor types (metal film, deposited carbon, wire-wound, tin oxide) is available within this range for selection. (See representative types in Table 3-2.)

In addition to these vendor-supplied types, there is the class of *screened-ink resistors* that are manufactured along with other parts directly on the (wafer) substrate in one of the two-dimensional approaches that will be discussed subsequently. For all practical purposes, such deposited flat parts on a common substrate can be considered to have no third dimension, and packaging capabilities of such parts could perhaps be expressed as parts per square inch; such refinement, however, is not truly necessary since, in the ultimate calculation of parts densities of functional assemblies built by such in situ deposition of parts, the number of parts on the substrate (both deposited and attached) is merely divided by the total outline volume occupied by the assembly to get the parts density of the finished circuit function. As a matter of information, however, such deposited ink resistors are capable of as low as $\pm 5\%$ tolerance in production, a resistance range of 5 ohms to 100 megohms, and a dissipation rating of about $\frac{1}{8}$ watt;* although the resistor inks and geometric patterns are selected to suit the resistance value desired, a $\frac{3}{16}$- by $\frac{1}{16}$-in. configuration would be typical of a "small" deposited resistor used in a microelectronic circuit system employing in situ formed parts.

Variable Resistors. Manually adjustable parts such as potentiometers, tuning devices (trimmer capacitors or inductors), and switches are necessary complements for adjustment or control purposes when microelectronic construction is being considered on an equipment level. Table 3-3 provides some perspective on the available miniature resistors in this class of parts, which does not have high population use. It is expected that these units will find prime use as individually mounted (panel) elements, rather than as elements assimilated physically into the microelectronic assembly proper.

Although the size of the variable resistor is large compared to the high-population fixed resistors and capacitors, their comparatively low level of use tends to offset their apparent depressing effect on attainable size reduction. Nonetheless, this class of parts is one that requires

* Metal film, metal oxide, and ceramic resistors are also 2-D types manufactured in situ on a substrate shared with other parts. Tolerances of the best of these are as low as $\frac{1}{2}\%$ (or better), with resistance ranges as high as 1 megohm.

TABLE 3-2. TYPICAL FIXED RESISTORS

Type	Profiles, actual size	Watts rating	Parts/ft^3 x 10^6	Remarks
General purpose		$\frac{1}{10}$	39	Composition type, uninsulated; RC-06 (MIL-R-11D), ±20% life stability, 10 ohms - 22 megohms
		$\frac{1}{8}$	3.5	Composition type, insulated; RC-08 (MIL-R-11D), ±20% life stability, 10 ohms - 22 megohms
		$\frac{1}{4}$	1.1	Composition type, insulated; RC-07 (MIL-R-11D), ±20% life stability, 10 ohms - 22 megohms
Stable utility		$\frac{1}{4}$	1.1	Tin oxide type; RC-07 (MIL-R-11D), ±5% life stability, 50 ohms - 150 kilohms
Precision		$\frac{1}{10}$	12	Film type; RN-55 (MIL-R-10509D), ±2% life stability, 200 ohms - 100 kilohms
		$\frac{1}{20}$	0.57	Wire-wound type; approximately ±2% stability, 1 ohm - 1 megohm; inductively wound

TABLE 3-3. TYPICAL MINIATURE VARIABLE RESISTORS

Type	Profiles, actual size	Life Stability, %	Watts at 70° C	Parts/ft³ x 10⁶ *	Remarks
General purpose	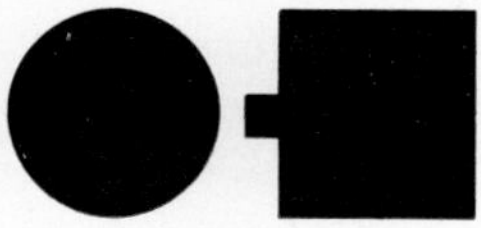	20	$\frac{1}{2}$	0.014	Carbon composition; 100-ohm-5-megohm ranges
Stable utility	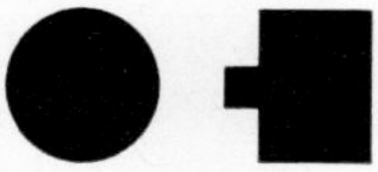	5	$\frac{3}{4}$	0.04	Cermet; 50-ohm - 50-kilohm ranges
Precision		2	1	0.06	Nichrome wire; 50-ohm - 20-kilohm ranges
		2	$\frac{1}{2}$	0.09	Nichrome wire; 50-ohm - 10-kilohm ranges

* Body volume only.

additional development attention to narrow down the gross size differences between it and other miniature parts.

The variable resistor for either panel or trimming control is just one of the many discrete parts that, at this time, has no equivalent in the thin-film or in the solid-state system. Accordingly, it is an essential complement to *all* microelement systems in applications requiring this type of electrical function.

Other parts which similarly have no available counterparts as yet in the newer technologies include: electric switches and relays (for applications requiring complete ohmic connects or disconnects), quartz crystals, coils and transformers,* and capacitors in the multimicrofarad ranges. Accordingly, until such time as technological advances in thin films and solid state permit, these discrete parts will be used as "outboard" elements whenever they are needed in equipment design. In this sense, the variable resistor and other tables on the following pages are useful in providing the necessary parts "backup" wherever the other (integrated) systems of construction need such discrete-parts capabilities.

* Very limited coil capabilities presently exist in the thin-film systems.

Fixed Capacitors. The large capacitances (0.1 μf and higher) for transistor circuit usage find best expression in the tantalum dry and wet series of parts available from many vendor sources. Table 3-4 illustrates some of the tubular (metal) form factors that are common. Molded and dip-protected styles are also being marketed at this writing with some incremental improvement in the packaging capabilities of this class of parts. These tantalum parts, in general, have dissipation factors of less than 5% and leakages of the order of 0.04 μa/volt; rated operating voltages are generally limited to 35 volts maximum.

Where the dissipation factors or leakages are inadequate for an application requiring coupling or filtering capacitors up to about 0.1 μf, the ceramic types of Table 3-5 represent a higher plateau of capability in these respects but with lower packaging densities.

Nonpolar types require substantial sacrifices in the packaging capability since two polar types must be used within the part envelope for such a special function.

The smaller capacitances for coupling, filtering, bypass, and precision applications are accommodated by the families of ceramic, glass, and mica capacitors typified in Table 3-5. Here again, some variations of these types (and others such as polyester film) are available, with incremental gains in packaging capability in some instances.

A more substantial capability for precision performance in transistor

TABLE 3-4. TYPICAL FIXED CAPACITORS (ELECTROLYTICS)

Type	Profiles, actual size	Capacitance		Parts/ft^3 x 10^6	Remarks
		μf-volts, max.	Range, μf		
Electrolytics		4	0.01 - 2	2.3	Tantalum wire
		10	0.05 - 10	0.9	Tantalum wire
		44	1.2 - 6.8	0.56	Solid tantalum
		1100	2.5 - 11	0.08	Wet-slug tantalum
		2350	30 - 47	0.029	Wet-slug tantalum

TABLE 3-5. TYPICAL MINIATURE FIXED CAPACITORS (COUPLING AND PRECISION)

Type	Profiles, actual size	D-c volts	Capacitance	Parts/ft³ x 10⁶	Remarks
Coupling and bypass		25	0.1 μf max.	0.16	Ceramic (multilayer)
		100	0.001	1.72	Ceramic
		100	0.047	0.1	Ceramic
Precision		300	1 - 300 pf	0.37	Glass (CY-10, MIL-C-11272B)
		300	220 - 1200	0.13	Glass (CY-15, MIL-C-11272B)
		300	12 - 300	0.17	Mica (1 dip protection)
		500	10 - 510	0.06	Mica (CM-15, MIL-C-5B)
		100	1 - 50	1.0	Ceramic, NPO

circuits is expected from the ceramic class of parts (illustrated by the one NPO unit in the table) with due recognition of the lower operating voltages. Production of an expanded line of such miniature ceramic types in other temperature characteristics is expected by 1963.

Variable Capacitors. The variable capacitors, most frequently used in trimming (tuning) operations in fixed tuned r-f circuits, are represented in Table 3-6 in the popular 1- to 10-pf range; trimmers up to 100 pf and

TABLE 3-6. TYPICAL MINIATURE VARIABLE CAPACITORS (TRIMMERS)

Type	Profiles, actual size	Capacitance range, pf	Parts/ft³ x 10⁶	Remarks
Air		0.8 - 10	0.07	Military quality; 0 ± 20 ppm/°C stability
Glass		0.5 - 7	0.16	Commercial grade

larger are also available for more special applications with almost proportionate decreases in packaging capabilities. Though other form factors exist, the two shown reflect the capacitance-size relationship typical of this class for precision and noncritical usage, generally in linear circuits.

Like the other mechanically actuated parts (potentiometers, switches, relays, etc.), the variable capacitor is a comparatively low-use item, so that its limited packaging density is not so depressing as the table suggests. It, too, is not suited particularly for inclusion in small modules such as cordwoods, although the lengths of the lower-valued trimmers (7 pf or so) are available in ⅝-in.-long (or less) bodies capable of being housed in cordwood designs; some form of protection against encapsulant leakage into the mechanical system, however, would have to be provided.

A particularly small ceramic trimmer (on a 0.31- by 0.31-in. wafer base) is available in the uniformly shaped parts inventory (Table 3-13) in trimmer ranges up to 18 pf.

Inductors and Transformers. Inductors find moderate use in linear circuits, so that the range of 20,000 to 40,000 parts/ft^3 packaging capability in some cases can represent significant limitation to achievement of good size reductions (low frequency circuits in particular). The difficulties are basic, in that the types of phenomena and materials available to physically express this reactance function L are limited. Closed-loop magnetic materials with high saturation densities and low losses, and very fine wire windings, yield part sizes that at best could be considered just tolerable for microelectronic purposes. It is difficult, also, to design circuitry so as to eliminate the inductor, although the unique characteristics of the transistor have been used in many instances to obviate the need for an impedance changing or output transformer. Positive reactance is seldom used in digital circuitry.

For r-f coils, the use of powdered iron and ferrite cores is almost mandatory up to about 50 Mc to obtain minimum size; the toroidal form of these materials is invariably used for low-level (signal) circuits. Table 3-7 illustrates one popular toroid size of magnetic material for r-f coil and transformer applications up to about 50 Mc.

For transmitter-antenna matching purposes where r-f power is to be delivered, a surprisingly small ferrite transformer can be designed because of the inherently low losses and high coupling possible with these materials. For example, for a 5-watt r-f power-handling capacity, such transformers with packaging capability of about 10,000 parts/ft^3 are estimated as possible for frequencies as high as 100 Mc.

Transistors and Diodes. Perhaps the most satisfactory of the discrete-component-parts family from the point of view of size are the transistors and diodes. In the decade or so of commercial and military design and production, these active devices have undergone phenomenal size reduc-

TABLE 3-7.

Type	Profiles, actual size	Parts/ft^3 x 10^6	Remarks
Audio transformers		0.055	Up to 500 mw output depending on impedance and DC current levels; frequency response = ±1 db, 300–3000 cps
		0.090	Same as above except for poorer response at low frequency and limited to higher impedances and lower DC currents
Pulse transformers		0.041	Blocking oscillator or circuit coupling applications
		0.120	Same as above except slower response time
RF chokes (cylindrical shape)		0.200	L = 0.1 to 1000 μh; self-resonant frequency 6 to 500 mc; Grade 1, Class B, MIL-C-15305B
RF coils and transformers (toroidal)		0.460	Toroidal coils are limited to low signal levels. Special coils (e.g., for low frequency) have densities as low as 0.10 x 10^6 parts/ft^3

tions as reflected by the packaging capabilities shown in Table 3-8. Within these shape factors, almost every type of transistor and diode useful for microelectronic fabrication can be found. Various silicon and germanium semiconductor technologies can be used for microelectronic discrete devices operating at frequencies up to 1,000 Mc.

The fact that some of the transistor styles are single-ended (terminations are axial and on one side) does not inhibit their usage in the cordwood system; the two end boards of a cordwood design can each accommodate these transistors placed flatly on the boards with sufficient headroom after assembly so as not to cause interference (in other words, head-to-head placement of transistors still can be compatible with the length requirement of cordwood parts).*

A matter of some concern in transistor usage involves the hermeticity of the device. Recent advancements in surface passivation of silicon have made possible the unequivocal acceptance of dip-protected silicon diodes, but some resistance to use of similarly dip-protected silicon

* See Sec. 3-3 for details of the cordwood concept.

transistors is still apparent in industrial and military operations; no such reservation exists with regard to the use of metal-can enclosures. In any case, Table 3-8 reflects only those form factors with which satisfactory performance has been shown,* whether the type illustrated is can-sealed or dip-protected.

Relays and Switches. The relay is identified with the class of outboarded parts (e.g., potentiometers, trimmers, manually operated switches, etc.), and as yet is not truly compatible with the microelectronic dimension.† Like the potentiometer, the relay is an occasionally

* For United States Army applications.

† Some qualification must be expressed here since one manufacturer has experimentally packaged a SPDT electromagnetic relay in a TO-5 can (packaging capability of 62,000 parts/ft³). This density probably represents the near limit of the electromechanical capability from practical manufacturing considerations.

TABLE 3-8. TYPICAL MINIATURE TRANSISTORS AND DIODES

Device	Profiles, actual size	Parts/ft³ x 10⁶
Transistors		12
		11
		1.3
		0.8
		0.6
		0.43
Diodes		33
		20
		18
		18
		11

TABLE 3-9. TYPICAL MINIATURE RELAYS

Type	Profiles, actual size	Parts/ft^3 x 10^6	Characteristics			
			Maximum contacts	Contact rating, amperes	Operating sensitivity, mw	Mode
Sensitive		0.035	SPDT	1	100	Nonlatching
Sensitive		0.032	SPDT	$\frac{1}{4}$ max.	40	Nonlatching
General purpose	Same as above	0.032	SPDT	1	300	Nonlatching
General purpose		0.014	DPDT	2	300	Nonlatching
Sensitive	Same as above	0.014	DPDT	$\frac{1}{4}$ max.	100	Nonlatching

used item in most applications, and for that reason its comparatively large size is usually not too serious (see Table 3-9).

The recent new subminiature reed relays are of particular interest, combining excellent reliability performance with small size. The reed types, available in single- and double-pole double-throw styles, are identified by the $\frac{1}{4}$ amp/contact rating. They are also available in higher multipole combinations.

The so-called "solid-state" relays (consisting of semiconductors and conventional parts packaged in one envelope) have been omitted from this tabulation since design experience of the past few years indicates a preference of designers for treating such a relay as part of the circuit complex, rather than as a vendor-supplied item. Solid-state relays have a proper place in microelectronic circuitry where the finite open-circuit impedances of these devices can be tolerated.

The relay is a necessary complement in any microelectronic system that needs a positive (ohmic) type of connect and disconnect. As such, it is an essential part of the systems when the remote-switching function is needed.

The three basic electromechanical switches universally used in electronic designs are typified in Table 3-10. In general, these parts are panel-mounted, and only the body volumes (behind the panel volume) have been considered in ascribing packaging capabilities to each type. The future outlook for any marked decrease in size of these parts is pessimistic at this time because of the problems of manufacturing the small linkages and contact surfaces, and the related electrical problems of maintaining a low-ohmic contact resistance, reasonable life, and ade-

Table 3-10. Typical Miniature Switches

Type	Profiles, actual size	Rating: amperes/ d-c volts/ a-c volts	Parts/ft³ x 10^6	Remarks
Toggle:*				
Single Pole		5/30/125	0.044	Available with center off and either or both sides momentary or maintained
Double Pole		5/30/125	0.018	Same as above
Rotary:†				
Sealed		2/30/125	0.018	Max. 1 pole, 10 position or 2 pole, 5 position
Unsealed		2/30/125	0.014	Max. 12 positions/section; up to 5 sections
Sensitive:‡				
Sealed		5/30/125	0.018	
Unsealed		7/30/125	0.044	

* MIL-S-3950A.
† MIL-S-3786A.
‡ JAN-S-63.

TABLE 3-11. TYPICAL MINIATURE RF CONNECTORS

Type	Profiles, actual size	50-ohm cable	Parts/ft^3 x 10^6*	Remarks
Plug		RG-196/U	0.067	Up to 5 kmc; screw-on coupling; not waterproof
Receptacle	Same as above	....	0.067	Mate for above; mated length $1\frac{3}{8}$ in.
Plug		RG-58/U	0.021	Up to 10 kmc; bayonet coupling; environment resistant (UG-1366)
Receptacle		....	0.022	Mate for above; UG-1471; mated length $1\frac{9}{16}$ in.

* Body volume only.

quate current-carrying capabilities. Being very low use items in electronic circuit design, their comparatively large size involves a size penalty that must be paid generally only once in an equipment design.

Switches, too, are complementary discrete parts to the thin-film and solid-state systems which do not have equivalents of this ohmic connect-disconnect function.

Connectors. The two connector families typified by Tables 3-11 and 3-12 (matched r-f connectors and subassembly plug-in types) are unique in that they are terminal connections for multiple-module subassemblies rather than for individual attachment to other parts or even to single functional packages (modules).

The need for such input-output terminal devices is essential in every construction philosophy if maintenance, repairability, and manufacturing feasibility of the finished equipment are to be retained. It seems unreasonable to project any complex equipment design in the image of a solid monolith of discrete parts (say, one 5-in.-diameter cordwood) or a complete receiver in a single block of silicon;* the many serious problems of manufacturing such a monstrous, nonrepairable complex suggests the more tenable approach of fabricating the circuit functions individually (modules), then grouping a number of such functions together to make up a subassembly, and finally integrating the subassemblies into assemblies, or into equipments. It is for subassembly terminations that these parts are being considered here.

* Conceptually not impossible, but of doubtful feasibility.

The tables are self-explanatory with regard to capabilities. It must be recognized that if an unmatched r-f connector can be used in an application, several much smaller configurations are presently available for use. These smaller connectors approximate the diameter of the cables they terminate.

The plug-in connector family is for use with printed wiring in the several discrete-parts systems to be described. The limitations of such connectors with regard to size is related to the limitations of terminal design on the printed-circuit boards. This subject of practical printed-wiring-board design for microelectronics and the effects on spacings and size of subassembly connectors is treated fully on the following pages.

Dimensionally Uniform Parts. The prior review touched only the systems of random-shaped parts available for discrete-parts assembly.

TABLE 3-12. ULTRAMINIATURE PRINTED-CIRCUIT CONNECTORS (UPC SERIES)

Type designation	Contacts/in.	Available pin and length combinations		Volume of mated pairs, in.³/in. length	Application
		Pin	Length, inches		
UPC 150-MR UPC 150-MS UPC 150-FS	11.5	7 12 17 23 28	0.8 1.2 1.6 2.0 2.4	0.125	Single-sided printed wiring
UPC 100-MR UPC 100-MS UPC 100-FS	16.5	9 17 25 33 41	0.8 1.2 1.6 2.0 2.4	0.125	Multilayered printed wiring
UPC 100A-MR UPC 100A-MS UPC 100A-FS	24.5	13 25 37 49 61	0.8 1.2 1.6 2.0 2.4	0.150	Multilayered printed wiring

NOTES:

1. All connectors are rated at 100 volts between adjacent terminals; 1 amp carrying capacity per contact.

2. MS (Male-straight) is used for right-angle mounting of subassemblies to printed wiring. MR (Male-right angle) is used for parallel mounting of subassemblies to printed wiring. FS (Female-straight) mates with the MS or MR male connector of the corresponding series.

TABLE 3-13. UNIFORMLY SHAPED PARTS
(MICROELEMENTS FOR MICROMODULES)

Type of part	Profiles, actual sizes		Parts/ft^3 $\times 10^6$	Remarks
		Mils (t)		
Fixed resistors:				
Utility		21 max.	3.4 max.	Cermets; 10 ohm–1 megohm; up to ½ watt/wafer; 4 resistors/wafer max.; 2 % tolerance; temp. coefficient 300 ppm/°C
Precision		11	6.4	Metal films; 10 ohms–1 megohm; up to ½ watt/wafer; 4 resistors/wafer; 1 % tolerance; temp. coefficient 200 ppm/°C
Capacitors:				
Electrolytics		82–138	0.2–0.13	Up to 470 μf-volts; dissipation factor <5 %
Coupling and bypass		11–100	1.6–0.18	50 volts up to 0.02μf, 25 volts above; ±15 % and +30–50 % types available; dissipation factor 1.5 %; 1,000 pf–0.5 μf
Precision		11–100	1.6–0.18	50 volts; 10–5,000 pf; dissipation factor 0.01 % at 1 Mc
Variable (trimmers)	310 t	60	0.3	50 volts; Q = 500; 1.5–5 pf, 3–10 pf, 2–18 pf available
Inductors:				
Coils	310 Dim. in Mils	82	0.2	Inductance up to 1.5 mh; temp. coefficient 50 ppm/°C up to 100 μh; up to 10 mh possible; Q = 50–90 (toroidal ferrite)
Transformers		138	0.13	Pulse transformers available as microelements; audio transformers available in micromodule dimensions only
Diodes		31–70	0.58–0.26	General-purpose, ultrafast-switching, rectifier, zener, and variable-capacitance types available
Transistors		47–87	0.38–0.21	High-frequency power (approx. ½ watt at 70 Mc), medium- and high-speed switching, and other types available
Quartz crystals		49	0.37	0.005 % stability; 7–70 Mc
Other:				
Ceramic filters		80	0.22	Individual resonator disks mounted in wafer recesses
Solid-state circuits		50	0.36	Digital circuits on wafer
Thin-film circuits		21	0.85	Experimental only (limited to digital circuits)

Such parts are available in mass production volume, and, in most cases, from multiple sources of supply.

In the several parts-assembly techniques to be discussed, a basic and unique requirement is demanded of the discrete parts, namely, the discrete parts must be of uniform geometry to permit efficient integration into functional circuits.

Tables 3-13 and 3-14 describe two such disciplined geometry systems: the *microelement family* of parts for the micromodule system and the *pellet family* of parts for the pellet system of assembly. The packaging capabilities of these families of parts have been included in the panoramic display of Fig. 3-3. The availability of the microelements from vendor sources is fairly extensive at this writing (over 40 suppliers), and is expected to expand in the next two years. The availability of pellet components is somewhat more limited at present, but growing user interest suggests that additional vendor sources (supplementing the two present prime sources) will develop. Some of the technological problems and gaps remaining to be solved in both these parts systems are

TABLE 3-14. DISCIPLINES GEOMETRY PARTS (PELLET SYSTEM)

Type	System 1		System 2		Remarks
	Profiles, actual size	Parts/ft^3 x 10^6	Profiles, actual size	Parts/ft^3 x 10^6	
Resistors:	$\frac{1''}{16}$		0.03"		
Composition		3.5		29	10 ohms - 5 megohms, 0.1 watt
Film		3.5		7.3	Approx. 50 ohms - 0.3 megohms
Wire-wound				2.9	
Capacitors:					
Electrolytics		0.6		1.2	System 1: 60 μf-volts (12 μf max.) System 2: 20 μf-volts (4 μf max.)
Ceramics		3.5		7.3	Up to 8,200 pf, 50 volts
		0.6		1.2	Up to 4,700 pf, 50 volts
Diodes				19	
Transistors		5.5		11.4	
Inductors		0.6		1.2	

evident from a survey of Tables 3-13 and 3-14. The uniform geometry, as mentioned previously, has the added attribute of being adaptable to easy machine handling, an inevitable process step in a matured microelectronic system (whether in the discrete-part or integrated-circuit philosophy).

2. Terminations and Interconnections

A printed-wiring board in a microelectronic application may be used for:

1. Integrating discrete components into functional circuits (modules).
2. Integrating modules into subassemblies (the latter generally being the plug-in type using a subassembly connector).
3. Integrating subassembly connector receptacles (back-plane wiring).

In these applications, certain general considerations are worthy of note. To permit maximum efficiency of part or module placement into printed-wiring boards, and to facilitate their production, it is necessary to use certain disciplines relating to layout of parts; for this purpose, standard grids have been devised for the location of the mounting holes. United States industry standards today recognize 0.025 in. as the basic grid increment.* In early applications of printed wiring when relatively large parts were used, a coarse grid spacing of 4× (100 mil) for the terminals of parts was found to be adequate. However, the introduction of transistors and the subsequent families of miniature passive parts necessitated printed-wiring-board layouts on a 2× (50-mil) grid to improve packaging efficiency. The more recent use of miniature discrete parts to build larger functional circuit aggregates (modules) has generated interest in the use of the intermediate 3× (75-mil) grid for spacing of the leads of these modules and of connectors for microelectronic subassemblies.

Other factors which must be carefully considered in designing the printed-wiring interconnections, particularly for microelectronic usage, are: (1) Adequacy of the amount of copper around the terminal holes necessary to assure a good solder joint and adequate adhesion to the laminate, (2) adequacy of the electrical spacing between terminal areas and conductors to preclude dielectric breakdown, and (3) the rather severe tolerances that must be imposed on the artwork and processing.

Single- and double-sided printed-wiring boards have been available for many years, but more recently multilayered boards[2] have been introduced, having as many as six layers (Fig. 3-4), to permit highly compact interconnection designs needed for high-density packaging systems.

* In other words, the finest standard grid is 0.025 in. with permissible terminal-lead spacings of parts of 0.025, 0.050, 0.075 in., etc. These incremental spacings are sometimes identified as multiples of the basic 0.025-in. grid (1×, 2×, 3×, etc.).

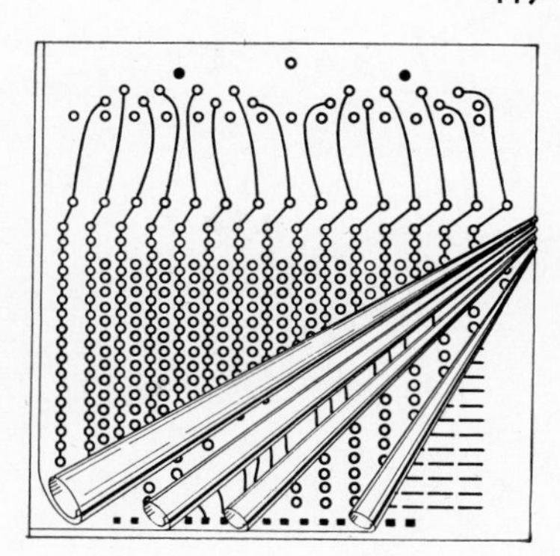

FIG. 3-4. Multilayer printed wiring. A multilayer board may consist of two or more layers of etched copper-foil circuitry on epoxy resin–impregnated glass-fiber base, each layer insulated from the others, and laminated under heat and pressure into a rigid board with accessible terminations to each level of circuitry. Problems of registry and interlayer connection require considerably more care and skill than for the conventional single- or double-sided etched-wiring board.

Through very extensive usage and perfection of processes, materials, and techniques, printed-wiring boards now provide substantial microelectronic interconnection capabilities.

Tables 3-15 and 3-16 show the recommended terminal-area spacings* for single- and double-sided printed-wiring boards. Tables 3-15 and 3-16 are for the following two cases, respectively:

Case 1. Where all terminations are adjacent to one another with *no conductors between.*

Case 2. Where *one conductor is allowed between* adjacent terminations.

These spacings were developed with due consideration to all the layout factors mentioned above. The data in these tables are particularly useful in establishing the pattern designs and the practical limitations for such high-termination-density† parts as connectors, cordwood modules, micromodules, etc., as will be discussed subsequently.

It should be noted from these tables that, for 100-volt (maximum) circuits, the least allowable spacing between adjacent terminations is 0.150 in. for the single-sided and 0.100 in. for the double-sided boards.‡

The more complex geometry of multilayer boards and the related termination density and crossover capabilities are shown in Table 3-17. Here, for 100-volt maximum working-voltage circuits, the minimum allowable terminal-area spacings (allowing subsurface crossover conductors between adjacent terminals) is 0.075 in.§

The composite picture of the interconnection density capabilities for all printed-board systems is given by Table 3-18. The maximum

* United States Army practice.

† Termination density is defined as the number of component-part terminations per unit area.

‡ The UPC series of subassembly connectors previously discussed (Table 3-12) is based on these maximum allowable termination densities.

§ Connectors based on this high-termination-density capability are being tooled for production at this writing.

TABLE 3-15. TERMINAL-AREA SPACINGS FOR SINGLE- AND DOUBLE-SIDED PRINTED WIRING (NO THROUGH CONDUCTORS)

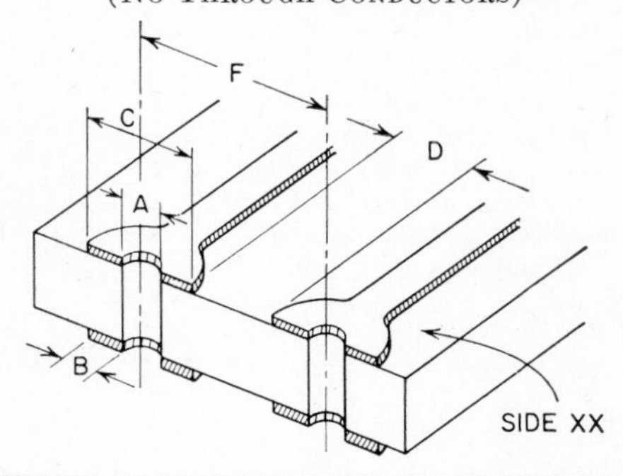

Dimensions, inches	Single-sided printed wiring, side XX		Double-sided printed wiring	
	300 volts	100 volts	100 volts	50 volts
A: Mounting-hole diameter, max.	0.060	0.045	0.035	0.020
B: Annular-ring width, min.	0.040	0.025	0.015*	0.015*
C: Minimum terminal-area diameter (A + 2B)	0.140	0.095	0.065	0.050
D: Minimum electrical spacing (coated)	0.030	0.020	0.020	0.015
E: Tolerances	0.030	0.035	0.015	0.010
F: Recommended mounting-hole spacing (C + D + E)	0.200	0.150	0.100	0.075

* Reduced annular ring made possible by use of plated through-hole for increased solder-joint strength.

theoretical termination densities are quite high; however, it is impractical to consider printed-wiring-board terminations only (no conductor interconnection freedom). Accordingly, a column of the estimated *practical termination* densities has been included in Table 3-18; these figures reflect the needs of average circuitry for ohmic connections between various terminations. For example: double-sided printed-wiring boards are capable of a theoretical termination density of 100 terminations/in.^{2} based on a 0.1-in. terminal spacing; however, a practical limit of only about 18 terminations/in.^{2} can be expected in a typical layout integrating discrete parts. This limitation is determined, first, by the part dimensions and, second, by conductor routing restrictions. It is conceivable that practical termination densities as high as 50 per in.^{2} in the double-sided board could be achieved in some instances when integrating modules and making back-plane (connector bank) interconnections.

TABLE 3-16. LAYOUT DESIGN FOR SINGLE- AND DOUBLE-SIDED PRINTED WIRING

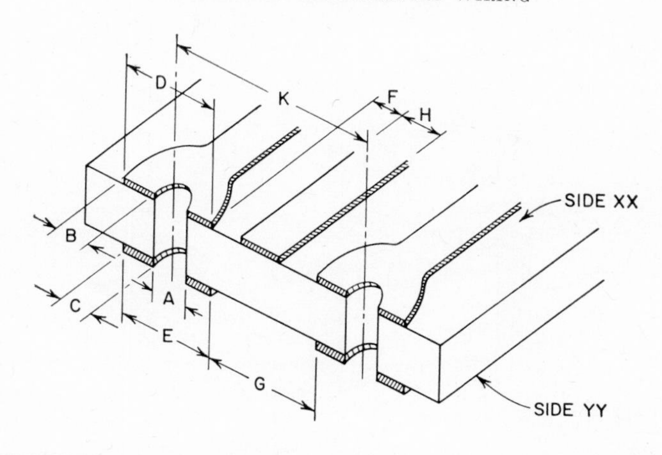

Dimension, inches	Single-sided printed wiring, side XX		Double-sided printed wiring	
	300 volts	100 volts	50 volts	50 volts
A: Mounting-hole diameter, max.	0.045	0.035	0.035	0.020
B: Annular-ring width, min., side XX	0.025	0.020	0.005*	0.005*
C: Annular-ring width, min., side YY	N.A.	N.A.	0.015*	0.015*
D: Minimum terminal-area diameter, side XX (A + 2B)	0.095	0.075	0.045	0.030
E: Minimum terminal-area diameter, side YY (A + 2C)	N.A.	N.A.	0.065	0.050
F: Electrical spacing, min., side XX†	0.030	0.020	0.015	0.010‡
G: Electrical spacing, min., side YY†	N.A.	N.A.	0.025	0.015‡
H: Conductor width, min., side XX	0.030	0.020	0.015	0.015
I: Tolerances, side XX	0.015	0.015	0.010	0.010
J: Tolerances, side YY	N.A.	N.A.	0.010	0.010
K: Recommended terminal-area spacing with one conductor (D + 2F + H + I)	0.200	0.150	0.100	0.075

* Reduced annular ring made possible by use of plated through-hole for increased solder-joint strength.

† Protective coating over circuitry.

‡ Minimum spacing available for 0.075 terminal-area spacing.

N.A. Not applicable.

TABLE 3-17. LAYOUT DESIGN FOR MULTILAYER PRINTED WIRING WITH CROSSOVER CONDUCTORS

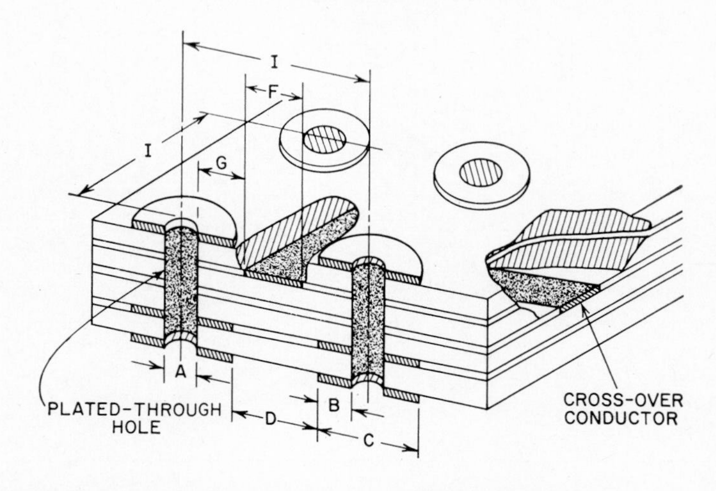

Dimension, in.	300 volts	300 volts	100 volts	50 volts
A: Mounting-hole diameter, max.	0.036	0.036	0.026	0.021
B: Annular-ring width	0.020	0.020	0.017	0.0025*
C: Maximum terminal-area diameter (A + 2B)	0.076	0.076	0.060	0.026
D: Electrical spacing between terminal areas, min. (coated)	0.030	0.020	0.011	0.020
E: Terminal-area tolerances (I-C-D)	0.044	0.004	0.004	0.004
F: Conductor width, min.	0.030	0.020	0.020	0.010
G: Electrical spacing between conductor and mounting hole, min. (coated)	0.030	0.015	0.011	0.005
H: Conductor-to-mounting-hole tolerances	0.024	0.014	0.007	0.009
I: Mounting-hole spacing (A + F + 2G + H)	0.150	0.100	0.075†	0.050†

* Reduced annular ring made possible by use of plated-through hole for increased solder-joint strength.

† The dimensions given for 0.150-, 0.100-, and 0.075-in. mounting-hole spacings apply for either the "plated-through" or "clearance-hole" multilayer fabrication techniques. Dimensions for the 0.050- and 0.075-in. mounting-hole spacings apply to the "plated-through" and "clearance-hole" techniques, respectively.

If one will accept minimal tolerances, minimal electrical spacing, and the use of plated-through holes, 0.075-in. terminal spacings can be used with double-sided boards. However, this close termination spacing is practically limited to the interconnection of three-dimensional modules or to back-plane wiring because of the general incompatibility of discrete-part dimensions with such small terminal spacings.

TABLE 3-18. COMPARISONS OF PRINTED-WIRING TERMINATIONS AND INTEGRATION CAPABILITIES

	Adjacent terminal spacing, inches	Minimum line width and electrical spacing, inches	Terminations/in.^{2} capability		Theoretical conductors/in.		Interconnection capability[a]
			Theoretical	Practical	Parallel	Crossover	
Single-sided	0.200[b](300 volts)	0.030	25	5	4	0	100
	0.150[b](100 volts)	0.020	45	10	7	0	315
Double-sided	0.100[b](50 volts)	0.020	100	18	10	10[c]	1,000
	0.075[b](50 volts)	0.020	177	40	13	0 6.5[d]	2,300 1,700
Multilayer (6 layers)	0.150[e](300 volts)	0.020	45	25	14	14	2,260
	0.100[e](300 volts)	0.020	100	50	20	20	4,000
	0.075[e](100 volts)	0.020	177	88	26	26	9,200
	0.050[f](50 volts)	0.005 for spacing 0.010 for line	400	200	40	40	32,000
References	Table 3-15				Table 3-16	Table 3-17	

[a] Interconnection figure of merit = termination X (parallel + crossover conductors/in.).
[b] Production capability.
[c] Crossover capability for 50 terminations/in.^{2}
[d] Crossover capability for 88 terminations/in.^{2}
[e] Pilot production capability.
[f] Development model capability.

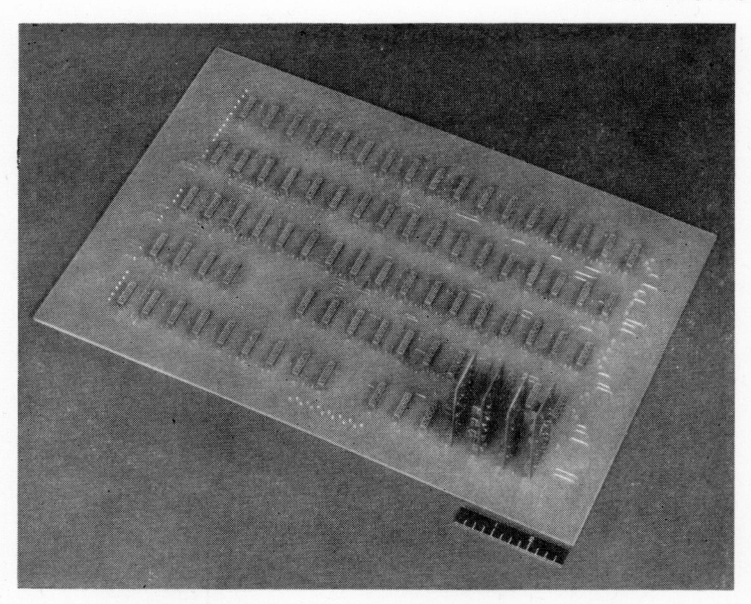

FIG. 3-5. A conventional single-sided printed-wiring board used for back-plane wiring of subassembly connectors for a reader-reperforator. See Fig. 3-6 for wiring pattern.

Multilayered printed wiring offers the following particular advantages:

1. Substantially higher conductor interconnection capability.
2. Broader tolerances for part location.
3. Considerably higher electrical rating for small conductor spacings; this attribute is derived from the fact that multilayered printed wiring is fully encapsulated.

With regard to back-plane wiring of banks of connector receptacles, terminal densities approaching 50 per in.^{2} pose very serious problems in design of the wiring layout; at such densities, the interconnection capability of single- or double-sided printed wiring becomes inadequate for handling the complex of back-plane wiring, and the design is said to be *conductor-limited*. Multilayered wiring is mandatory in such a situation. Figure 3-5 illustrates the low-volume efficiency (wide spacing of receptacles) that results when the printed-wiring layout is conductor-limited. The upper part of Fig. 3-6 shows the back-plane wiring of Fig. 3-5; superimposed on part of the board is a multilayer redesign (six layers) of the same circuit. The same 82 receptacles are easily contained on the new board which requires only 30% of the area of the single-sided design. Ample space is available on each layer for the inter- and intra-receptacle wiring.

The progressive shrinkage in the sizes of discrete parts, modules, and related subassembly connectors, and the increasing use of solid-state and thin-film microassemblies will require terminal spacings of 0.075 in. (177 terminations/in.^{2}) and even 0.050-in. terminations (400 per in.^{2}) with a concurrent very high order of conductor density. These demands will negate the use of any printed wiring other than the multilayered type.

It is significant to note that for microelectronic integration by multilayered printed wiring (i.e., integration of cordwood, micromodules, and other types of functional circuits to be described), a maximum practical termination density of 88 joints/in.^{2} (50 volts maximum circuitry) is anticipated. The size of the solder joint, which up to this point has been tolerable, now begins to appear too large and too crude for any finer

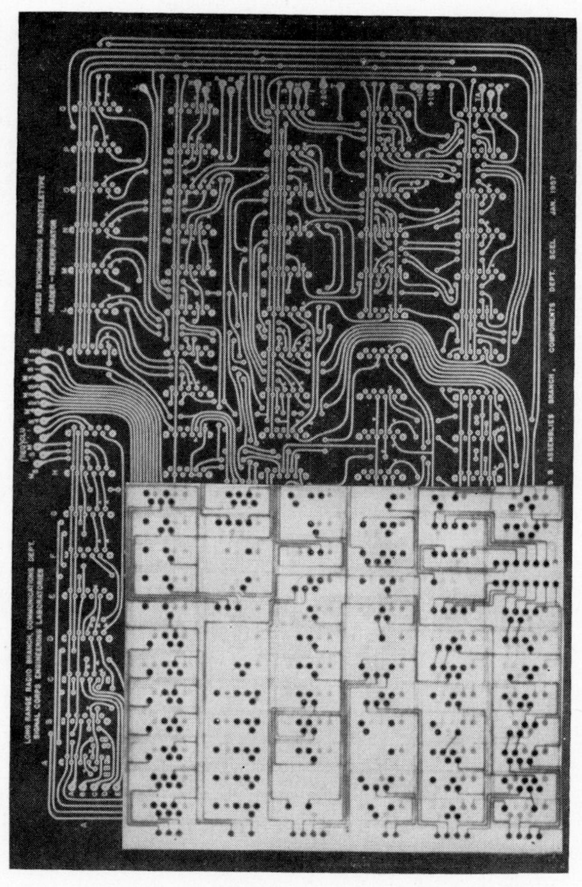

FIG. 3-6. The single-sided wiring pattern of the reader-reperforator of Fig. 3-5 compared to a six-layer design of the same circuit. Ample wiring area is available on each level of the multilayer design, as compared to the extremely constricted room for interconnecting wiring on the single-sided board. Successive layers are partially visible through translucent top layer. (*Internal Resistance Co.*)

termination capability. For all practical purposes, this density is considered the practical production limit for solder-joint printed wiring with present technology.

This limitation is one of the most cogent reasons for considering welded microjoints as the next generation technique to accommodate the microelectronics of the near future. Welding at these microdimensions is treated fully in the discussions of advanced interconnection technologies and techniques in Sec. 3-5.

The following paragraphs explore some of the special considerations related to using printed wiring in microelectronic designs.

Printed-wiring Boards for Component-parts Integration. The use of printed-wiring boards for *parts integration,* highly popular in the nonminiature or moderately miniaturized equipments (such as Fig. 3-2) has permitted practical parts densities in planar structures up to about 50,000 parts/ft.^{3}* Table 3-16 has shown that double-sided boards in a 50-volt (maximum) system may have line widths of 0.015 in. and spacings of 0.010 in., in which case a termination density as high as 40 joints/in.^{2} is attainable. However, in the general case of a practical printed-wiring layout, even when using the smallest discrete parts available, such densities are not even approached. When double-sided or multilayer boards are used for parts interconnection, it becomes all the more obvious that such planar-card circuits are component-size-limited and not *integrating-board-limited.* It is conceivable that progressive size reduction of discrete parts in their present conventional varied shapes may one day permit utilization of the termination capabilities indicated in Table 3-15; however, as a practical matter of the moment, it must be recognized that the microelectronic production capabilities of the planar-card-type structure for integrating parts is limited to about 100,000 parts/ft^3 maximum.

Although printed wiring, as referred to above, has been associated with the common etched-foil-clad laminate type of circuit, the same considerations are applicable to those discrete-parts-assemblies systems that use screened conductors on ceramics for integration. These systems, even using two sides of a wafer for conductors, are components-size-limited for the same reason as the etched-foil derived circuits.

Printed-wiring Boards for Module Interconnection. The next most common usage for printed-wiring boards in microelectronic construction is for the integration of 3-D circuit modules into equipment subassemblies (Fig. 3-7). However, as the parts density of the modules increases (as it no doubt will as microelectronic technologies grow), it will be found that even double-sided printed wiring will become both termination- and conductor-limited. Multilayer wiring provides a ready solution in such

*In some extreme cases of select circuits, densities as high as 100,000 parts/ft^3 in a module have been realized.

FIG. 3-7. Integration of modules into subassemblies. Double-sided printed-wiring boards are used to integrate about 30 functional circuits (micromodules in this instance) on each subassembly. The two subassemblies are then nested together to form one "booklet" of circuitry for the logic section of a microminiature computer. The 61-pin connectors (0.1-in. pin centers) and double-sided boards are adequate in this design. If the approximately 30 modules on each board were put in a tight arrangement of module next to module, a substantial shrinkage in the area of each board would be accomplished, but a 0.075-in.-pin-spacing connector would have to be used, and the constricted wiring area would have demanded multilayering of the wiring board. The arrangement illustrated avoids these connector and printed-board complications. (*RCA.*)

cases up to the point where 0.050-in. spacing is still adequate to accommodate the number of terminations on the module.

Printed-wiring Boards for Subassembly Connector Integration. The final general use for printed-wiring boards is as a subassembly integrator; the subassembly-connector receptacles are mounted on the board which provides all interconnecting wiring (e.g., Figs. 3-5 and 3-6). This is the most stringent application for printed wiring since it requires, in a majority of cases, the connection of two or more conductors to a single terminal. Accordingly, the conductor capability of single- or double-sided printed wiring is quickly exceeded, and the use of multilayered printed wiring becomes necessary. A production capability now exists to provide layered printed wiring for 0.150-, 0.100-, and 0.075-in. terminal spacings. At this writing (1963), a 0.050-in. terminal-spacing layered printed-wiring capability, projected for use in the near future, is under development.

Other Techniques. *Wire-wrap* is another technique for individual-part, circuit-module, or subassembly integration, which has a good interconnection capability. Multiple turns of wire are wound tightly on a pin-type terminal possessing at least one sharp edge. The use of a convenient mechanical tool for the wrapping operation permits uniform

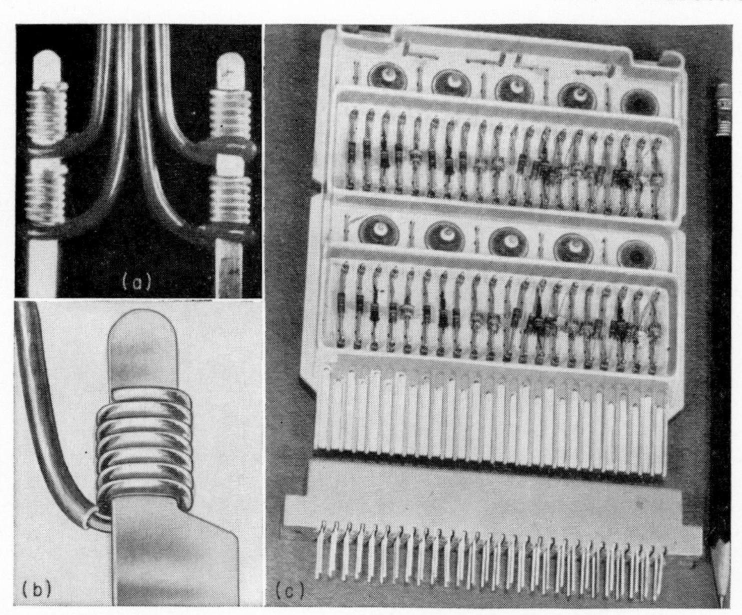

FIG. 3-8. Wire-wrap termination technique: (*a*) Double connection, typical of usage in back-plane wiring of connectors; (*b*) a typical single connection of six turns; (*c*) an application of wire-wrap at the component-part level.

and dependable gastight joints to be made quickly. The absence of heat, flux, and solder splash avoids damaged connections and short circuits. Figure 3-8 illustrates this technique.

Table 3-19 summarizes the wire-wrap termination and interconnection capability and provides a basis for comparison of over-all "connection efficiency" of this technique with that of printed wiring as shown on Table 3-18. The wire-wrap interconnection capabilities (or figures of merit) shown are per square inch of electronic chassis area. Whereas printed-wiring-board terminations require about 0.100 in. thickness over-all,* the wire-wrap technique uses terminals from 0.500 to 0.750 in. high. This 5 to 7.5 times poorer volumetric efficiency of wire-wrap should be recognized. However, of possibly greater significance is the ease and flexibility with which wire-wrap can be accomplished for fabrication of small lots of modules or equipments as compared to multilayer printed wiring. Artwork and special tooling are at a minimum for wire-wrap, and the consequent cost advantage holds up to the point of moderate (small lot) production requirements. Extensive laboratory characterization has verified the reliability of wire-wrap when established design criteria are rigidly followed.

* Based on 0.062 in. thickness of a typical multilayer board and 0.038 in. height of solder fillets above the board.

Table 3-19. Wire-wrap Termination and Interconnection Capability

Adjacent terminal spacing, inches	Terminal dimensions		Wire diameter over insulation		Turns per pin	Termination density, terminations/in.^{2}		Conductor density, conductors/in.^{2}		Interconnection capability*
	Cross section, mils	Pin length, inches	B&S wire size no.	Wire diameter inches		Theoretical	Probable	Parallel	Cross-overs	
0.200†	20 × 60	0.75	24	0.060	7	25	21	25	25	1,250
0.150†	45 × 45	0.75	26	0.050	7	45	38	35	35	3,150
0.100‡	22 × 25	0.75	28	0.048	7	100	85	50	50	10,000
0.075§	25 × 25	0.5	30	0.043	7	177	150	65	65	23,000

* Interconnection capability (a figure of merit) = terminations/in.^{2} × (parallel + crossovers per inch).
† Automatic production capability in 1962.
‡ Handtool production capability in 1962.
§ Prototype of handtool production capability (1963 or 1964).

The use of special solder and welding techniques (both resistance and electron beam) will be considered later in the discussions of module fabrication where these techniques provide a unique microelectronic integration capability.

3-3. MODULAR (3-D) CIRCUITRY

The preceding pages have described the diversified types of discrete parts (of both random and uniform shapes) that make up the *miniature-parts inventory*, from which microelectronic circuits can be constructed. The following two objectives are generally considered most important in selecting a technique for integrating such discrete parts:

1. Achievement of the maximum volumetric efficiency (smallest size) of the finished subassembly
2. Reduction of assembly costs to a minimum with no sacrifice in the reliability* of the assembled circuit

The planar printed-wiring approach to integration of parts, as stated before, has inherent limitations that make achievement of high parts densities (beyond 50,000 parts/ft^3) very difficult. Furthermore, efficient stacking of such flat assemblies in an equipment is severely limited because the minimum separation between the boards is determined primarily by the thickest or highest component on the boards. Also, the vacant space that must be allowed above the component-part leads and above the terminations is reflected as waste volume for the full thickness of each planar assembly. It follows that the more efficient utilization of the vertical dimension (as is done in three-dimensional cordwood modules) will permit higher parts densities to be attained with random-sized parts. Further increases in circuit packaging efficiency are attained in such 3-D systems by selecting parts of uniform (or similar) lengths so that the fullest use is made of the selected height of the module. As this type of single-dimension control is augmented by additional geometrical restrictions, families of parts of firmly disciplined geometry may be generated (e.g., microelements for micromodules). Such disciplined geometries permit significant reductions of the waste space between parts in assembly and a major step forward toward more efficient integration of parts into modules; the disciplined shape of the resulting modules then permits efficient combinations of such modules into subassemblies and these, in turn, into equipments. A concomitant attribute of these disciplines is the excellent adaptability of uniform geometry parts to

* That is, the assembly process should cause no degradation in reliability over that expected from a more conventionally assembled circuit.

machine handling for automatic assembly of modules, and even of modules into subassemblies.

The many considerations related to volumetric efficiency, to the various concepts of assembly of discrete parts, to the methods of interconnections, and to geometric disciplines currently in use are described on the following pages.

1. Random-geometry Techniques

Three-dimensional modules (Fig. 3-9) may be comprised of parts of various sizes and shapes. The natural transition from the circuit diagram to physical realization suggests a complex interconnection of parts randomly oriented in three dimensions to achieve the module structure. Such an elemental and undisciplined approach to construction of a functional assembly, particularly with extremely small parts, frequently yields very poor volumetric efficiency, and creates considerable assembly difficulties in production. However, by careful consideration of the packaging operation and by better utilization of the volume, it is possible to produce a relatively high-density equipment, using miniature conventional parts. Figure 3-10 shows an example of such an equipment—Mariner A magnetostrictive delay-line memory developed for Jet Propulsion Laboratory. The total power (including all input-output logic and associated electronics) is less than 1 watt. As such structures receive more attention with regard to size reduction, and especially to better producibility and maintenance, a modular approach must be considered.

Random Orientation. In one early form of module for the variable-time fuze, randomly oriented conventional parts were held for intercon-

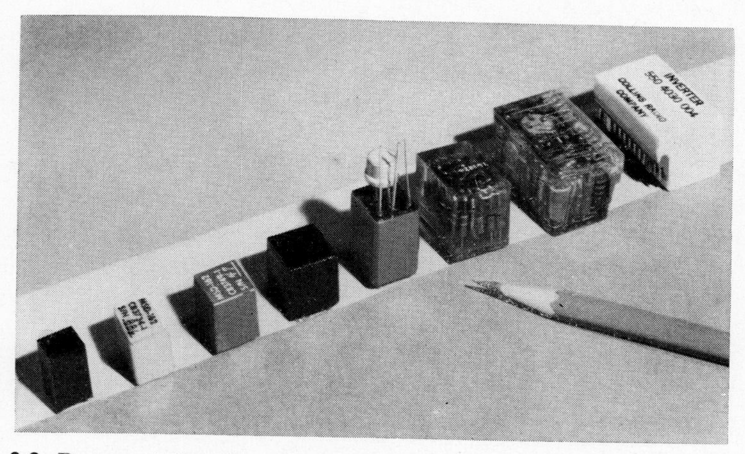

FIG. 3-9. Representative three-dimensional modules utilizing discrete component parts.

FIG. 3-10. Random-parts assembly. The interconnection of randomly oriented miniature parts in three dimensions, as illustrated in the above hand-wired assembly, yields an assembly that has parts overlaying other parts, low volumetric efficiency (in this case, about 2,000 parts/ft^3 in the underchassis areas), with severe demands on skills and care in the assembly operation. (*Computer Control Co.*)

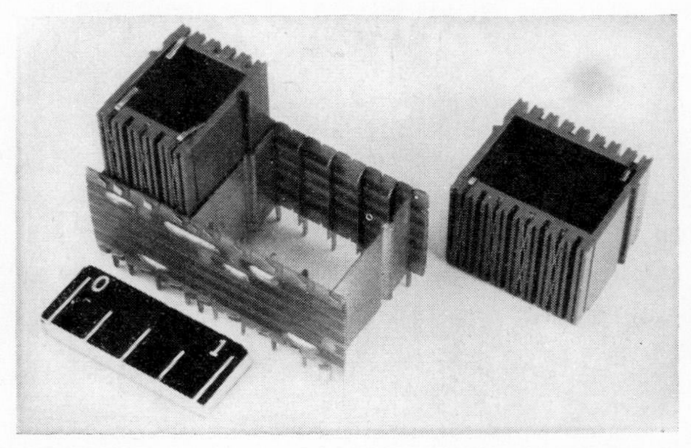

FIG. 3-11. Module shells with slide-in contacts for integration into larger subassembly shells. (*AMP-MECA.*)

FIG. 3-12. Module shells with plug-in terminals for ready removal from printed-wiring integrating boards. (*Elco Mod-U-Con.*)

nection by a molded polyethylene preform. After the leads were connected together, a complete polyethylene encapsulation provided rugged support. More recently, molded epoxy shells have served as containers in which randomly oriented parts can be potted. One approach, the AMP MECA system, also provided external slide-in contacts so that several such modules could be readily plugged in and removed from a larger subassembly (see Fig. 3-11). Another manufacturer, Elco, provides their Mod-U-Con module form already terminated in plug-in type individual contacts (see Fig. 3-12).

Parallel Orientation. The problems of assembling randomly oriented parts into modules on a production basis are considerably simplified if the axes of all the cylindrical parts are maintained parallel to each other. This configuration, which for obvious reasons has received the picturesque designation *cordwood* (Fig. 3-13), simplifies the placement of the parts for assembly. Also, the leads are readily available for interconnection at the common end planes. The manner of interconnecting these leads has also influenced the design of cordwood-type modules. The Ordnance Division of Eastman Kodak Company made an early application of printed wiring for interconnecting the leads of component parts stacked in cordwood fashion. Fixtures were provided to facilitate threading the

FIG. 3-13. Cordwood module. Uniform orientation achieves efficient packaging of conventional parts if these are of essentially uniform lengths. Module length and width may be varied to accommodate circuit size; the height of this module is approximately ½ in. (*Sprague Electric Co.*)

component leads into corresponding holes in the end plates. Where space was not at such a great premium, other manufacturers have used terminals located at notches around the edges of the plates.[3] The parts are then assembled with their leads in these notches by simple edge-soldering of the modules. More recently, considerable attention has been given to interconnecting the leads of cordwood modules by means of resistance welding to nickel ribbons (Fig. 3-14).

Each of the foregoing cordwood approaches requires careful attention in design to prevent undue complication of the interconnections at either end of the module. To minimize expense, crossovers must be studiously avoided. The over-all height of the module is, of course, determined by the length of the longest component. In some cases, short components such as transistors have been efficiently included in module designs by placing them on opposite end plates in such a manner that they are assembled back to back. Just as in an equipment, no size advantage is gained by placing short modules among tall ones, so similarly there is no point in paying

FIG. 3-14. Welded cordwood module. Strong high-temperature joints may be achieved if all process variables are maintained under very strict control. (*Elco.*)

a premium for short components to be scattered among the longer ones of a given module.

Though the above factors bearing on the size and producibility of cordwood modules are quickly recognized in the design process, other considerations highly important to the reliability of such modules are often overlooked. Excessive tensile and compressive forces may appear alternately at the terminals of the component parts as an encapsulated cordwood module is cycled through its extremes of operational temperatures. The differential between the temperature coefficients of expansion of the encapsulant and of the various materials employed for the parts and leads may readily result in detrimental forces being applied to the part termination. Damage in terms of failed parts, such as diodes and resistors, has been experienced. The high reliability considered to be inherent in the use of established conventional pigtail parts does not necessarily apply when those parts are subjected to environments more severe than those for which they were designed and tested. Rigid encapsulation, far from being a panacea to permit the use of submarginal parts, may instead introduce a particularly hazardous environment, especially in the case of military equipments operating over wide temperature extremes. In many designs, the force created by the excess expansion of the encapsulant (which may easily exceed 0.002 to 0.003 in.) is applied against the relatively large areas of the end plates; the force is then transmitted through the part lead to the most critical point—the junction between the lead and the active element within the part. Much of this force is also applied at the joint between the part lead and the interconnecting circuitry. This situation has been a factor in stimulating the search for techniques such as welding, which has higher intrinsic strength than soldering. However, it should be noted that the basic problem of excess stress on the components is not removed by this approach. Actual stress measurements shown by curve *A* in Fig. 3-15 indicate that the cyclic forces on a copper lead (the minimum situation) may readily reach values of the order of 14 lb. Though this force is not necessarily enough to break a noticeably high percentage of most parts in a few tem-

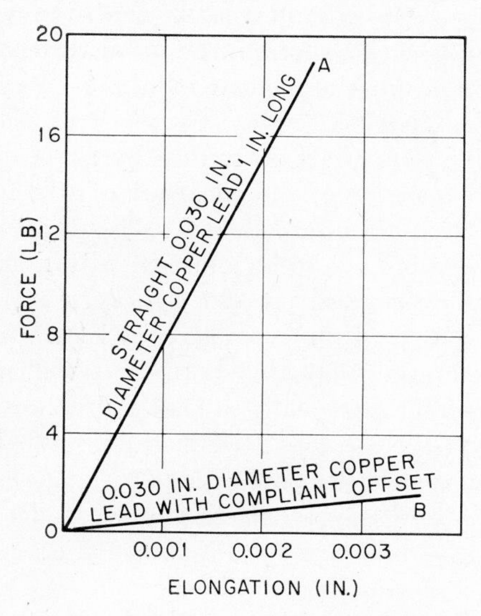

FIG. 3-15. Force on component-part leads in cordwood modules.

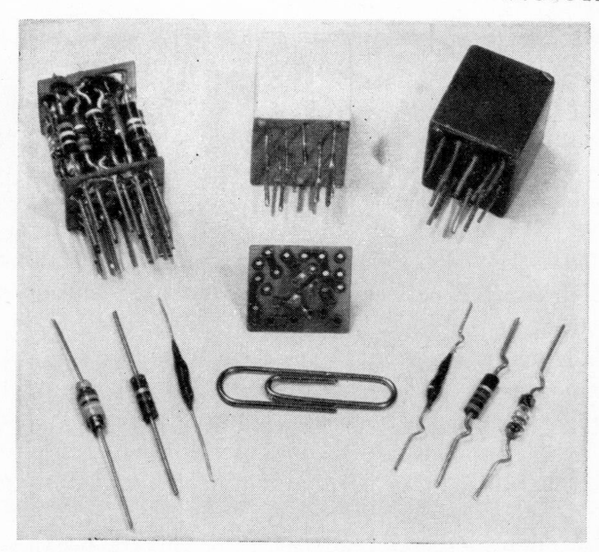

FIG. 3-16. Modified cordwood modules. Compliance introduced into cordwood design limits stress on parts due to differential expansion. (*Republic Aviation.*)

perature cycles (or else the problem would have been more widely recognized), it has, in at least one Army application, completely negated the high-reliability objective and markedly affected the production yield of a variety of module designs. The obvious solution was to provide compliance so that the expansion of the encapsulant would not cause undue force to be transmitted to the part termination. This was done as shown in Fig. 3-16 by introducing a kink into each of the component leads. Curve *B* of Fig. 3-15 shows that such compliance was adequate to reduce by at least an order of magnitude the effect of expansion of the encapsulant; the force transmitted to the part termination was now reduced to less than one pound. However, to permit this compliance to be effective within the rigid encapsulating material, it was necessary to precoat the entire structure of parts, leads, and end plates with a thin coating of a Dow Corning DC-271 silicone resin. Tests showed that this permitted the encapsulant to expand independently of the part leads so that the values of curve *B* could actually be realized in the encapsulated module. While these modifications greatly reduced the force to which the part itself was being subjected, no data were available to establish that even under such moderated stress environment the inherent high reliability of conventional parts might still not be suffering degradation.

To establish that the foregoing modifications[4] were indeed adequate as a general solution, some 480 modules were constructed of the nine

TABLE 3-20. MODIFIED CORDWOOD-MODULE OPERATIONAL LIFE-TEST DATA

Module type	Quantity	Solder Joints	Transistors	Diodes	Resistors	Capacitors	Failures
Read amplifier	22	1936	132	44	352	132	5*
Flip-flop	41	2706	82	328	410	164	0
Inverter	58	2610	116	116	464	116	0
2 Input gate	67	2077	0	603	201		0
3 Input gate	21	588	0	168	84	0	0
7 Input gate	10	400	0	160	20	0	0
Noninverter	42	840	84	0	42	84	1
Clock driver	10	510	30	30	80	50	0
Write amplifier	24	2040	96	96	96	48	0
Total*	273	11,771	408	1521	1397	562	1

* Read amplifier failures resulted from incorrect assembly of electrolytic capacitors (reversed polarity). These data were eliminated from the totals and from further analysis.

different types indicated in Table 3-20. After assembly, precoating, and encapsulation, the modules were checked against their individual operational test specifications.[5] As part of acceptance testing, each module was then given a preconditioning thermal shock test of 5 cycles from −65 to +85°C, in accordance with method 107, condition A, of MIL-STD-202B. This weeded out some modules which were found to have manufacturing defects. The units were then divided into several groups. One group was subjected to a range of environmental tests including shock (50*g*), vibration (10 to 55 and 55 to 2,000 cps), and moisture resistance. No failures resulted. The remaining 295 modules were operated in typical digital circuits in an 85°C ambient for 3,750 hr. The modules were monitored throughout the test. The six failures are indicated in Table 3-20, which also notes the fact that 5 of the 22 read-amplifier modules failed because the electrolytic capacitors had been assembled in reverse polarity. All data on these 22 modules were censored from the totals of Table 3-20 and from further analysis. The remaining significant failure was a short-circuited ceramic capacitor in the non-inverter module. The statistical analysis of these results (see Table 3-21) indicates demonstrated failure rates for the various classes of parts which are consistent with those experienced in conventional unencapsulated printed-wiring-board assemblies. Beyond merely verifying that the modified cordwood module design was indeed adequate to avoid degradation of parts reliability, the data obtained to date actually suggest a potential for a markedly enhanced reliability in such a structure. Accordingly, the modules of Table 3-20 are being subjected to extended

TABLE 3-21. RELIABILITY ANALYSIS OF OPERATIONAL LIFE-TEST DATA ON MODIFIED CORDWOOD MODULES

Component type	Quantity in sample	Failures in sample	Total unit hours $\times 10^6$	At 75% confidence	
				Demonstrated minimum MTBF	Calculated maximum failure rate, % per 10^3 hr
Transistors	408	0	1.53	928×10^3	0.108
Diodes	1,521	0	5.7	3.94×10^6	0.26
Resistors	1,397	0	5.24	3.77×10^6	0.27
Capacitors	562	1	2.11	764×10^3	0.13
Printed-wiring solder joints	11,771	0	44.10	31.3×10^6	0.004

life tests for an additional 15,000 hr to obtain data upon which a valid improved reliability estimate may be based. Results are expected in July of 1964. (The long-time factor is another indication of the problems involved in establishing assured high-reliability assembly techniques.)

Of course, not all cordwood module designs are solidly encapsulated. However, if the parts are terminated to rigid end-plates, differences in expansion between leads and parts of various constructions must still be accommodated. Designs which employ thin end plates without rigid encapsulation provide the necessary compliance through the flexibility of the end plates. However, unless such designs are given secondary

FIG. 3-17. Welded cordwood with plastic honeycomb support. Honeycomb facilitates introduction of compliance by use of flexible end encapsulation. (*Hughes Aircraft.*)

protection, they are subject to damage by handling during equipment assembly and maintenance. Protection may be provided by a molded shell. Also, the parts assembly may be ruggedized by a "wash"-type encapsulation which coats all surfaces but does not fill large voids between parts, leads, and end plates. This results in reduced weight without some of the temperature rise, moisture entrapment, and materials incompatibilities which may be introduced by foamed encapsulants. Another design of cordwood module construction by the Hughes Aircraft Company (Fig. 3-17) employs a rigid honeycomb to support the bodies of the individual parts. Interconnection is by resistance welding with nickel ribbon. Undue stress on the part termination is avoided by completing the end encapsulation with a semiflexible encapsulant.

2. Uniform-geometry Techniques

From a size-efficiency point of view, the cordwood-type module encountered limitations occasioned by dissimilar lengths and diameters of cylindrical parts, and by the relatively restricted interconnection capability of the two end planes. Also, as the size of parts is reduced, the proportion of the volume required by the structural and protective material and leads of the part, as opposed to the volume of the active element material, is increased. Hence, greater emphasis must be given to achieving geometric uniformity and connection compatibility between the various parts and their interconnecting media. Absolute three-dimensional control of parts geometry, such as the specification of a uniform cube, is not practicable. The variety of parts, types, characteristics, ratings, etc., involves a wide range of basic volumes to accommodate active elements ranging from thin resistive films to bulky inductors or electrolytic capacitors. Hence, at least one dimension must be a variable if minimum size for each element is to be achieved. Further, the geometry should be selected to facilitate assembly for the range of circuit applications likely to be encountered (see Sec. 3-2 under Dimensionally Uniform Parts).

Wafer Form Factor. The foregoing requirements are most effectively met in elements intended for assembly into three-dimensional modules by establishing these elements in wafer form of uniform rectangular shape. The thickness of the wafers can be made greater as required to accommodate the individual elements. Such elements may then be stacked one over the other to assemble modules of uniform cross-sectional area. Restriction of element terminations to the periphery of the wafer permits most effective utilization of wafer area for the element itself. Interconnection between elements is then accomplished by vertical conductors along the side faces of the wafer stack or module. Considerations of maximum flexibility in the termination and use of individual wafer

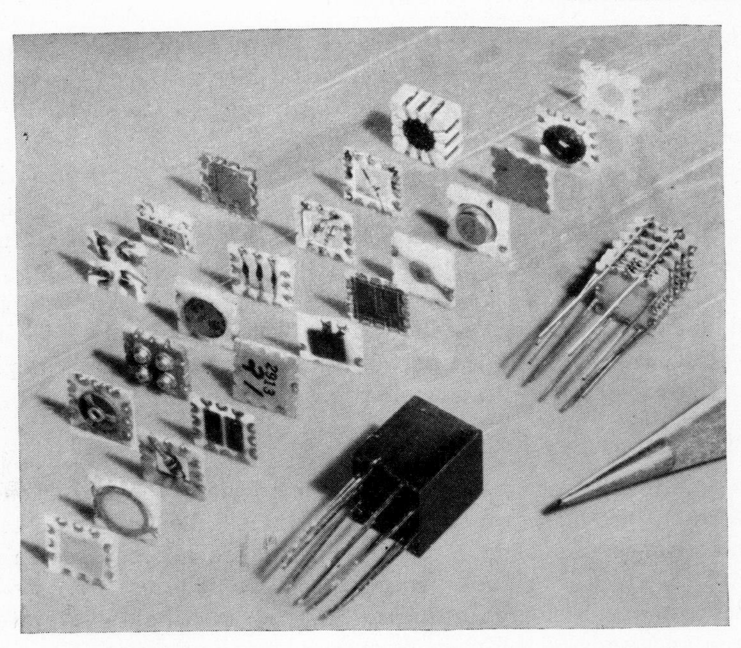

FIG. 3-18. Microelement family of disciplined geometry parts. Resistors, capacitors, transistors, diodes, coils, pulse transformers, trimmer capacitors, and quartz crystals, all in a 0.31-in.-sq. form factor of varying thickness, make up the microelement inventory at this time. At the extreme left are three experimental wafers embodying four pellet transistors in one wafer, a solid-state logic circuit on a second wafer, and a thin-film logic circuit on the third. A machine-assembled stack of microelements is at the upper right, and a finished epoxy encapsulated micromodule is in the center.

elements suggests that the rectangular wafer shape be further restricted to a square. This also permits most convenient and efficient stacking of numbers of adjacent modules into subassemblies and equipments. The remaining criterion for ensuring that these modules of wafer-type elements will permit maximum parts density to be realized when assembled into subassemblies and equipments is that the modules, themselves, should be of essentially uniform height. This becomes a matter of module design and involves balancing the number of circuit functions and elements within different modules in consideration of the various element thicknesses.

The particular size to be selected for the square element shape is influenced by the minimum area required to accommodate the range of circuit-element parameters, such as those normally encountered in transistor circuits, and by the area required to provide for the element terminations and interconnections to adjacent wafers. A further consideration is that completed module structures are conveniently integrated into subassemblies by use of printed-wiring boards containing one or more layers of circuitry. Such multilayer printed-wiring interconnecting media strongly suggest that the module terminations be coordinated with the rectangular grid layouts customarily associated with printed wiring. In selecting wafer size, it is also pertinent that consideration be given to accommodating multielement networks on a wafer, to the extent that the availability or the projected capability of advanced element technologies are indicated.

The micromodule program[6] (Fig. 3-18), begun in 1958, had as its objective a demonstrated parts density of 600,000 parts/ft³ to achieve a tenfold size reduction of military equipment. It was considered essential that the micromodule concept be applicable right from the start to the design and production[7] of a wide range of transistorized military equipment, of at least comparable performance and reliability to that attained with conventional parts. Since cost of electronics is of great interest, the approach was to provide significant cost savings as well.

Figure 3-19 shows the outline dimensions for a standard microelement and micromodule.[8] Three riser wires, or leads, are spaced 0.075 in. apart along each side of a 0.300-in. square. The leads are located on a basic

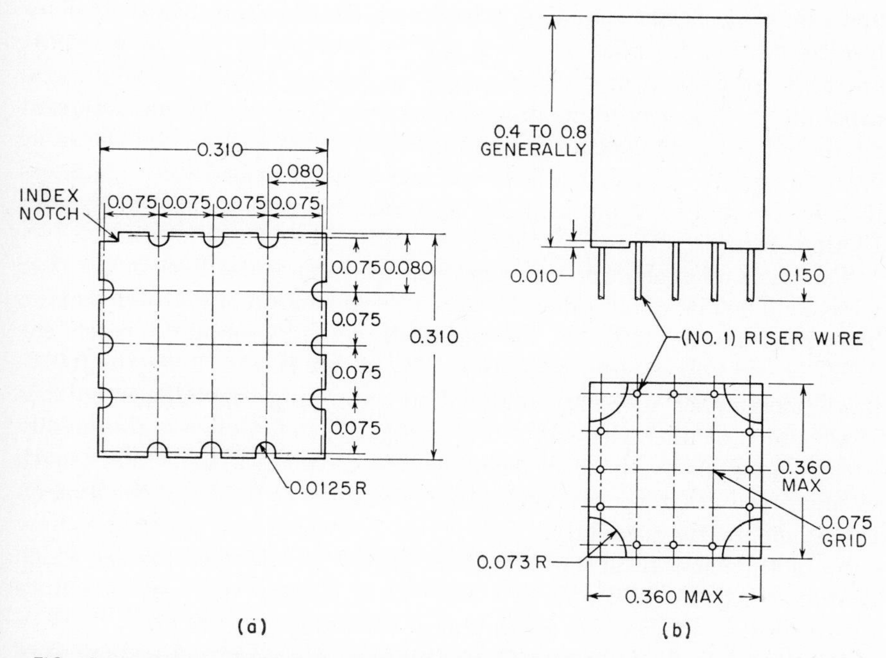

FIG. 3-19. Micromodule standards: (*a*) Outline dimensions of standard microelement wafer. Substrate thickness is 0.010 in. (*b*) Outline dimensions of an assembled and encapsulated micromodule.

FIG. 3-20. Encapsulation of a micromodule stack. The assembled and tested micromodule stack is inserted into an epoxy shell (shown at left in cross section) with riser wires threading through perforations at the bottom. The shell is filled with an epoxy formulation to the requisite height (center cross section) and cured, and the funnel head is then cut off. Subsequent marking and final testing completes the terminal processing.

0.075-in. grid. Corresponding holes in a printed-wiring board used for interconnection between modules can be located by step and repeat operation of drilling or punching facilities having a basic 0.150-in. grid capability. The over-all module dimension of 0.360 in. permits adjacent micromodules to be located on 0.4-in. centers with all leads on the same 0.075-in. grid. Adequate allowance is made for reasonable tolerances in hole size and location, lead size and location, module skewness, positioning, etc.

The basic microelement wafer extends past the center lines of the riser wires by 0.005 in. on all sides to an over-all dimension of 0.310 in. square. This provides for 0.025 in. of encapsulation all around to reach the 0.360 in. over-all module dimension. As shown in Fig. 3-20, the 0.016-in.-diameter riser wires are soldered to appropriate metallized notches in the stack of microelements. Subsequently, the module is thoroughly encapsulated, as by the shell technique shown in the figure. The length of micromodules generally varies from about 0.4 to 0.8 in., depending on the elements they contain. Individual microelement substrates may vary in thickness from 0.010 to 0.030 in. or so. Microelements which contain relatively bulky devices may have recesses or holes to reduce the over-all thickness to a maximum of the order of 0.100 in.

Micromodule Performance Capabilities. A generalized statement of micromodule design and application capabilities[9] is that they will provide small size, full performance, and high reliability for portable, vehicular, guided-missile, and projectile equipment. The specific service environments for which the micromodule has demonstrated full capability are shown in Table 3-22. Functional circuit capability is determined primarily by the active devices employed. Specific designs have involved digital repetition rates and communication frequencies approaching tens and hundreds of megacycles, respectively. In dense assemblies of hundreds and thousands of modules, power densities of the order of 100 mw per circuit function may readily be utilized. By application of more elaborate thermal design techniques, efficient cooling may be provided for individual stages dissipating a watt or more of power.

The range of circuit performance achievable with micromodules is given in Table 3-23, which lists the principal parameters of some 30 modules selected from over 300 designs. Figure 3-18 and Table 3-13 indicate the variety of classes of microelements which are available to the circuit designer. New types and increased ranges of parameters continue to be added. The ranges of parameters in which individual types of microelements are available are discussed in Sec. 3-2 under Dimensionally Uniform Parts.

Micromodule Reliability. Figures 3-21 and 3-22 show two equipments which demonstrate the feasibility of the application of micromodules: a 50-Mc f-m helmet radio transceiver designated AN/PRC-51 (see Fig. 3-21), and a portable Micropac computer designated AN/TYK-9 (see Fig. 3-22). The Micropac was designed more or less after a jeep-mounted mobile version. These two equipments, completed in 1962, met their operational specifications. To obtain reliability data covering these module designs and certain improvements and advances (such as multielement wafers) which had been introduced into the program, quantities of modules from these equipments were also subjected to full environmental and operational-life testing. Many of the modules were operated for over 5,000 hr. An analysis of the life-test results is given

TABLE 3-22. MICROMODULE ENVIRONMENTAL CAPABILITY

1. *Operating temperature of microelements:*
   a. −55 to 85°C
   b. −55 to 125°C
2. *Vibration:* 2,000 cps at 10*g*
3. *Shock:* 15,000*g*, 4 msec rise time and 8 msec total duration
4. *Spin:* 20,000 rpm, 1 in. radius
5. *Storage:* −65 and 71°C
6. *Altitude:* to 150,000 ft
7. *Temperature cycling:* −55 to 85°C or −65 to 125°C dependent on required maximum temperature of operation. MIL-STD-202, method 102A, test condition *A* or *C* as applicable.
8. *Salt atmosphere:* MIL-STD-202, method 101A
9. *Moisture resistance:* MIL-STD-202, method 106

TABLE 3-23. RANGE OF MICROMODULE DESIGNS
Communications Modules, AN/PRC-51

R-f amplifier (XM-995):
Center frequency, 51 Mc. Bandwidth at 6-db points, 2.3 Mc. Power gain, 20 db. Input, 50 ohms

Mixer (XM-997):
Center frequency, 51 Mc. Bandwidth at 6-db points, 320 kc. Conversion gain, 2 db. Input, 1,000 ohms. Output load, 50 ohms

Oscillator 61.7 Mc (XM-996):
Output frequency, 61.7 Mc. Output, 150 mv. Output load, 220 ohms

H-f and i-f amplifier (XM-998):
Center frequency, 10.7 Mc. Bandwidth at 6-db points, 300 kc. Power gain, 34 db. Input, 50 ohms

Converter (XM-999):
Oscillator frequency, 10.245 Mc. Conversion gain (including filter), −8 db. Oscillator output, 150 mv. Input, 10.000 ohms

I-f filter (XM-1000):
Center frequency, 455 kc. Bandwidth at 6-db points, 50 kc. Insertion loss, 1.0 db. Output load, 1,000 ohms

First and second i-f module (XM-1001):
Operating frequency, 455 kc. High-frequency response, −6 db, 1 Mc. Gain, 38 db. Input, 360 ohms

Third i-f and limiter module (XM-1002):
Operating frequency, 455 kc. Gain, 38 db. Limiting, 0.7 volt + 0–1 db for input 10 to 100 mv. Input, 360 ohms

Discriminator (XM-1003):
Center frequency, 455 kc. Audio, 1 kc output for 0.5 volt input ±8kc. Deviation, 30 mv. Distortion at 1 kc, ±15 kc. Deviation, 5%

Audio amplifier (XM-534):
Power output, 2 mw at 1 kc. Voltage gain, 40 db. Frequency response, 3 db, 500–5,000 cps. Distortion, 10%. Squelch voltage, 1.45 volts

Squelch (XM-533):
Frequency response at 1.0 volt input, 1.3 volts d-c at 3 kc, 2.45 volts d-c at 10 kc. Output load, 15,000 ohms

Amplifier, tone oscillator (XM-523):
Oscillator output, 2.0 volts at 1,000–2,000 cps. Amplifier output at 4.0 mv input, 1.0 volt at 500 cps, 2.3 volts at 1 kc, 2.5 volts at 5 kc. Distortion, 10%

Transmitter oscillator (XM-521):
Center frequency, 8.5 Mc. Voltage output, 2.1 volts

Tripler (XM-524):
Center frequency, 25.5 Mc. Conversion power gain, 8.5 db. Efficiency, 90%. Input impedance, 3,000 ohms

Intermediate-power amplifier (XM-525):
Center frequency, 51 Mc. Power gain, 8.5 db. Efficiency, 35%. Input impedance, 150 ohms

Phase modulator (XM-522):
Center frequency, 8.5 Mc. Output, 1 volt Zener 6.2 volts. Distortion at 1 kc and ±4 kc deviation, 20%

Power amplifier (XM-526):
Center frequency, 51 Mc. Power gain, 5.5 db. Efficiency, 55%. Input impedance, 50 ohms

Doubler (XM-1067):
Center frequency, 51 Mc. Conversion power gain, 3 db. Efficiency, 15%. Input impedance, 900 ohms

TABLE 3-23. RANGE OF MICROMODULE DESIGNS (*Continued*)
Micropac Modules, AN/TYK-9

Double standard gate (XM-707):
Standard power module. One 2-input gate and one 3-input gate. Provision for diode-cluster connection at 3-input gate

Double low-power gate (XM-709):
Low-power module. One 4-input gate and one single-input gate. Output of 5-input gate connected to single-input gate. Provisions for diode-cluster connection to single-input gate

Low-power flip-flop (XM-718): Low-power module. Two low-power gates connected to form a flip-flop

Standard flip-flop (XM-752):
Standard power-drive module. Two gates connected to form a flip-flop

One-shot multivibrator (XM-719):
One-shot multivibrator for generating a short-duration control level

Nixie driver (XM-720):
Amplifier for actuating Nixie display lamps

Neon driver (XM-721):
Amplifier for actuating neon display lamps

Input-line amplifier (XM-722):
Input-line amplifier to raise input-line signal to required level

Diode cluster (XM-726):
Two groups of 5 diodes per group. Used to extend input capability of gate modules

Shift-register driver (XM-754):
Shift-register driver for triggering groups of shift-register modules

Input-output sampler (XM-742):
Input-output strobe generator

Gage amplifier (XM-732):
An AND circuit for the strobe pulse and the memory-plane output signal

Output amplifier (sense amplifier) (XM-733):
Power amplifier stage of the sense amplifier

Decoder gate (XM-738):
An AND gate used to select the memory address lines

Memory address switch (XM-788):
A saturating switch used in conjunction with the regular switch to select the memory address lines

Strobe inhibit gate (XM-740):
An AND gate for the memory strobe signal and the strobe enable level

in Table 3-24. The table also shows the calculated values of mean time between failure (MTBF) in hours for a 10-element micromodule and the failure rate calculated on an average per element basis in terms of per cent per 1,000 hr. The particular confidence levels, representing the degree of desired trust or assurance in a given result, used for comparison are not standardized in industry. Hence, the calculations are shown for 90, 60, and 10% confidence levels.

To facilitate reliability comparisons with other given or projected situations, tabular values are given in Table 3-25, which has application to the analysis of any reliability life-test data. Knowing the total test time in unit-hours and the number of failures (catastrophic and degradational) which were experienced, the maximum failure rate corresponding to the selected confidence level may readily be calculated. Of course,

FIG. 3-21. A tactical field radio (micromodule development model). A hand-held f-m transmitter consisting of 7 modules (left) and a helmet-mounted double-conversion receiver consisting of 11 modules demonstrate the feasibility of applying the micromodules concept to communication equipment. Frequency 51 Mc; r-f power output, 100 mw; receiver sensitivity, 2μv, 10 mw output (with noise-operated squelch). Transmitter, 4.2 in.^{3} over-all, 14 oz (with batteries). Receiver, 3.8 in.^{3} over-all, 8 oz (with batteries). The micromodule chassis portion of the receiver is only 5 in. long by ½ in. wide by ¾ in. high (about 2 in.^{3}).

relative comparisons should take into account any differences in specification limits, accelerated testing stresses, etc.

Data available to date indicate comparable MTBF between micromodules and Minuteman parts under equivalent operational conditions.

FIG. 3-22. Micropac computer. A general-purpose digital computer (1.6 Mc clock frequency; random access memory, −40 to 52°C operation), consisting of approximately 1,650 micromodules comprising 25,625 components (8,400 diodes, 2,950 transistors, 10,800 resistors, 3,050 capacitors and 425 miscellaneous parts). Two plug-in memory units and a 350-watt power supply occupy two-thirds of the computer case (height, 12½ in.; width, 17¾ in.; depth, 21 in.).

TABLE 3-24. MICROMODULE LIFE-TEST ANALYSIS
(July 1962)

	No. of modules	Element hours	Failures (catastrophic and degradational)	MTBF in hours for 10-element module at confidence level of			Element failure rate, % per 1,000 hr at confidence level of		
				90%	60%	10%	90%	60%	10%
Communications:									
Initial phase	179	9,494,000	11	57,200	75,350	121,000	0.175	0.133	0.0825
AN/PRC-51 helmet radio	88	1,579,900	0	57,500	173,620	1,504,200	0.146	0.05	0.0061
Estimated present capability	267	11,074,000	3*	165,000	265,200	635,000	0.061	0.038	0.016
Digital:									
Initial phase	233	24,456,000	3	366,000	586,000	1,410,000	0.027	0.017	0.0071
AN/TYK-8 MICROPAC field computer	171	11,078,900	2†	205,160	357,380	1,007,170	0.049	0.028	0.01
Estimated present capability	404	35,535,000	3*	534,000	851,000	2,030,000	0.0187	0.012	0.0049

* Eight communications and two digital failures from the initial program were deleted from these totals, on the basis that corrections for these types of failures were instituted in the equipment modules and that these failures have not reappeared.

† Two catastrophic failures: (1) Open joint on diode microelement wafer; (2) increased forward resistance of zener diode.

TABLE 3-25. CALCULATION OF FAILURE RATE

Based on the number of failures, find the value of $nt\lambda$ in the table corresponding to the desired confidence level.

To find the mean time between failures (MTBF), divide nt (the summation of unit hours of test) by the value of $nt\lambda$.

To find the Failure Rate (λ) in units per unit hour, divide $nt\lambda$ by nt. To convert this failure rate to per cent per thousand hours, multiply by 10^5.

VALUES OF $nt\lambda$

No. of failures	Confidence level, per cent					
	50	60	70	80	90	95
0	0.695	0.915	1.20	1.61	2.30	3.00
1	1.68	2.02	2.44	3.00	3.89	4.74
2	2.68	3.10	3.62	4.28	5.3	6.3
3	3.67	4.18	4.76	5.5	6.7	7.75
4	4.67	5.25	5.9	6.7	8.0	9.15
5	5.65	6.3	7.0	7.9	9.25	10.5
6	6.65	7.35	8.1	9.1	10.6	12.8
7	7.65	8.4	9.2	10.2	11.8	13.2
8	8.65	9.45	10.3	11.4	13.0	14.4
9	9.65	10.5	11.4	12.5	14.2	15.7
10	10.65	11.5	12.5	13.6	15.4	17.0

3. Interconnection Problems

The various discrete component parts which comprise an equipment are generally assembled into several successive levels of structures (modules, subassemblies, assemblies, units) to build the complete equipment. This involves many mechanical, thermal, electrical, operational, and maintenance considerations. The electrical interconnections consist of the conductors and joints which create the current paths between the component parts. Many joints occur within the structures of the discrete component parts or wafer microelements. However, this section is concerned only with those interconnections which occur between parts within 3-D modules, between modules in subassemblies, and between subassemblies in an assembly. For purposes of discussion, the first category is treated as fixed interconnections while the latter is treated in terms of connectors which facilitate removal and replacement of subassemblies. Particular attention is given to those aspects which involve critical limitations of one sort or another and, hence, frequently give rise to what are referred to as "interconnection problems."

Within 3-D modules, the interconnection problems involve primarily (1) the form and manner of joining of the component-part terminations to the interconnecting and terminating conductors, (2) the mechanical design of the module such that the physical integrity of the circuit is maintained, and (3) the thermal design considerations within a module as influenced by its external environment. Most of the considerations relating to item 2 were discussed previously and will not be repeated here.

Cordwood-type modules generally employ either soldering or resistance welding. For soldering the part leads to printed-wiring end plates, the two end plates are commonly supported parallel to each other in jigs separated ½ to 1 in. farther apart than their final positions. The parts have long leads which are threaded through the appropriate terminal holes in each of the end plates. The end plates are then brought together to the final spacing; the leads are trimmed to extend about ¹⁄₁₆ in. through the end plates; and the connections from the leads to the printed wiring are accomplished by dip soldering. If weldable printed wiring is used, such as the AMP tab-type circuitry, the lead ends are resistance-welded to the circuit-terminal tabs. Frequently, welded cordwood modules are assembled, using punched Mylar sheets for the end plates. The part leads extend through about ⅛ in. and are then interconnected by resistance-welded nickel ribbons. Other variations include directly joining the leads of the parts to each other at either end of the module, and substituting molded eggcrate or honeycomb structures for the end plates as the means of positioning the parts for interconnections.

The wide acceptance of the use of solder is based on its relatively noncritical requirements with respect to time, temperature, lead materials, and form factor. It does require clean surfaces for good connections, and is limited for high-temperature applications to the softening point of the solder, about 183°C. Resistance welding, on the other hand, produces a weld nugget which has high strength up to the melting point of the lead materials and resulting alloy. However, the localized high temperatures needed to effect a resistance weld introduce several requirements which become particularly critical for the small conductors and terminations involved in microelectronics. For one, the required electric energy must be dissipated at the weld area. This necessitates careful control of all resistances in the weld circuit and, hence, surface cleanliness, lead resistivities, contact pressures, even electrode dynamic characteristics, all play a part. The optimum conditions vary for different lead materials, lead dimensions, and lead coatings. Hence, many different welding machine adjustments, or "weld schedules" as they are called, are required to accommodate the wide variety of component parts which are commonly used.[10]

Furthermore, variations in batch to batch of components, if not reflected in modification of the weld schedules, may result in marginal or unreliable joints. Solutions to these problems are being sought through standardization of part leads. However, the preferred leads for welding are not in all cases those which are optimum from the point of view of part performance and reliability. Resistance welding becomes increasingly difficult as spacings become smaller and tolerance requirements tighter in microelectronic assemblies.

Interconnection of microelements into micromodules is accomplished by soldering of riser wires into metallized notches. The notches are surrounded also by metallized terminal areas to support a fillet of solder at each joint to effect a connection of uniform size. All surfaces are pretinned to facilitate reliable soldering. Over 250 million joint-hours in micromodule life tests without a failure have demonstrated a failure rate of less than 0.0004% per 1,000 hr. Looking to the day when more interconnecting conductors may be required within such a module, electron-beam welding of 0.002- by 0.010-in. copper ribbons to the edge of the wafers would provide an additional 24 leads (see Sec. 3-5 under microassemblies). Also, development of modified forms of resistance welding may provide similar capabilities with assured high reliability.

An interesting aspect of the interconnection of 3-D modules, in general, is thermal design. The problem is well illustrated in connection with the thermal classification of micromodules. These are classed as either 85 or 125°C designs, on the basis of their maximum allowable internal hot-spot temperatures. The 85°C class, for example, requires the use of microelements rated for at least 85°C, and the use of the encapsulation procedure established for this class. Thermal classification is based on the internal temperature since, to remove the heat being dissipated within the module while not exceeding the maximum allowable internal temperature, the exterior of the module must be cooled to a temperature somewhat lower, depending on the amount of heat being dissipated. Further, in microminiature equipment, the precise thermal environment for a module becomes difficult to determine. Hence, an 85°C class of micromodule, unlike an 85°C rated component part, is not intended to provide rated performance in a still-air ambient of 85°C, but rather is to be cooled adequately so that the surfaces of the internal microelements do not exceed the 85°C temperature.

For the immediate future at least, the design of 3-D modules is looked upon as being an inherent part of equipment design, using generally available standard microelements, rather than being a matter of specialized complex component design. On this basis, it is expected that modules will be custom-designed for specific applications, rather than selected from a catalog of standard modules by the equipment designer. To ensure high reliability of the resultant custom-designed modules, two more-or-less obvious thermal requirements must therefore be met. First, microelements should be of adequate temperature rating, and, second, no microelement should be hotter than the internal-temperature capability of the module encapsulation. This second point has, perhaps, some not quite so obvious implications. If several microelements, each operating within its rating, are included in a single module, their combined power dissipation may readily result in excessive internal module temperature. Thus, the module must be designed on the basis of its total power dissipation and the manner and degree to which it is to be cooled. In conventional hand-wired and print-wired assemblies, the component-part hot spots and the internal and external temperatures of the thin conformal coating over the parts are generally one and the same temperature for all practical purposes. However, with the order of magnitude reduction in size achievable with wafer-type modules, the power density in terms of watts per cubic inch is correspondingly increased. The internal and external temperatures of the module may therefore be significantly different to the degree that power is being dissipated by the module.

Also, for a given module (internal) temperature classification (say, 85°C), considerably more power may be dissipated in particular applications by increased cooling, to reduce the exterior temperature of the module. Thus, if certain thermal considerations of module design are left to the equipment designer, the latter is free to design on the basis of standardized microelement ratings and assembly procedures. This is much as he has always done with conventional equipment which utilized standard component parts and hand or printed wiring. Once a module design is completed, arbitrary ambient temperatures may be specified for test purposes, with the understanding that these are test equivalents to the actual thermal environment that the module will see in the equipment design. Except in the case of nondissipative modules, such test temperatures would always be below the module (internal) temperature classification.

Interconnections between 3-D modules are most commonly accomplished by use of single- and multilayer printed wiring. Also, as with assemblies of individual parts, leads of modules may be interconnected by hand soldering with hookup wire or by resistance welding with nickel-ribbon conductors. As the number of terminations per square inch increases at the bottom (usually) of the module, requirements on the printed wiring or other interconnecting medium become increasingly severe for smaller hole sizes, reduced terminal-area diameters, and closer conductor spacing. The need for cross connections, or conductor crossover capability, also increases markedly, particularly for digital applications. Details of these requirements and the availability of several levels of multilayer printed-wiring capability are discussed in Sec. 3-2.

Interconnections between Subassemblies of 3-D Modules. For most equipment of reasonable complexity, the dictates of producibility and maintainability call for construction in which the subassemblies can be

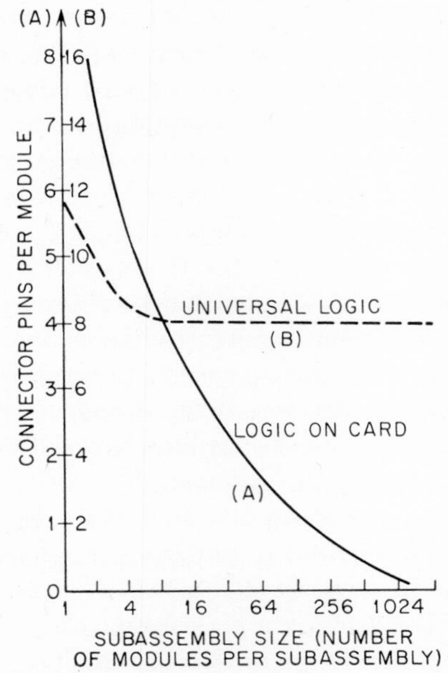

FIG. 3-23. Connector pin per module requirements versus booklet size.

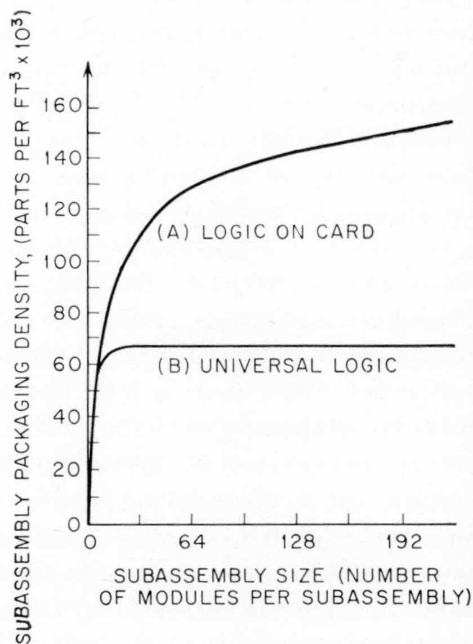

FIG. 3-24. Subassembly packaging density versus subassembly size.

readily removed. This necessitates the use of connectors. The size of the individual subassemblies, and hence of the connectors, is a matter affecting the over-all size, initial cost, and ease and cost of maintenance.

The analysis of this problem as applied to a 1650-module logic section of a computer* provides a case in point.

Figure 3-23 shows the number of connector pins required per module as the number of modules per subassembly is increased. Curve *A* is for designs aimed at achieving maximum parts packaging density in the equipment. To this end, the circuit was optimized for each card; hence the designation "logic on card." Curve *B*, on the other hand is for a "universal logic" approach wherein the over-all design is based on the minimum feasible number of different types of modules and of different types of subassemblies. This approach is highly advantageous from the maintenance viewpoint. However, as shown in Fig. 3-23 the number of pins per module for a practical 64-module subassembly is 8 for universal logic versus only 1.6 for the selected logic on card. The greater number of connector pins means greater volume of connectors in the finished equipment. The number of modules per subassembly also has a marked effect on the over-all packaging density of the complete logic section. The

* Micropac Field Data Computer, RCA, Army Micro-Module Program, Contract DA36-039 sc-75968.

average module had an individual-parts density of about 400,000 parts/ft.[3] Figure 3-24 indicates the resultant density when the modules are packaged in subassemblies within the computer logic section. The limiting effect of universal logic on packaging density (68,000 parts/ft^3) is clearly shown in curve *B*.

3-4. FLAT (2-D) CIRCUITRY

Section 3-3 discussed how the limitations of nonuniform component parts were overcome by use of improved modes of three-dimensional packaging, including the development of families of parts of disciplined geometry. Improved geometric control of part geometry, consistent with the technology of the part itself, may likewise be applied to component parts intended for assembly in thin flat-type modules. The prime requirement on parts for such 2-D modules is conformance with the maximum-thickness limitations. The high proportion of volume taken up by pigtail leads and their terminations strongly suggests the need for more direct interconnection of parts. Approaches taken include recessing of leadless cylindrical parts into the substrate wafer, use of very thin wafer parts mounted on the substrate, and re-forming parts into short cylinders, which are then inserted flush into cylindrical holes in a substrate of thickness equal to the height or thickness of the cylinders.

1. Random-parts Techniques

Given an assortment of small parts of reasonably small thickness dimension, these may readily be mounted in the conventional manner on a small printed-wiring board. Figure 3-25 shows such a flat circuit fabricated on a 0.7-in.-square, 0.05-in.-thick epoxy glass wafer.[11] The component parts are soldered to solder-plated printed wires on one side of the wafer. Printed wires on the wafer terminate in 16 tabs on the wafer's peripheral edges. Some reduction in over-all effective thickness of the subassembly is accomplished by dispensing with the leads and fillets of solder extending $\frac{1}{32}$ in. below the board. Also, removal of all circuitry from the under-

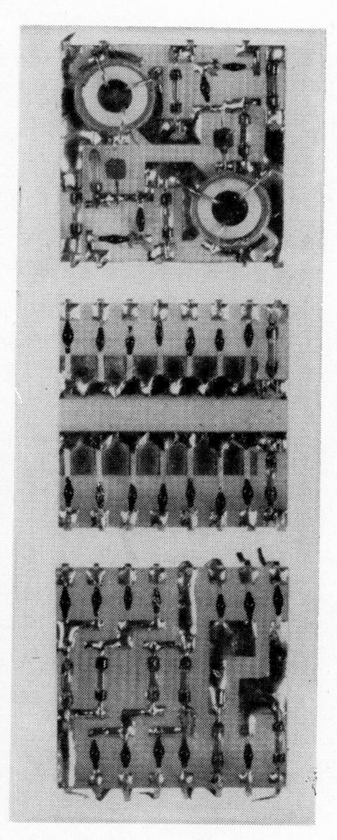

FIG. 3-25. Random 2-D subassembly. (*Arma Div.*)

FIG. 3-26. Electron-beam-welded 2-D subassembly. Conventional parts are welded to metallized conductors on a ceramic wafer. Uniform spacing of termination points permits rapid indexing of the assembly under the beam during the welding operation. (*Hamilton-Standard.*)

side assists in insulating between adjacent boards, with resultant further space savings. Figure 3-26 shows such subassemblies. The part leads are attached by electron-beam welding. The metallized ceramic substrate provides adequate bond strength for direct support of the individual component parts.

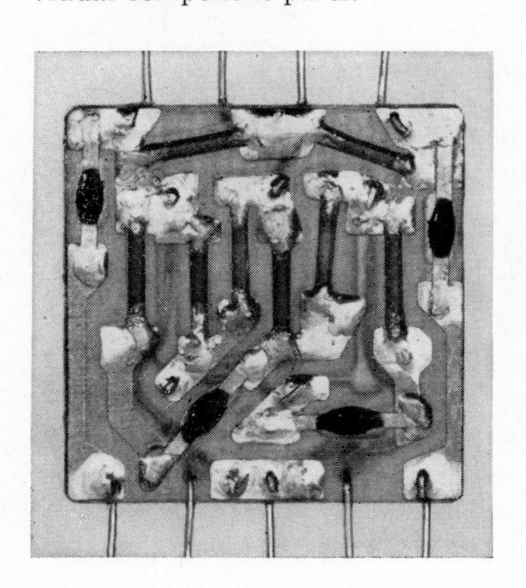

FIG. 3-27. A 2-D subassembly using conventional (leadless) parts and soldered terminations. (*Cleveland Metal Specialties.*)

Split-tip welding, a newly developed modification of resistance welding, employs a thin electrode which consists of two halves laminated together lengthwise with insulation between. Over-all tip diameters down to about 0.001 or 0.002 in. are compatible with microelectronic dimensions. A wide range of applicability is achieved by precise control of current pulses.

A further gain in total thickness is achieved by forming holes or recesses in the substrate, into which the component-part bodies are placed. Interconnection space may be saved by substituting terminal areas for the part leads. Figure 3-27 shows such leadless cylindrical parts soldered directly to corresponding terminal areas on the substrate. Figure 3-28 illustrates an approach where the substrate is eliminated; here the parts are mechanically supported and electrically connected by welding to a preformed sheet-metal base plate. After encapsulation, the module leads are cleared by trimming off the supporting bar. Since the component parts are fully protected to start with, the encapsulant is required only to perform a relatively noncritical mechanical support function.

In seeking parts which naturally lend themselves to use with flat substrates, screening techniques may also be used which have been developed over the past decade or more.[12] Fired metallization on ceramic has

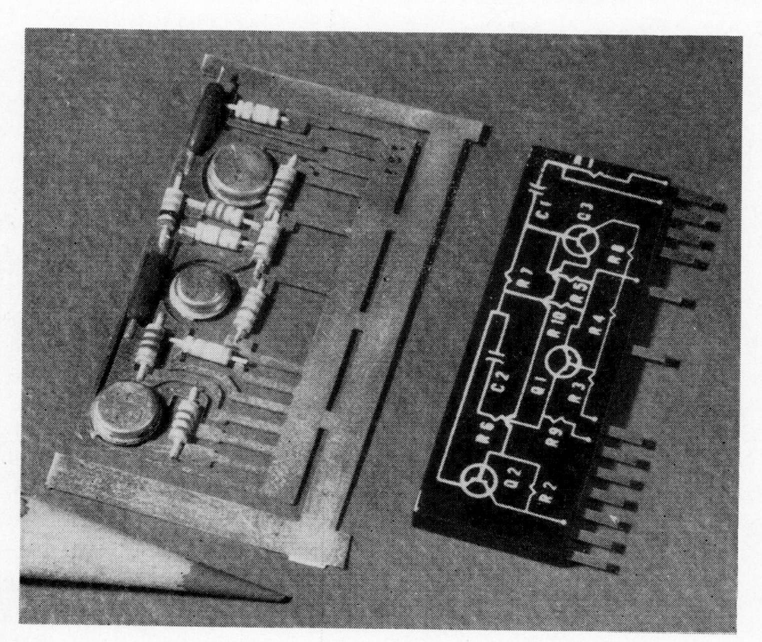

FIG. 3-28. A 2-D subassembly using conventional parts resistance-welded to a preformed metal matrix. Terminals are readily accessible for controlled welding. Encapsulation and removal of integrating bar (bottom of assembly) provides a rugged module. (*Collins Radio.*)

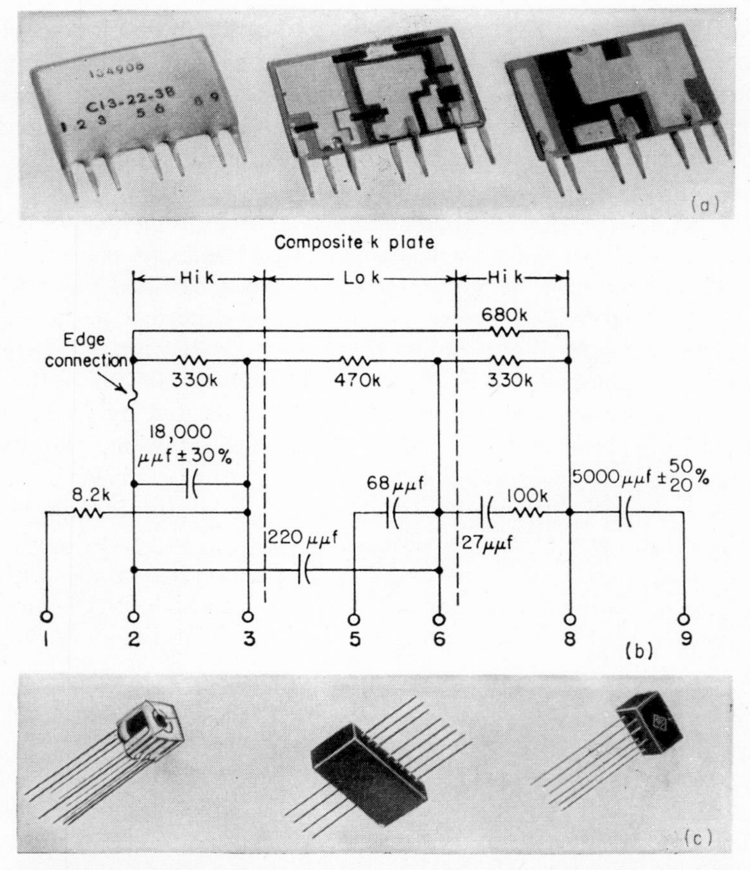

FIG. 3-29. A 2-D ceramic subassembly using deposited passive parts. Screen deposition of resistor and silver inks provides considerable flexibility in design of *RC* networks, which find extensive use in commercial radio and TV electronics in the United States. (*a*) and (*b*): Fairly complex *RC* network using a composite substrate having sections of different dielectric constant. (*c*) Three-dimensional stacking of planar deposited circuits with active devices permits efficient usage of deposited circuits in parallelepiped shape. Module parts densities of the order of 1 million/ft^3 are attainable in practical designs. (*Centralab, Division of Globe Union.*)

long been accepted for conductors and capacitor electrodes, and sprayed or screened deposition of stable resistive formulations provides a wide range of capability. These capabilities are exploited in the passive network shown in Fig. 3-29. The substrate in this case serves as the capacitor dielectric. It contains sections of high and low dielectric constants to suit the circuit requirements. For most general application, active devices (transistors and diodes) are needed also. Figure 3-30 shows an approach in which photolithographic techniques are used to provide connections to the semiconductor devices mounted in the sub-

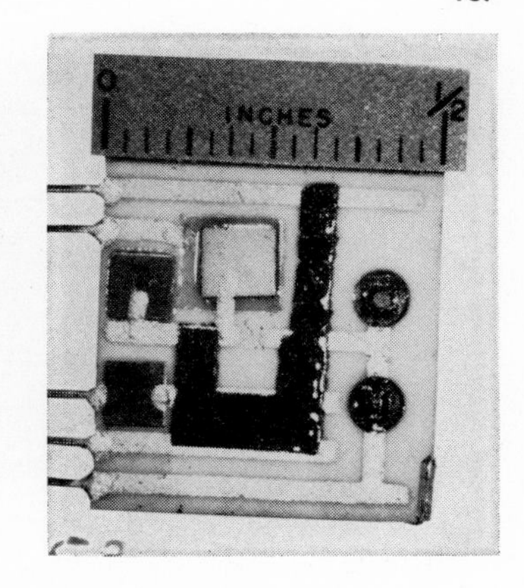

FIG. 3-30. A 2-D ceramic subassembly with deposited and attached parts and uncased microminiature semiconductor devices. Direct connections from semiconductor terminations to wafer circuitry are made by photo-lithographic techniques. (*Diamond Ordnance Fuze Laboratories.*)

strate.[13] In this illustration, small-value capacitors are fastened on top, while a larger unit is recessed into the substrate.

For high-performance high-reliability applications, hermetic sealing of transistors is generally considered to be a necessity, to avoid possible surface contamination. The transistor shown in Fig. 3-31 has its leads soldered directly to terminal pads on the substrate, is supported by its leads, and is mounted in a hole cut in the glass substrate. Conductive, resistive, and capacitive elements have been deposited on the glass by thin-film techniques which will be discussed more fully in the following chapter. Hermetically sealed semiconductor devices also are available in flat planar form such as those shown in Fig. 3-32. By mounting these units parallel to but slightly away from the substrate, useful substrate surface area is retained for resistors and conductors. Techniques have been established whereby the flat ribbon leads may be readily soldered to the thin-film terminal areas.[14]

FIG. 3-31. A 2-D thin-film circuit using hermetically sealed transistor. (*International Resistance Co.*)

FIG. 3-32. A 2-D planar thin-film computer subassembly using 56 outboarded transistors simultaneously reflow-soldered to thin-film-circuit terminal areas. (*International Business Machine Company.*)

2. Uniform-geometry Techniques

The range of component-part technology reviewed in the preceding section would appear to support a fairly broad circuit-fabrication capability. However, it should be noted that as one departs farther and farther from use of available parts of established performance and relia-

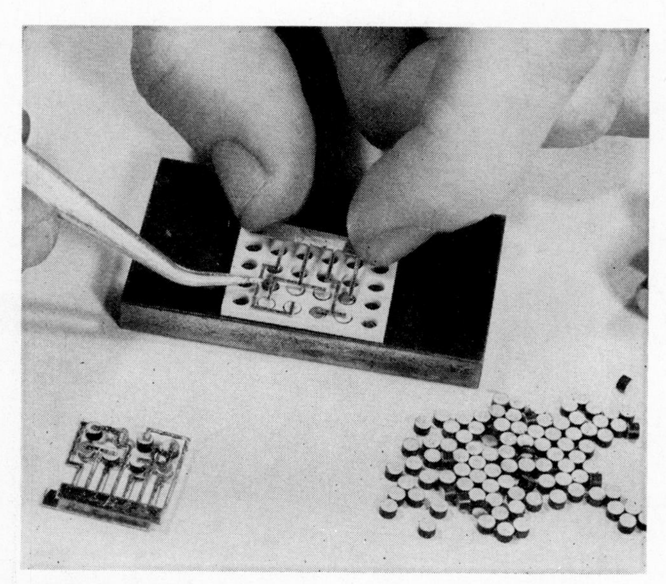

FIG. 3-33. A 2-D assembly system using discrete parts of uniform shape (pellet system). Circular parts of any reasonable diameter but with thicknesses restricted to 1/16 in. are readily inserted into recesses. Surface terminations are interconnected by soldering of a preformed matrix (or by using other interconnecting media), followed by testing and encapsulation. See ~~Fig.~~ Table 3-14 for list of pellet parts. (*P. R. Mallory.*)

bility, one assumes the responsibilities, risks, and difficulties of assuring that specially modified parts or those fabricated in situ are indeed fully adequate for the job. The slightest trouble encountered in this regard may quickly negate any potential savings in cost projected for the do-it-yourself basis.

Families of pellet-type parts of uniform height (see Sec. 3-2) have been made available which can be inserted into holes in prepunched substrates as in Fig. 3-33.[15] In one form, circuit interconnections are accomplished by use of conductive cements (Fig. 3-34). The pellet parts are also available with flat tabs, extending from the top and bottom surfaces. These tabs may then readily be welded or soldered to adjacent land areas on the substrate. Pellet-type parts are made in two general heights, 0.062 and 0.030 in. While the height must be kept constant for assembly in a given substrate, the diameters may vary, as for some capacitance values, up to 0.250 in. In addition to a range of resistors and capacitors, there are inductors, thermistors, diodes, and hermetically sealed transistors also available in this form factor. Since the parts are prefabricated, they are subject to the same types of vendor quality-control procedures that normally apply to conventional parts (see Sec. 3-2).

3. Interconnection Problems in 2-D Circuitry

Essentially the same techniques are used to interconnect 2-D modules into a subassembly as are used for the interconnection of 3-D modules into subassemblies and assemblies. The individual 2-D modules may be assembled parallel to and one over the other like a stack of printed-wiring boards. Alternatively the individual circuits such as shown in Fig. 3-25 may be mounted coplanar on a common printed-wiring mother

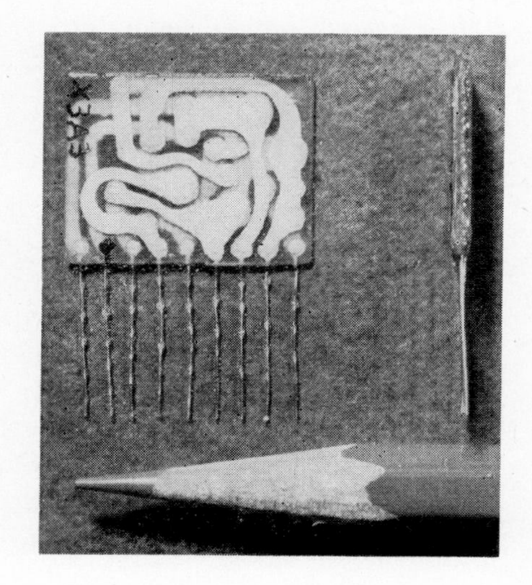

FIG. 3-34. A 2-D pellet module showing screened conductors (prior to encapsulation). (*Hughes Aircraft Co.*)

FIG. 3-35. Area integration of 2-D circuits (using discrete parts) into a subassembly for a computer. Printed wiring is used for this planar integration technique which provides adequate surface area for the interconnections. This planar-integration approach is an alternative to vertical integration of ceramic circuit substrates. (*American Bosch Arma Corp.*)

board (see Fig. 3-35). Interconnections from the modules to terminal holes in the plug-in mother board are by soldered wire leads. Figure 3-36 shows a complete computer package assembled in this way.[11]

Within 2-D modules, the variety of types and form factors of parts gives rise to a corresponding variety of interconnection schemes. For small subassemblies constructed of tiny pigtail parts conventionally soldered on printed-wiring boards, the principal interconnection difficulties lie in providing small terminal areas and close conductor spacings. With regard to reliability, however, the individual solder joints remain virtually unstressed. The small parts are normally held to the board by the conformal coating, and the compliance of the bends in the component-part leads absorbs any relative motion due to differential expansion between the parts and the board.

Provision for limiting such stresses in microminiature 2-D circuitry, however, becomes increasingly difficult as size is reduced. At the same time, the stress due to differential expansion between two directly coupled dissimilar members is constant, regardless of their length. Furthermore,

FIG. 3-36. Stacking of area-integrated subassemblies in a computer package. Back-plane wiring of the male connectors at the bottom is done with harness wiring. (*American Bosch Arma Corp.*)

differential expansion between dissimilar materials, if concentrated at the area of a small bridging electric connection, can introduce severe cyclic elongation and contraction of the connection as the temperature changes. Figure 3-37 shows how the problem can arise. A discrete part (say, an alumina-based resistor) is inserted into a cavity in a Pyrex substrate. The space between the part and the cavity walls is filled with a resin bridging material, over which the electric interconnection is deposited. In such a case, a reasonable match exists between the part and substrate, except for the expansion and contraction of the resin filler, which is squeezed in and out as temperature changes.

A more severe situation exists when a portion of a brass lug or terminal

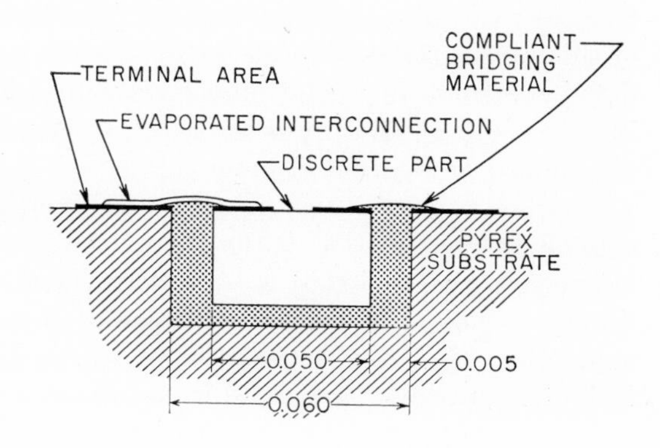

FIG. 3-37. Bridging of conductors from a substrate to a recessed part. (*Melpar.*)

of, say, 0.250-in. width is substituted for the part. The temperature coefficients of expansion of the several common materials are, per degree centigrade:

Pyrex glass	1×10^{-6}
Microsheet glass	3×10^{-6}
Silicon	5×10^{-6}
Alumina	7×10^{-6}
Brass	18×10^{-6}
Epoxy (unfilled)	$45\text{–}65 \times 10^{-6}$

Thus, the differential coefficient between Pyrex and alumina is 6×10^{-6} per °C, and between Pyrex and brass 17×10^{-6} per °C. For a 180°C change in temperature (−55 to 125°C), the differential expansions across the 0.050-in. alumina and the 0.250-in. brass, respectively, are as follows:

0.050-in. alumina vs. Pyrex −0.00006 in.
0.250-in. brass vs. Pyrex −0.0007 in.

These numbers look quite small. However, if the brass terminal should be located slightly off center, the gap to the substrate on one side may be but 0.001 in. One-half the differential expansion appearing across such a gap will result in a cyclic contraction and elongation of the interconnection of 35% as the temperature is cycled from −55 to 125°C. Even the alumina part in a Pyrex substrate would generate a 3% contraction and elongation. These figures are further magnified by the expansion and contraction of the filler mentioned previously. No casual application of solder, plating, conductive cement, etc., should be expected to withstand such service for long. Certainly even hookup wire should not be subjected to many cycles of even a small fraction of one per cent cyclic elongation if reliability is to be assured. Careful matching of materials may indeed mitigate this problem. However, this imposes marked restraint on the microcircuitry designer's freedom. At the time of this writing (1963), intensive efforts are underway, directed toward a definitive solution. The objective is to assure that the much touted inherent high reliability of the joints internal to microelectronic modules will indeed be realized in practice.

Another type of interconnection frequently used within 2-D modules also bears mention. It is the thermocompression bond commonly made between a fine (0.5 mil) gold wire and the surface of a semiconductor device, terminal part, or gold-plated circuit terminal area. Precise control of this joint is essential in semiconductor device manufacture; lack of such control may lead to weakness, which may be apparent after

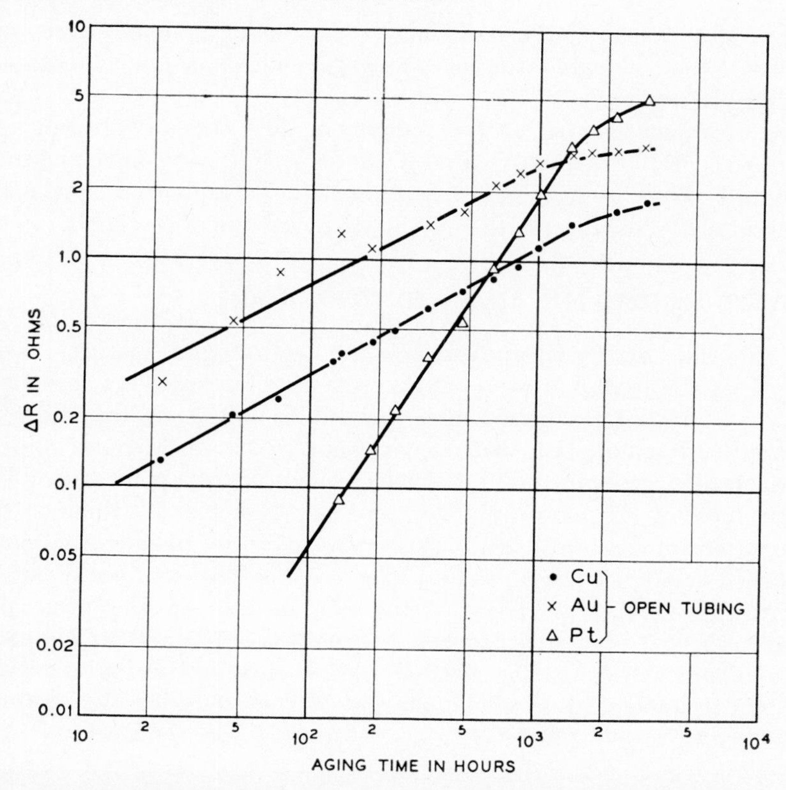

FIG. 3-38. Aging characteristics of bonded wires (aged at 200°C).

extended high-temperature aging. Figure 3-38 shows aging data* obtained for thermocompression bonds, using three types of fine wire bonded to an aluminum layer, which had been evaporated into a *p*-type germanium substrate. The structure was mounted in a tubulated type TO-28 transistor encapsulation. Samples were run sealed in room air and also in open tubulated cans. Figure 3-38 is for open tubing. The sealed units had much shorter aging life than those left open to air.

The curves show significant increases in joint resistance of the order of an ohm within 100 to 1,000 hr of aging at 200°C. Data for silver wire were also taken but were not plotted since open-circuit failures occurred generally within the first ten hours. This failure mechanism of thermocompression bonds as a result of thermal aging involves a reduction in mechanical strength of the bond as well as a change in electrical resistance. Even at lower temperatures, this failure mechanism proceeds at a rate sufficient to seriously limit the ultimate life of the connection. Thermocompression bonds should, therefore, be used with caution in microelec-

* Bell Telephone Laboratories, Signal Corps Contract DA36-039 sc-88931.

tronic circuitry, particularly so when it is noted that a commonly accepted value for excessive degradation of a conventional solder joint is not one ohm, but one milliohm.

Also, ultrasonic cleaning, as used in many module assembly procedures, has been reported to cause failure of thermocompression bonded leads in certain types of semiconductor devices. The unsupported leads may be excited into resonance, with consequent overstress of the bond.

3-5. HYBRID SYSTEMS AND ADVANCED TECHNOLOGIES

As thin-film and semiconductor-circuit technologies mature, their range of capability and areas of applicability may be expected to broaden greatly. In the interim period, their capabilities frequently may best be exploited in combination with more established techniques. Such hybridization may be introduced at many levels of equipment assembly. In fact, most of the more fully developed technologies continue to be incorporated into hybrid assemblies as new advances enhance the more established capabilities also. Thus, effective and efficient hybridization occurs at the module level (both 3-D and 2-D), as well as at the subassembly assembly and equipment levels. An analysis of the probable impact upon equipment design of the successive levels of hybridization as microelectronic technological developments unfold is of considerable interest.

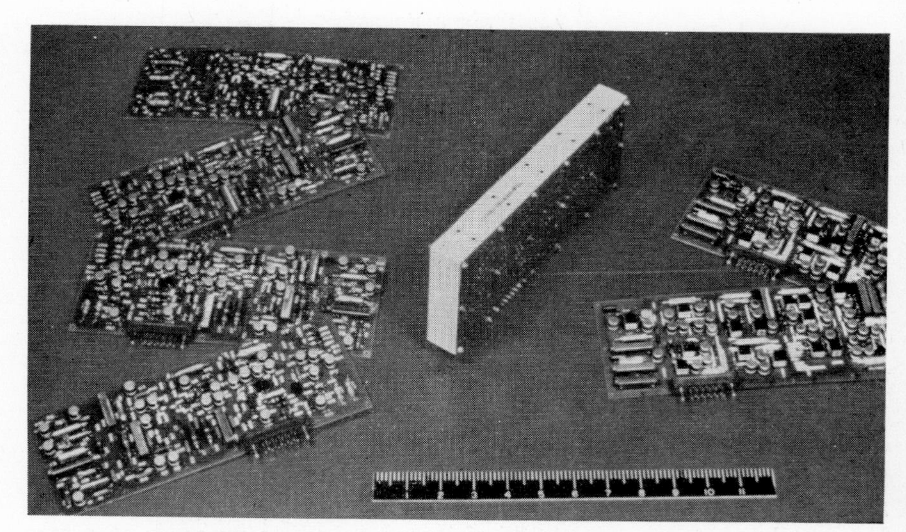

FIG. 3-39. An illustration of hybrid use of conventional parts and functional circuits. Four circuits boards (left) made by conventional printed-wiring and miniature-parts technique could not be accommodated in the specified two-board assembly format (center). By utilizing micromodules for selected circuits, the necessary two-board design of identical circuitry was possible. (*Courtesy RCA/Navy.*)

FIG. 3-40. A hybrid subassembly of cordwood and conventional parts. By outboarding certain parts, it was possible to limit this board to only three different types of modules. (*Republic Aviation.*)

1. Hybrid Subassemblies

The printed-wiring-board subassembly has for years been a prime medium for integration of new components into operating equipment. Hence, it is not surprising to find micromodules hybridized into printed-wiring subassemblies,[16] as at the right side of Fig. 3-39, for the same reason that the other components are there—they are needed by the equipment. In this instance, the requirements on an existing portion of an equipment (center) were doubled. This would have necessitated four boards (left) in place of the original two if it were not for the high-density micromodules. Prime savings were realized by avoiding the repackaging of the entire equipment to meet the increased requirements.

Similarly, cordwood modules have been hybridized with conventional parts to accommodate different circuit requirements from board to board in a large digital equipment (Fig. 3-40). The fewest number of types of different modules were designed and then a few conventional parts were outboarded on the individual printed-wiring boards as required.

2. Hybrid 3-D Modules

The cordwood type of 3-D module is particularly well suited to integrating a wide variety of pigtail-leaded components. However, the newer thin-film and semiconductor circuit devices are not especially

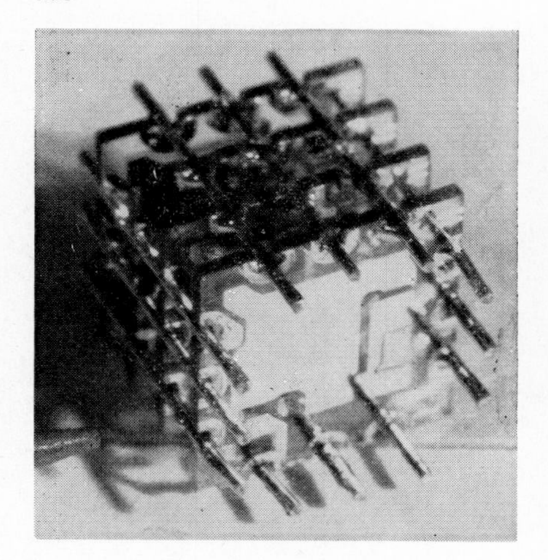

FIG. 3-41. A three-dimensional (micromodule) assembly of thin-film-circuit wafers. Each wafer is an interconnected *RC* array to which tiny active devices are added by thermocompression bonding. When techniques are developed for fabrication of thin-film active devices in situ on such a thin-film circuit, hybridization by addition of discrete active devices could be eliminated. (*Experimental—Servomechanisms, Inc.*)

amenable to this form factor, and hence have not found general acceptance within cordwood modules. The multiwafer-type modules, such as the micromodule, on the other hand, are compatible with the newer approaches, which may be employed on some or all of the wafers. Figure 3-41 illustrates thin films incorporated in a four-wafer micromodule comprising two flip-flops. Completed eight-wafer modules built with four such flip-flops are 0.65 in. high and have 64 parts. Tiny microdiodes and transistors compatible with the parts-density capability of the thin films are used to achieve the parts density of 1,700,000 parts/ft^3 in these modules.

FIG. 3-42. An integrated semiconductor logic circuit on a microelement wafer. Semiconductor hybridization of such circuits with thin films and other microelements is readily accomplished within the micromodule. (*Texas Instrument.*)

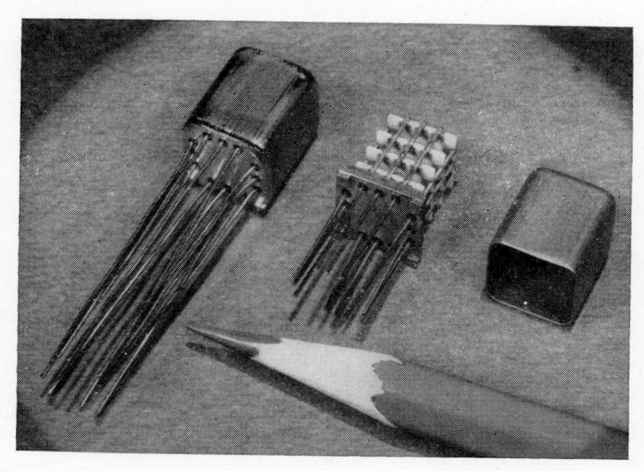

FIG. 3-43. An electron-beam-welded micromodule. Riser wire connections and seal of hermetic can are made by electron-beam welding. (*Experimental—Hamilton Standard.*)

Figure 3-42 shows a hermetically sealed semiconductor circuit mounted on a microelement wafer. Micromodules made of three such wafers are 0.250 in. high and contain 24 parts in the following circuit functions: 1 flip-flop, 2 gates, and 2 emitter followers. The module-parts density is 1,360,000 parts/ft^3. This might readily have been doubled if this were an all-flip-flop module.

Various advanced interconnection capabilities discussed in Sec. 3-4 may likewise be utilized on microelement wafers as applicable. Similarly, as shown in Fig. 3-43, welding techniques may be employed for hermetic sealing and interconnection of wafers of the micromodule when such techniques have been established to have production or reliability advantages.

3. Hybrid 2-D Modules

Two-dimensional modules naturally lend themselves to hybridization of different techniques. Several of the illustrations in Sec. 3-4 could strictly be classed as hybrids. The interconnection problems encountered within such modules are, to a large extent, influenced by the dissimilar materials involved. For hybrid 2-D modules using thin-film techniques for conductors, resistors, and capacitors on glass substrates, another type of problem may occur where the thin films meet the terminal areas for external connections to the module. These external connections are usually in the form of a wire or ribbon conductor, and in use are subject to possible application of relatively large forces. Hence, the terminal area must be strongly bonded to the substrate, a requirement

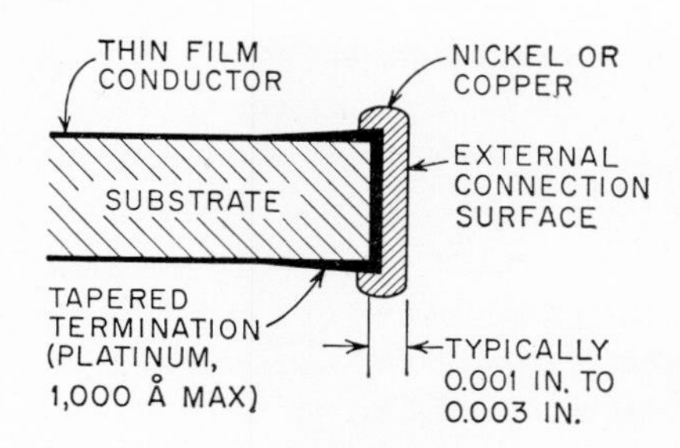

FIG. 3-44. Substrate termination for thin-film to heavy metallization.

which generally suggests a firing process to provide a thickened terminal area. The thin films may be of the order of 100 A thick whereas the termination is preferably of the order of 0.001 in., or 2,500 times as thick. Hence, the thickness of the terminal area must be gently contoured so as not to be too abrupt for the thin film, or else reliability of the connection will be compromised. Figure 3-44 shows a cross section of this critical area.

4. Microassemblies

It becomes apparent, as the circuit functions on a small conventional subassembly are reduced first to a 3-D-module form factor, and then further to a small 2-D wafer module, that an assemblage of such 2-D

FIG. 3-45. Microwelding by electron beams. Metallized ceramic substrates are interconnected by 0.002- by 0.01-in. copper ribbons on 0.025-in. centers. This technique could permit extremely compact interconnection of microcircuit wafers into microassemblies. (*Hamilton Standard.*)

wafers may no longer represent just an equipment subassembly. It may quite possibly represent a major assembly of the equipment. To this new composite, the term *microassembly* has been applied.[17] A conceptual model of a microassembly is shown in Fig. 3-45. The interconnection requirements for such a microassembly are several times greater than for the 3-D modules discussed in Sec. 3-3. Interconnections are by electron-beam welding of 0.002- by 0.010-in. copper ribbons on 0.025-in.-centers.[18] Of course, an operating unit would require encapsulation or hermetic sealing for environmental protection.[19] Since it undoubtedly would have to be removable, appropriate connectors will be required. At the time of this writing, such connectors are under development.

The electron-beam welded microconnections provide a reliable, serviceable, and producible means of effecting interconnections within the microassembly. Figure 3-46 shows a close-up of these connections. In the models shown, the many copper-ribbon conductors on 0.025-in. centers are welded to the continuously metallized edges of 0.030-in.-thick substrate wafers. (In an equipment, discrete terminations would be pro-

FIG. 3-46. Close-up (150 magnification) of electron-beam welds of copper ribbons to a metallized ceramic substrate. Ribbons are 10 mils wide (see also Fig. 3-45).

vided.) The wafers shown in the photograph are stacked on edge on 0.050-in. centers. The wafers are separated by resin, which is recessed 0.005 in. back from the edges. Welds can also be made to 0.010-in.-thick wafers on 0.025-in. centers. This is a termination density of 1,600 connections/in.^{2} On a volume basis, 64,000 of these interconnections would occupy only one cubic inch, or about one-tenth the volume required for the same number of soldered micromodule connections. Substrates used are of high-density alumina and Pyrex glass, Corning no. 7740. These are considered to be adequately representative of the range of substrate material characteristics likely to be encountered in microassemblies.

For the alumina wafers, a strong bond between the metallizing and the substrate is obtained with molymanganese coating (duPont no. 7619), dried at 160°C for 15 min and then fired at 1450°C for one hour in a hydrogen atmosphere. The coating is applied so as to be 0.0005 to 0.0008 in. thick after firing. Too thick a buildup on the peripheral edges results in a convex coating, making it difficult to assure full area of contact with the flat-ribbon conductors. This coating, while highly adherent, provides too little metal at the surface to ensure a good weld. This need is met by overplating the molymanganese with 0.001 to 0.002 in. of nickel or copper. The nickel plating is annealed in dry hydrogen at 1000°F for one hour to relieve the residual plating stresses. The ribbon conductors are 0.002- by 0.010-in. hard-drawn copper. This is of reasonably low resistance and is considered most practicable for fabrication.

For the purpose of microwelding a beam current of 0.4 ma is accelerated by a voltage of 90 kv in a vacuum of 0.1 μ. This beam is applied for 5.3 msec to each weld, to yield an average energy input of 0.19 watt-sec/weld. To assure even distribution of heat over a given weld area, the beam is focused to a spot approximately 0.003 in. in diameter. The beam is then programmed to sweep sinusoidally at 1,100 cps across an area 0.009 in. wide. The resultant weld area exceeds that of a circle 0.005 in. in diameter. Since the entire ribbon is melted, it must be kept free of tension during the welding process. This is accomplished by special fixturing which supports a band of 20 ribbons in parallel and ensures accurate positioning.

Mean value of resistance of the welded microjoint is 0.5 milliohm. The chance of the resistance of any joint exceeding even a value of 2 milliohms has been determined by extensive tests to be less than 1 in 100,000.

With regard to joint strength, the ribbon conductor is the weakest link. The copper is partially annealed by the welding process, so that it fails in tension at a mean value of 330 g. The weld strength and adherence to alumina substrates are well in excess of 500 g.

The joints withstand the full gamut of shock (including 15,000g acceleration), vibration, temperature cycling, and 20 cycles of thermal shock (−55 to +200°C) with no degradation of reliability.

REFERENCES

[1] O. B. King, Miniaturization in Missiles and Satellites, chap. 4, pp. 44–64, in "Miniaturization," Reinhold Publishing Corporation, New York, 1961.

[2] Layered Printed Wiring, Final Report, Contract DA36-039 sc-78941, International Resistance Company.

[3] J. Ritter, A Versatile Modular Packaging Concept, *Proc. 5th Natl. Conf. Military Electron.*, June 26–28, 1961, pp. 44-47.

[4] USAERDL (Electronic Components Department) Technical Guidelines, Process Requirements for Soldered Printed Wiring Cordwood Modules (Tentative), dated Sept. 25, 1961.

[5] USAERDL (Electronic Components Department) Technical Guidelines, Test and Evaluation of AN/USD-4 Cordwood Sub-module, dated Feb. 20, 1961.

[6] S. F. Danko, The Micromodule Approach to Microminiaturization, "Electronics Reliability and Microminiaturization," vol. 1, pp. 65–72 (or by same author, The Micro-module: a Logical Approach to Microminiaturization, *Proc. IRE*, vol. 47, pp. 894–904, May, 1959).

[7] J. P. Morone, Jr., and S. M. Stuhlbarg, A Mechanized Micro-module Assembly System, *Electron. Packaging and Prod.*, vol. 2, no. 3, pp. 10–14, May-June, 1962.

[8] Signal Corps, Technical Requirements for Micro-modules, No. SCL-7700, issued by U.S. Army Electronics Research and Development Laboratory, Fort Monmouth, N.J. [Specification MIL-M-55183(EL) is scheduled to be issued in 1963.]

[9] Micro-module Production Program, Quarterly Reports 1 through 21 covering period Apr. 1, 1958, to June 30, 1963, and Final Report covering Apr. 1, 1958, to June 30, 1963, RCA, Contract DA36-039 sc-75968.

[10] Welded Electronic Circuits, Space Technology Laboratories, *Missile Design & Develop.*, March, 1959, pp. 50–52, 82, 87.

[11] E. Keonjian and J. Marks, Micro-computer for Space Application, *Proc.* 1962 *Space Computer Eng. Conf.*, PGEC-IRE, Anaheim, Calif.

[12] A. S. Khouri, Packaged Electronic Circuits · · · , *Elec. Mgf.*, vol. 64, no. 4, pp. 164–166, 1958.

[13] T. A. Prugh, J. R. Nall, and N. J. Doctor, The DOFL Microelectronics Program, *Proc. IRE*, vol. 47, pp. 882–894, May, 1959.

[14] Planar Integration of Thin Film Functional Circuit Units, Quarterly Reports 1 to 3, Contract DA36-039 sc-87246, International Business Machines Corporation.

[15] F. Z. Keister, Microminiature Packaging Using Dot Components, *Electron. Packaging and Prod.*, vol. 1, pp. 54–58, September-October, 1961.

[16] D. S. Elders, Applications of High Density Electronic Packaging Systems, *Proc. 6th Natl. Conf. Military Electron.*, June 25–27, 1962, pp. 221–226.

[17] R. A. Gerhold, Integration of Microcircuitry Into Microassemblies, *IRE Trans. on Military Electron.*, vol. MIL-5, no. 3, pp. 227–233, July, 1961.

[18] Modular Interconnections for Microassemblies, Final Report, Contract DA36-039 sc-85347, Hamilton Standard Division, United Aircraft Corporation.

[19] Modular Interconnections for Microassemblies: Phase II, Final Report, Contract DA36-039 sc-87301, Hamilton Standard Division, United Aircraft Corporation.

BIBLIOGRAPHY

Hurowitz, Mark: Reliability of Welded Electronic Components, *Proc. 6th Natl. Conf. Military Electron.*, June 25–27, 1962, pp. 310–318.

Keonjian, E.: Microminiature Electronic Circuitry, 1959 *WESCON Convention Recd.*, part 6.

Church, S. E., and M. Geroulo: Cellular Packaging Improves Reliability, *Electron. Packaging and Prod.*, vol. 2, no. 3, pp. 15–16, May-June, 1962.

Keonjian, E.: Microminiature Computer Full Adder, AIEE Winter General Meeting, New York, 1960.

Suran, J. J.: Circuit Considerations Relating to Microelectronics, *Proc. IRE*, vol. 4a, pp. 420–426, February, 1961.

Harper, C. A.: Electronic Packaging with Plastics, *Electron. Packaging and Prod.*, vol. 2, no. 3, pp. 45–49, May-June, 1962.

Darnell, P. S.: Future of the Component Parts Field, *Proc. IRE*, vol. 50, no. 5, pp. 950–954, May, 1962.

CHAPTER 4

THIN-FILM CIRCUITS

By Rudolf E. Thun, William N. Carroll, Charles J. Kraus, Jacob Riseman, and Edward S. Wajda

4-1. INTRODUCTION

1. Application of Thin Films in Electronic Assemblies

In present usage, the term *thin film* is applied to coatings up to a thickness of a few microns ($1\ \mu = 10^{-4}$ cm $= 10^4$ A). Electronic component parts are deposited on a supporting substrate, consisting of a glassy or ceramic dielectric, less frequently of a single-crystal semiconductor wafer, and in a few instances of a polished metal plate. The various fabrication methods for thin films are shown in Table 4-1. Historically, films have been introduced into the manufacture of such component parts to save space and cost: The first products were paper or Mylar capacitors with evaporated metal plates and various types of resistors, mainly in the form of evaporated Nichrome, sputtered tantalum, and chemically deposited tin oxide coatings.[1–10] More recently, evaporated films have been introduced into the manufacturing process for transistors and diodes; contact areas and doping materials are deposited commercially in this fashion. However, all these applications of deposited films are outside the area of modern microelectronic developments and will not be discussed in detail.

Microelectronics has spurred three major film approaches. The first applies to linear as well as digital circuitry and duplicates conventional electronic components and devices such as resistors and capacitors in film form. *RC* film circuits with attached bulk diodes and transistors are in commercial production, and large *RC* networks containing more than 50 fully interconnected circuits are fabricated in the laboratory. Research on film diodes, gates, and amplifiers is intense and yields encouraging initial results with regard to the functional characteristics obtained. The search is carried far beyond the conventional junction device to such principles as field effects, plasmas in solids, and a wide range of electron

TABLE 4-1. FABRICATION METHODS FOR THIN FILMS IN ELECTRONIC APPLICATIONS

Method	Application	Explored applications
Vacuum evaporation	*RC* networks, resistors, capacitors, auxiliary method in semiconductor-device fabrication	Integrated circuits, semiconductor devices, metal-dielectric sandwich tunnel devices
Reactive vacuum evaporation	Oxide dielectrics, resistors, lands, interconnections	Semiconductor devices
Cathode sputtering	Capacitor dielectrics, resistors, insulation	
Electroplating	Reinforcement of conductors	
Chemical plating	Reinforcement of conductors	
Vapor decomposition	Tin, indium oxide resistors; Ge, Si, GaAs *npn* layers	Integrated circuits
Printing	Resistors, conductors	Integrated passive circuits
Spraying	Protective coatings	
Alloying	Auxiliary method in semiconductor-device fabrication	
Diffusion	Auxiliary method in semiconductor-device fabrication	
Thermal oxidation	Insulating layers, capacitor dielectrics	Auxiliary process in tunnel-device fabrication
Other techniques	Protective coatings	

and hole tunneling phenomena. Only inductive devices defy the film approach since the planar geometry of film network is incompatible with an effective coupling path of the magnetic flux. Low-Q film inductors up to a few microhenrys are possible, but more practical is the attachment of bulk microinductors and transformers to the film circuit or the choice of an unconventional solution to the inductor problem.[11]

The second approach (as well as the third) is computer-oriented. It was recognized early that the rotational switching mode in thin magnetic films is a considerably faster process than the wall-motion switching in ferrite cores, thereby permitting the design of a bistable memory element of improved speed. The development is mainly aimed at planar bit arrays which can be produced in a batch process and are well suited for high-density packaging. However, more recently film cores deposited on cylindrical glass rods or metal wires have found increasing interest because of their lower drive requirements and improved signal voltages.

In the search for a compatible design of the logic circuitry, the feasibility of magnetic-film parametrons has been explored, particularly in Japan. However, the various serious limitations of this approach have hindered, for the present, its serious consideration as a competitive computer technology.

The third major field of film electronic development exploits the phenomenon of superconductivity. The film cryotron, consisting of a simple pair of crossed or superimposed insulated lead and tin wires, is an ideal microelectronic switch which combines extremely low power dissipation with a favorable noise characteristic, sufficient gain, and miniature size. Since it represents a switch of very low potential cost, a more versatile computer organization may be facilitated by its use through the utilization of large associative memories. The cryotron is supplemented by the cryogenic sheet-film memory, which is a promising approach to the large-batch-fabricated random-access memory of relatively high speed. If the peripheral circuitry is not considered, cryotrons and sheet-film memory suffice for building large-scale computers. It is probably for this reason that the development of the superconducting tunnel amplifier seems to be less vigorously pursued. This device consists of two different semiconducting metal films, separated by a thin dielectric layer. Its operation is based on the relative shift of the superconducting band gaps in the two metals under an applied field which regulates the tunnel current through the dielectric.

In addition to these major film technologies, films play an important role in other microelectronic approaches. For instance, semiconductor integrated circuits are fabricated on semiconductor wafers of germanium, silicon, or gallium arsenide by transistor techniques, and are thus utilizing film deposition in various process steps. Moreover, design improvements regarding better performance, higher complexity, or lower cost often are possible by depositing film insulation, connections, crossovers, capacitors, or resistors over the basic array of semiconductor devices formed on the substrate wafer. However, this chapter will be restricted to a discussion of films deposited on an electrically neutral substrate since semiconductor circuits are treated in Chap. 5.

2. Comparison of Films with Other Microelectronic Approaches

Because of their unique performance characteristics, cryogenic and magnetic films do not compete directly with other forms of microelectronics. In the case of cryogenics, it must be seen if the film transistor is able to match the performance and price of the cryotron for large computer designs. A question still exists regarding the trade-off between the low-power dissipation and noise achievable with the cryogenic device and the additional expense of the necessary cryostat. Since the development of film semiconductor devices is very much in its infancy, it is impossible to give a conclusive answer to this question.

However, film circuits with attached chip semiconductors compete today with semiconductor circuits in linear and digital applications. Both approaches probably offer the same long-range potential with regard to reliability, component density, maintainability, and performance, and the manufacturing cost might thus very well be the decisive factor. The semiconductor-circuit approach is more naturally adapted to a functional integration at the circuit level. Film electronics, on the other hand, emphasizes the integral and functional interconnection of a larger matrix of repetitive basic circuits. However, the differences are not distinct enough to result in such a separation of applications. Indeed, semiconductor circuits may very well be combined with a basic film component and interconnection network.

More directly, film circuits are competing with printed or chemically deposited and etched multilayer panels containing resistor and conductor patterns. For the present at least, it seems that a better process control can be maintained in the film deposition of larger resistor arrays. In many applications, the combination of a first-level film-circuit package and a second-level package consisting of a laminated printed interconnection board today seems to offer the optimum packaging solution.

More specifically, film *RC* networks offer the following advantages:

1. Even for complex integrated functions, a lumped-parameter analysis is possible because of the discrete nature of the film components and interconnections, if a correct topography is chosen, i.e., if crosstalk is minimized.
2. An assembly unit can be selected which represents a more complex electronic function than a semiconductor circuit.
3. A free choice can be made of a topography for optimum heat dissipation. Semiconductor chips, for instance, can be mounted in areas not occupied by resistors.
4. The possibility exists of minimizing the connection pattern by freely selecting the optimum assembly level.
5. The film approach permits the choosing of the active element yielding the lowest cost or highest circuit performance, be it an individual diode or transistor chip or a more complex functional semiconductor block.

3. General Discussion of Film Materials and Fabrication Processes

Of the processes shown in Table 4-1, vacuum evaporation as well as cathode sputtering are used in the fabrication of electronic film networks. Both methods use almost identical vacuum plants. Usually, the process chamber is evacuated by an oil-diffusion pump backed by a mechanical pump. A valving system permits the "roughing out" of the chamber to a pressure of about 10^{-1} torr (mm Hg), where the diffusion pump becomes effective. It also isolates the pumping system from the chamber while the latter is being loaded with substrates and evaporant or sputtering material. Large diffusion pumps with high pumping speeds are preferred to shorten the pumping period. Cooled baffles between pump and chamber suppress the back-diffusion of pump oil, and act as auxiliary pumps. In the range below 10^{-3} torr, the chamber pressure is measured by ionization gages.

Evaporations are performed at vacuum below 10^{-5} torr. At these pressures, vaporized atoms or molecules reach a mean free path sufficiently exceeding the chamber dimensions. The vapor particles form thus a beam, extending conically from the evaporation source toward the film substrate. To minimize a contamination of the deposited film by occluded particles of the residual gases, chamber pressures in the 10^{-6}- and 10^{-7}-torr range are preferred.

The evaporant material is heated to achieve the required vapor pressure of 10^{-1} to 10^{-2} torr. Numerous types of evaporation sources are in use.

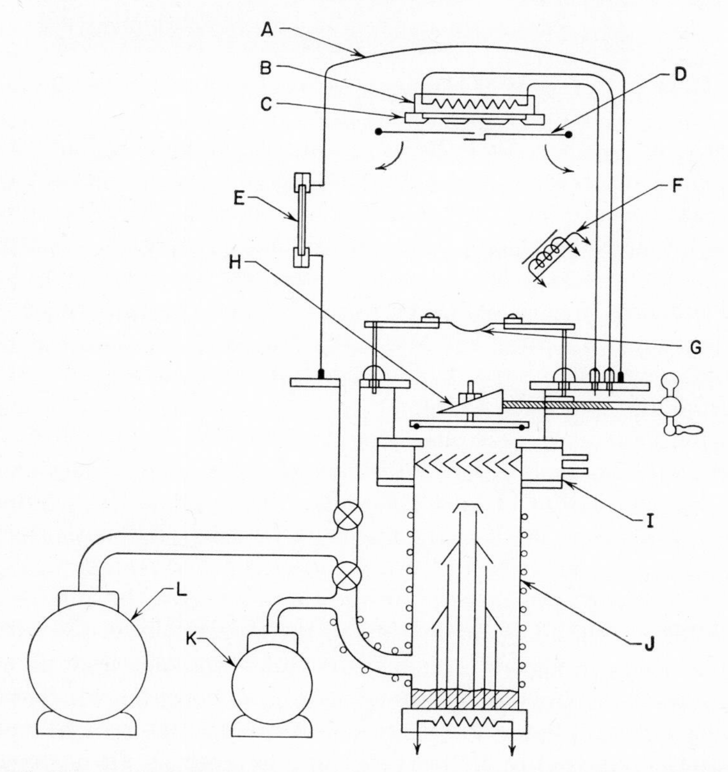

FIG. 4-1. Schematic diagram of evaporator for the deposition of film components: *A*, Bell jar; *B*, substrate holder with heater; *C*, mask; *D*, shutter; *E*, window; *F*, ion-gage-rate monitor; *G*, evaporation source; *H*, main valve; *I*, cooling baffle; *J*, diffusion pump; *K*, mechanical holding pump; *L*, mechanical roughing pump.

FIG. 4-2. Photograph of evaporator used in the electronic industry.

Electron-beam sources offer the advantage of minimizing film contamination, since the electron beam can be directed at the evaporant surface only, leaving the container or crucible relatively cold. Resistance-heated sources, on the other hand, have the advantage of greater simplicity. Often, the source crucible is heated indirectly by thermal radiation. Baffled crucibles with a restricted orifice narrow the cone angle of the vapor beam and minimize spitting. Figures 4-1 and 4-2 show a typical evaporation system. For a detailed description of vacuum and evaporation techniques, see Refs. 12–14.

Cathode sputtering takes place at a pressure of 10^{-2} to 10^{-1} torr. Two parallel plates are mounted at a distance of a few inches and connected with a d-c potential of about 3,000 volts. The cathode is covered with the sputtering material, and the anode carries the film substrate. The positive gas ions produced in the glow discharge between the plates are accelerated toward the cathode, where they vaporize the surface atoms or molecules. The films condensing at the anode side on the substrate surface contain, in general, a higher amount of gaseous impurities than evaporated films, because of the higher chamber pressure employed. By using an argon glow discharge, chemical reaction can be avoided. The reactive sputtering of metals in air, oxygen, or nitrogen, on the other hand, permits the deposition of various compounds often not obtainable by vacuum evaporation. The refinement of the sputtering technique during recent years has led to sputtering rates which are approaching the deposition rates achievable with vacuum evaporation. High-melting metals such as tantalum, molybdenum, or tungsten are sputtered easily. Figure 4-3 shows the schematic diagram of a sputtering chamber. Details of the technique are discussed in Refs. 15–18.

In the film electronics field, deposited films must fulfill many new and exacting requirements. In resistive, magnetic, or superconducting films, highly structure-sensitive properties must be utilized within close tolerances, demanding the accurate control of the deposition parameters. Other complications requiring a tight deposition control arise from the use of complex film structures and the adjacency of different materials of widely varying mechanical and chemical properties. As a consequence of these conditions, the development of a semiautomated evaporation process is not only a problem of cost reduction but a technical necessity.

The deposition of films in vacuum is basically a simple and extremely clean process, accomplishing the transport of matter by the purely physical phase transitions: solid $\rightarrow$ liquid $\rightarrow$ vapor $\rightarrow$ solid. In practice, however, various deposition parameters influence the contamination, structural order, and most microphysical properties of the deposit to a considerable degree. The most important of these parameters are: source material, source geometry, evaporation rate, chamber pressure, substrate temperature, substrate surface structure, and film thickness.[19–20] The parameters requiring independent and simultaneous control during deposition are: evaporation rate, chamber pressure, substrate temperature, and film thickness.

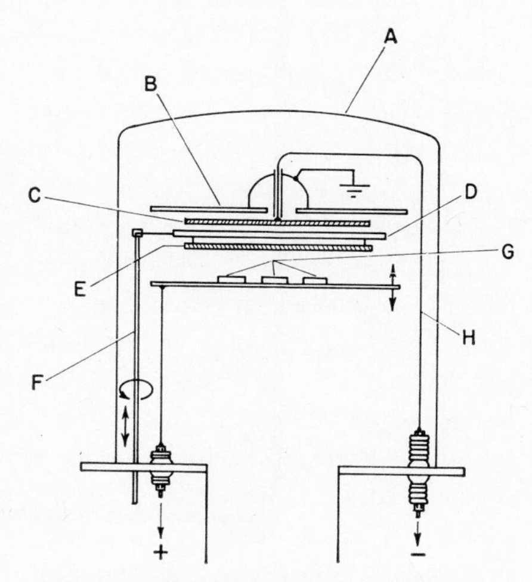

FIG. 4-3. Schematic diagram of cathode-sputtering chamber: *A*, Bell jar; *B*, grounded cathode shield; *C*, tantalum cathode; *D*, grounded gold-cathode shield; *E*, gold cathode; *F*, shaft to rotate gold cathode with shield; *G*, substrates; *H*, high-voltage lead.

Structure and properties of most film materials are not extremely sensitive to small variations of the substrate temperature. This insensitivity can particularly be observed in temperature regions around a property maximum or minimum. Resistivity minima, for instance, have been measured for various materials at easily obtainable substrate temperatures.[21] In most applications, it is thus sufficient to heat the substrate by attaching it to a resistance- or radiation-heated hot plate, and to measure the temperature with a thin-wire thermocouple mounted close to the substrate. The use of an automatic temperature-control loop is advisable.

The critical parameters in keeping the film properties within close tolerances are deposition rate and film thickness. An accurate method of determining the film thickness during deposition utilizes the crystal oscillator. The change of frequency is closely proportional to the mass deposited on the free-crystal surface. The rate is obtained by differentiating the output signal electronically.[22]

For production use, a more practical rate monitor is a modification of the Alpert ionization gage. It measures directly the rate, and the film thickness is obtained by integrating the signal. The principle of the ion-rate monitor is shown in Fig. 4-4.[23] The electrons from a hot tungsten filament are accelerated by an anode to an energy of about

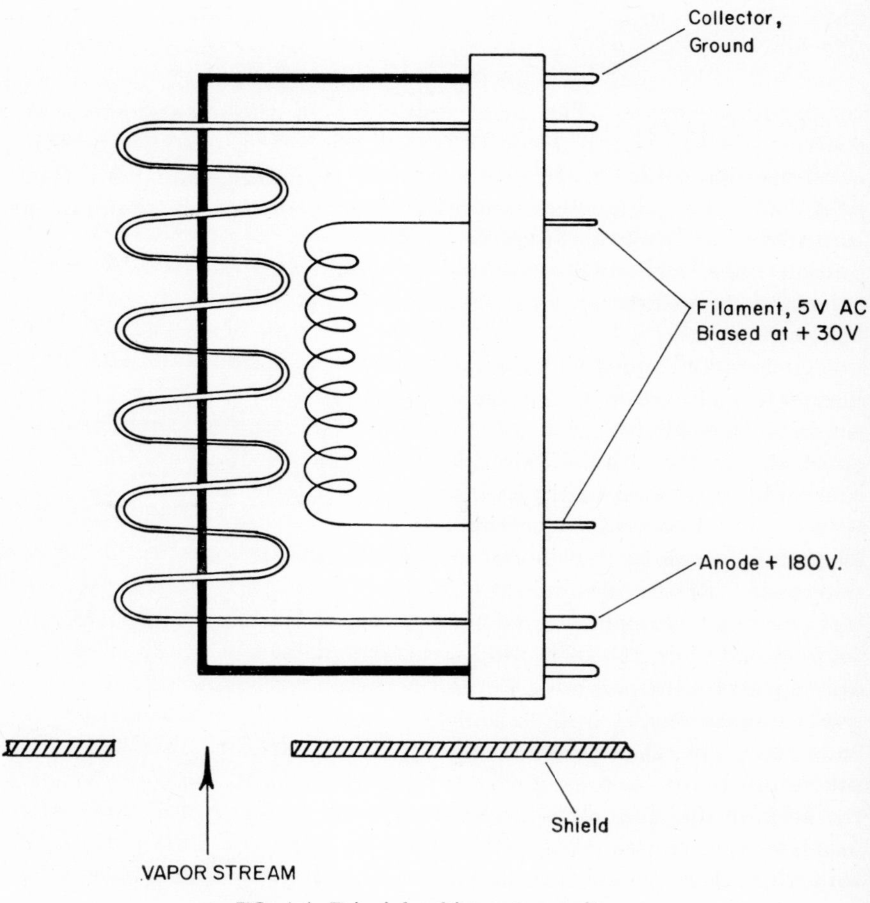

FIG. 4-4. Principle of ion-rate monitor.

150 ev. The vapor stream enters the gage parallel to the anode and is ionized by collisions with these electrons. The interception of the vapor ions by a collector yields a collector current proportional to the ion flow. Metal vapors simply may be deposited onto the collector. The rate measurement of dielectrics requires a collector heated to a temperature high enough to reevaporate the deposit.

With these rate-sensing elements, the power input of the evaporation source can be controlled automatically by means of a closed-loop system, as shown in Fig. 4-5.

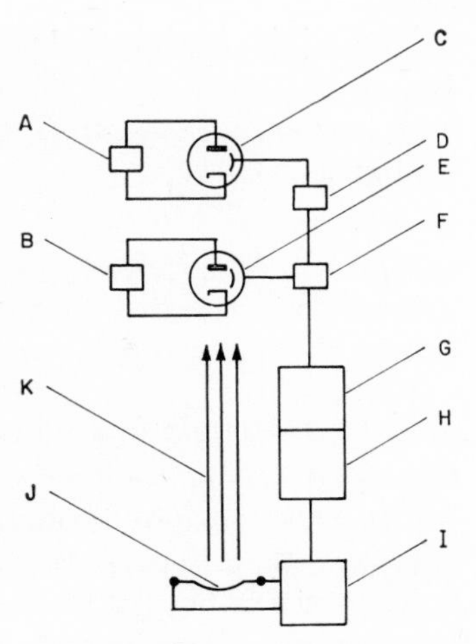

FIG. 4-5. Automatic rate control by ion-rate minotor: *A*, Vacuum-gage power supply; *B*, rate-monitor-gage power supply; *C*, VGIA vacuum tube; *D*, pressure-compensation network; *E*, rate-monitor gage; *F*, signal amplifier; *G*, L and N recorder; *H*, L and N controller; *I*, magnetic amplifier; *J*, source; *K*, vapor stream.

4-2. CIRCUITS WITH PASSIVE FILM COMPONENT PARTS

1. Film Resistors

In microelectronic applications, resistors or resistor networks are deposited, on planar substrates, in strips terminated by high-conductance lands. Current density and resistance of odd-shaped configurations have been analyzed,[24] but simple rectangular resistor forms are preferable. A range of resistor values of 1 to 100 can be obtained with a single resistor film thickness by varying the length-to-width ratio of the resistors from one-tenth of a square to ten squares. Depending on resistor material, film thickness, and structural order of the film, stable ohms per square values from about ten to a few thousand are within the capabilities of the present art.

The minimum linear dimension of a film resistor are determined by the masking and process accuracy available and the permissible resistor tolerance. Mechanical masks are usually fabricated to a line-width accuracy of ± 0.3 mil. A resistor width of 15 mils results, therefore, in a spread of $\pm 2\%$, owing to the inaccuracy of the mask. The maximum power load is limited by the permissible temperature rise at the resistor, and depends on such factors as the heat conductance of the substrate, the ratio of resistor area to total substrate area, the cooling mechanism used, and the ambient temperature. It is always advisable to perform an experimental and theoretical heat flow and temperature analysis of a chosen layout and packaging approach, but thermal loads of 0.25 watt/in.^{2} of substrate area and 20 watts/in.^{2} of resistor area have been realized with relatively simple cooling provisions.

The minimum length l and width w of a resistor are calculated from the given resistance R, ohms per square value $\mathcal{R}$, dissipated power P, and permissible power dissipation per square inch $\mathcal{P}$, by use of the formulas

$$w = \sqrt{\frac{P \cdot \mathcal{R}}{\mathcal{P} \cdot R}} \tag{4-1}$$

$$l = w\frac{R}{\mathcal{R}} \tag{4-2}$$

Various materials have been investigated as to their usefulness for deposited resistors. Elemental semiconductors such as germanium and silicon have not found widespread use because of the difficulty of achieving consistent dopant concentrations by evaporation techniques, and because of the limited temperature range for which a small and constant temperature coefficient of resistance can be obtained. Tin oxide is deposited by vapor-decomposition methods, which are not easily adaptable to masking techniques. The methods of evaporating compound semiconductors of group III-V or II-VI composition, while mastered in the laboratory, are not sufficiently well understood to permit the close-tolerance deposition of resistor arrays in a routine fabrication process. Nevertheless, large-band-gap semiconductors are potentially useful film-resistor materials.

For the present, however, metals, metal alloys, and cermets (mixtures of metals and dielectrics) are the resistor materials used in film networks.

According to Matthiessen's rule, the resistivity of a metal can be represented by

$$\rho = \rho_d + \rho(T) \tag{4-3}$$

The term ρ_d originates from the scattering of the conduction electrons at defects of the crystal lattice. It is nearly independent of the temperature over a restricted temperature interval, but it decreases irreversibly when the lattice defects are reduced by an annealing process. ρ_d can be considerably larger in films than in the bulk material since the condensation from the vapor phase tends to yield a high structural disorder.

With decreasing film thickness, ρ_d increases also, owing to an increasing influence of the electron scattering at the film surfaces, with a resulting reduction of the electron mean free path. This influence of the film thickness on the electrical resistivity has been treated theoretically for monovalent metals only.[25–28] However, the resulting thickness-resistivity curve describes also the behavior of the transition metals fairly well if the electron mean free path is chosen as an adjustable parameter (Fig. 4-6).

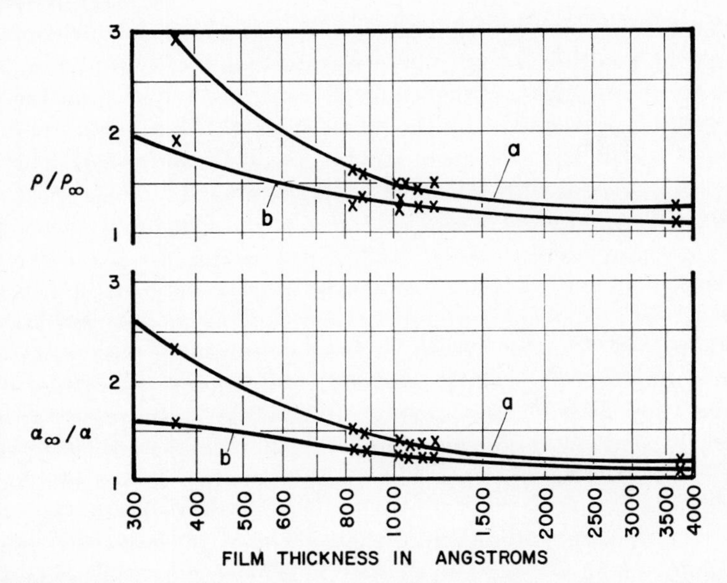

FIG. 4-6. Resistivity ρ and temperature coefficient of resistance α as a function of film thickness for nickel. ρ_∞, α_∞ = bulk values. a, unannealed films; b, annealed films.

The temperature-dependent term $\rho(T)$ is caused by scattering due to the thermal vibrations of the lattice. At temperatures T above 0.3θ (θ = Debye temperature), the resistivity of a metal varies nearly linearly with the temperature according to

$$\rho \approx \rho_0[1 + \alpha(T - 273.15)] = \rho_d + \rho_0\alpha(T - 0.15\theta) \tag{4-4}$$

where θ and T are temperature in degrees Kelvin, ρ_0 is the resistivity at 0°C, and α the temperature coefficient. Since $\rho_0\alpha$ changes but little with the structural order, a small relative change of the resistivity with temperature,

$$\frac{d\rho/dT}{\rho} = \frac{1}{\rho_d/\rho_0\alpha + T - 0.15\theta} \tag{4-5}$$

evidently requires a large ρ_d, or, in other words, a maximum concentration of lattice defects. Since highly disordered films are thermodynamically less stable, the requirements for a low α and a small drift of the resistance during the resistor life are difficult to reconcile with pure metals. The problem can be solved by using high-melting metals with a negligible atomic mobility at operating temperatures, or by choosing deposition conditions resulting in not fully metallic films. This can be achieved by sufficiently low deposition rates and high residual gas pressures to permit the enclosure of nitrogen and oxygen atoms (as low a rate as 10 A/sec, and a pressure of 10^{-5} mm mercury suffice in many cases) or, more effectively, by the deposition of metal-dielectric mixtures. It should be mentioned that purely metallic conductivity always results in a positive temperature coefficient of resistance. The negative coefficients frequently reported as the result of metal or alloy evaporations, therefore, permit the conclusion that these films contain a substantial amount of oxides or nitrides. Belser has discussed the electrical resistances of thin metal films before and after artificial aging by heating.[29]

The most frequently used film-resistor materials are nickel-chromium alloys chosen for their low temperature coefficient of resistance. Although work on nickel-chromium alloys has been reported amply in the literature,[30–37] more recently developed resistor alloys are often kept as trade secrets. With good SiO and organic overcoats, nickel- and chromium-based films show a reasonable stability, even under moderately adverse conditions, but their life expectancy under serious temperature cycling and extremely high humidity leaves much to be desired.

A better resistor material from the viewpoint of corrosion resistance is tantalum, which is usually deposited by sputtering. Maissel et al.[38–39] have given a description of an appropriate sputtering apparatus and of microcircuits utilizing tantalum resistors. The advantages of tantalum are twofold: It is a high-melting metal with a low atomic mobility at operational temperatures, and it forms a dense oxide layer at the surface, preventing the oxidation of the tantalum layer in depth. Maissel has shown that a uniform film thickness and resistivity distribution can be achieved over a significant part of the cathode area. The fabrication of extended tantalum resistor networks by cathode sputtering to reasonable tolerances seems thus possible, without the necessity of trimming individual resistors.

At 250°C, tantalum stabilizes in air in about one hour. Significant annealing does not take place at this temperature or, in any event, is masked by the formation of an oxide film at the surface, since the resistance increases during this period by about 50%. Thermal or, alternatively, anodic oxidation can thus be used to trim the sheet resistance after deposition to the accurate design value. Since the oxidation does not progress measurably below 150°C, one can assume that this temperature represents the upper operational temperature limit for tantalum resistors. Unfortunately, the amount of resistance increase during a 250°C air stabilization varies considerably within each deposition, and even more from run to run. It is also observed that these oxidized films show a considerably increased negative temperature coefficient of resistance and a poor high-frequency characteristic when compared to pure tantalum films.

Maissel[40,41] has shown that these effects are mainly caused by the diffusion of oxygen from the film surface and interior traps to the grain boundaries of the tantalum film. Maissel found that he could keep the oxygen out of the grain boundaries by doping the film with gold. These gold-doped and temperature-treated films yield extremely stable resistor films with a good high-frequency characteristic.

The fabrication of such resistors begins with the sputtering of a pure tantalum film. Typical deposition parameters are represented by an argon pressure of 60 μ, a current density of 0.5 ma/cm^2, and a cathode voltage of 3,800 volts. Under these conditions, the Crookes dark space extends about halfway between substrate and cathode. Deposition rates of about 11 A/sec. are observed.

To avoid the partial oxidation of the tantalum before the gold diffusion can be completed, the gold film is sandwiched between two tantalum layers. The first tantalum layer is kept thin ($\sim$1,000 ohms/square) to ensure good adhesion of the gold. The thickness of the gold should be about 7% of the total tantalum value. The diffusion can be carried out either immediately after film deposition or in a separate vacuum oven.

After 1,700 hr of life testing at 150°C, gold-doped and temperature-stabilized tantalum resistors representing a resistance range of 25 to 80 ohms/square exhibited a maximum resistance change of 0.5%. The temperature coefficients of resistance did not exceed the limits of $+50$ and -80 ppm.

Tantalum films share one limitation with other metal resistors: The usable sheet resistance range is relatively restricted. To extend this ohms-per-square range and obtain higher operational temperatures, various metal compounds and metal-dielectric mixtures (cermets) have been investigated as film-resistor materials.

Layer and his coworkers have developed various nitride, silicide, and oxide films deposited by evaporation.[42,43] Of the nitrides, chromium-titanium nitrides have exhibited, according to Layer, the most promise for resistor applications, and have been developed to the greatest degree. These films have been prepared in three steps. First, a 35% chromium–65% titanium alloy (by weight per cent) is evaporated at a vacuum better than 10^{-4} mm Hg and at a rate between 1 and 100 A/sec. Secondly, the film is heated in an ammonia atmosphere at about 1000°C for 10 min. The third step consists of heating the film for several days at a temperature of a few hundred degrees centigrade to stabilize the properties of the nitride film. At a temperature coefficient of resistance close to zero, Layer has obtained sheet resistivities between 100 and 1,000 ohms/square, varying with film thickness and deposition temperature. Half-watt resistors changed 1% in resistance during 1,000 hr of operation under full power at 150°C.

Depending on the substrate temperature during deposition, the films showed a wide sheet-resistance range from 100 to about 50,000 ohms/square. According to Layer, the films exhibited adequate stability under temperature storage, but data taken under power, temperature, and humidity seem to be lacking.

Beckerman and Thun have studied various metal-dielectric mixtures.[44] To avoid a change of film composition with film thickness and deposition time due to the different vapor pressures of the constituents, two different deposition methods have been used in this study. Most evaporations were made from simultaneously operated but separate sources for the metal and the dielectric. The sources are automatically controlled, through feedback loops, by evaporation-rate sensors. The evaporation from two separate sources was used to accomplish the rapid and flexible investigation of the various cermet compositions. Within each run, films were deposited on glass slides for X-ray analysis, thickness determination, and electrical measurements; on rocksalt for electron-microscope studies; and on carbon rods for spectrographic analysis. Figure 4-7 shows schematically the experimental arrangement. As a simpler source, more amenable to production use, a powder-fed flash-evaporation source was developed (Fig. 4-8)[44,45] The powder feed delivers a metered amount of the premixed cermet powder to a resistance-heated tantalum strip for instantaneous evaporation. This flash technique yielded homogeneous cermet films of excellent uniformity and reproducibility.

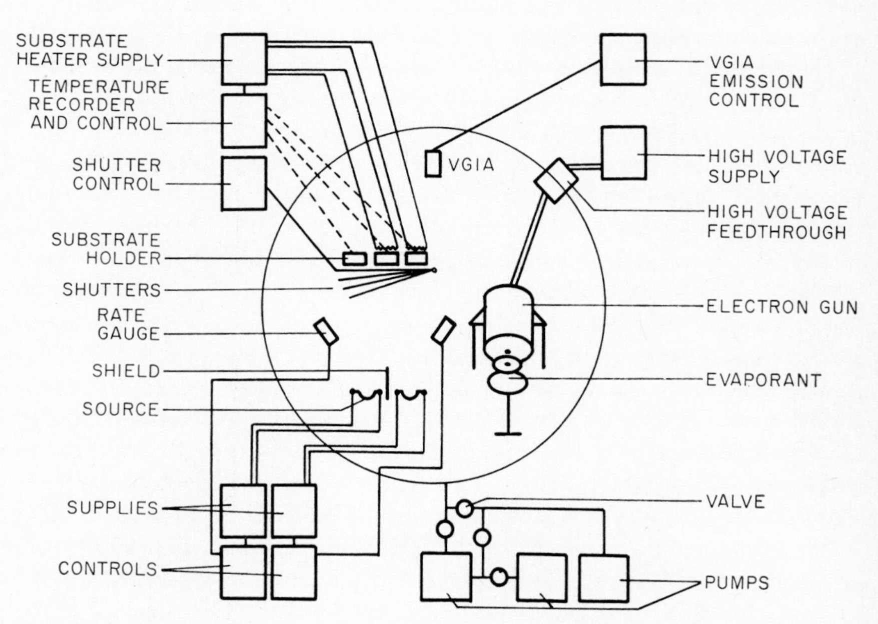

FIG. 4-7. Schematic diagram of evaporation system for the study of cermet systems.

The investigation of various metal-dielectric combinations revealed the following trends:

1. Noble metals such as gold resulted in relatively soft films, whereas transition metals with a good chemical affinity to the dielectric anion yield films which are hard and durable.

2. A crystalline dielectric-film material such as MgF_2 results in a cermet which shows a strongly nonlinear change of resistivity with decreasing metal content. Dielectrics as, for instance, SiO, which indicate their low atomic mobility by condensing in an amorphous state, on the other hand, lead to a reasonably linear composition-resistance relation, and thus are yielding more controllable cermet resistors.

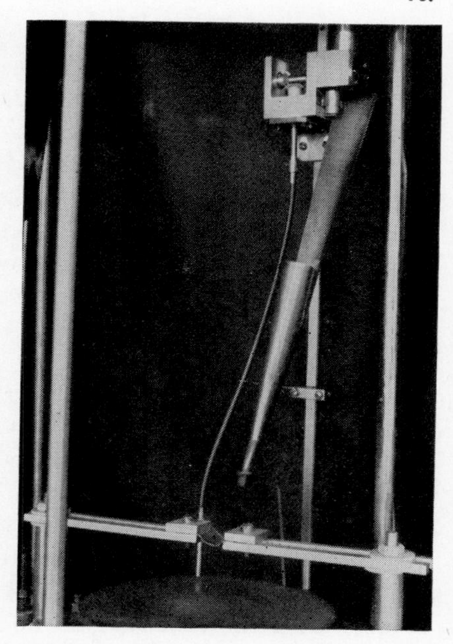

FIG. 4-8. Flash-evaporation source with powder feed.

3. The metal grain size and, in crystalline dielectrics, the dielectric grain size are always smaller in the mixture than in the corresponding

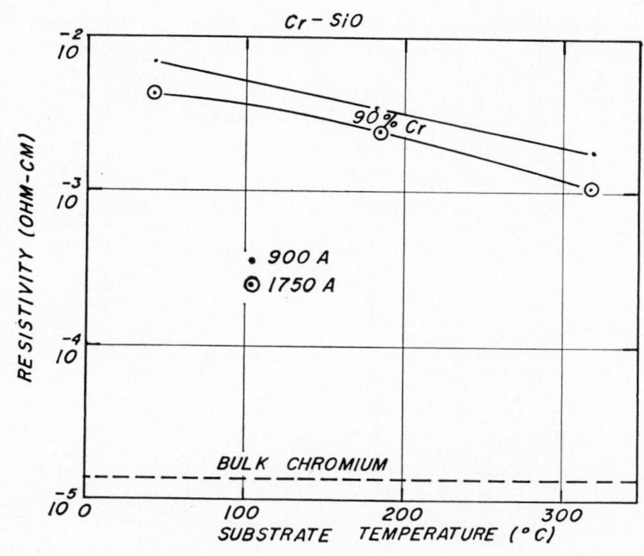

FIG. 4-9. Chromium–silicon monoxide resistivity versus substrate temperature.

pure metal or dielectric film deposited under identical conditions of substrate temperature and rate.

The best cermet system with regard to hardness, adhesion, uniformity, and stability found by Beckerman and Thun was the system chromium-silicon monoxide.[44] Figures 4-9 and 4-10 show the resistivity of this cermet versus substrate temperature and composition, respectively. Probably because of the higher rates applied, flash evaporation yielded films with a resistivity reduced by about one order of magnitude. A 70% Cr–30% SiO powder, for instance, gave a 300 ohms/square coating at 300 A thickness. Hardness and stability of the films were not impaired.

Beckerman and Bullard have reported on the fabrication and performance of 70% Cr–30% SiO resistors made by the flash evaporation of premixed 325- to 400-mesh powder.[45] The resistors were deposited on glass with a 5,000-A SiO underlay. The substrate temperature was 220°C, the rate 3 to 10 A/sec, and the pressure in the low 10^{-5}-torr range. The sheet resistance was monitored during deposition and stopped at 1.2 to 1.5 times the desired resistor value. After the application of gold lands with a 100-A chromium underlay and a protective SiO overcoat of 10,000 to 15,000 A thickness, the resistors were annealed in air or a reducing atmosphere to the resistor design value.

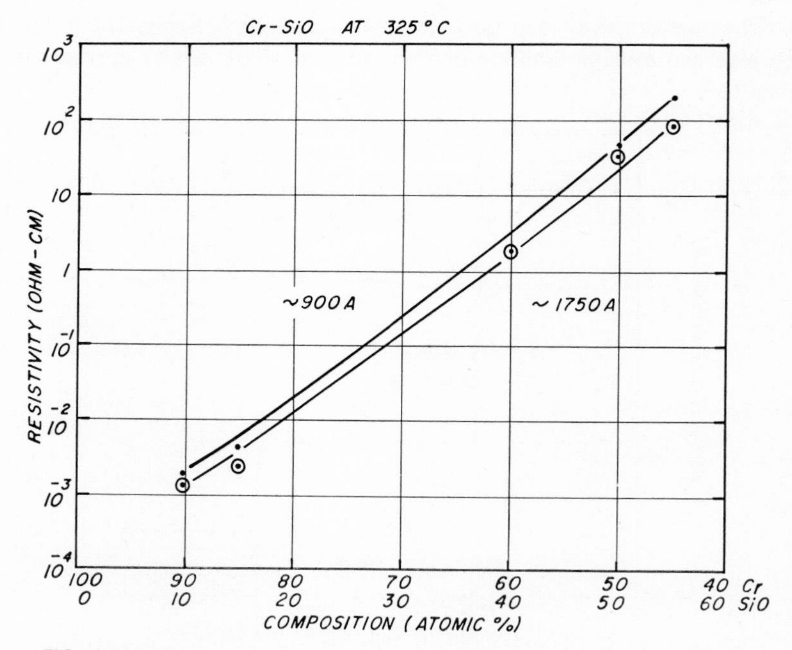

FIG. 4-10. Chromium–silicon monoxide resistivity versus composition.

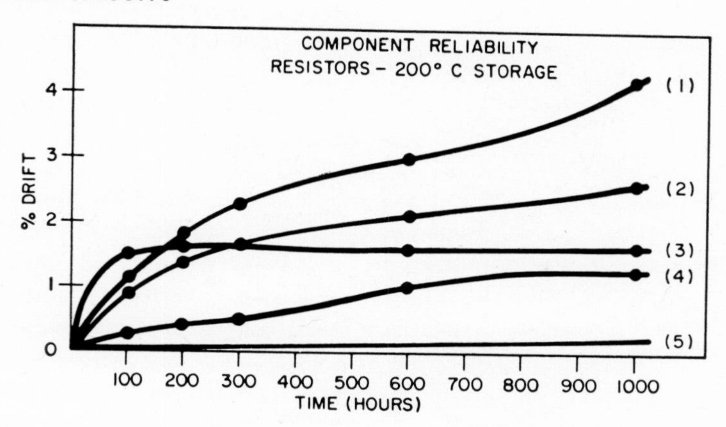

FIG. 4-11. Comparison of 200°C resistor storage-test results.

The simultaneous deposition of over 200 resistors with dimensions of about 15 by 200 mil on a 2.5- by 3.5-in. substrate resulted in a reproducible spread of only ±5% without trimming. A spread of ±2% can be achieved when mask dimensions are large with regard to dimensional tolerances. These Cr-SiO cermet films compared very favorably with metal-film resistors in various environmental tests. Figures 4-11 and 4-12 show the results of 200 and 300°C storage tests, and Figs. 4-13*a* and 4-13*b* the profile of the temperature-humidity cycle used and the corresponding test results for a resistor load of 11 watts/in.[2]

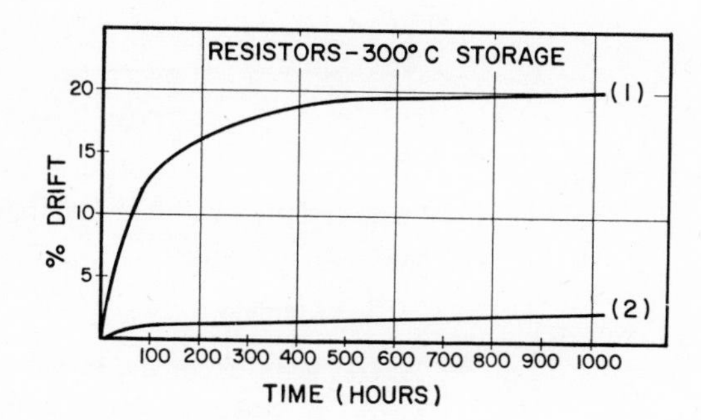

FIG. 4-12. Comparison of 300°C resistor storage-test results.

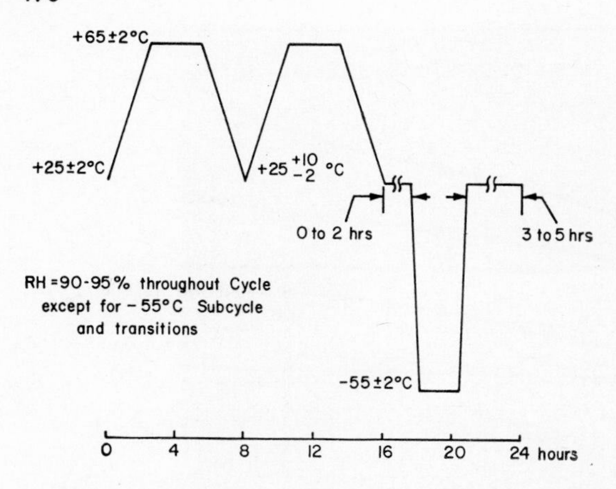

FIG. 4-13*a*. Resistor temperature-humidity cycle.

Recently, the sputtering of tantalum nitride resistors showing excellent stability has been reported by Gerstenberg and Mayer.[46] These authors have added to the argon-sputtering atmosphere small amounts of nitrogen (1 to 10% by pressure), which tend to override the accidental impurities, like oxygen and water vapor. The effect on the stability of the resistors

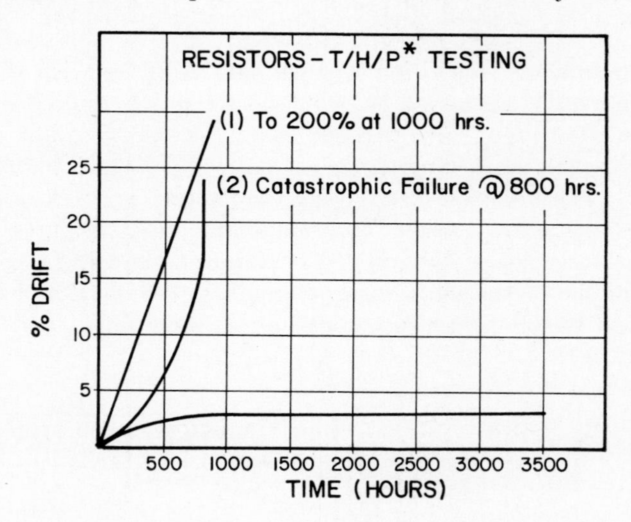

(1) Commercially Available Flat Film Assemblies
(2) Ni-Cr Alloy Films
(3) Cr-SiO Cermets

* T/H/P Tests according to Mil-Std 202, Method 106A, at power dissipations from 0 to 11 w/in^2

FIG. 4-13*b*. Comparison of temperature–humidity–power-cycling test for various resistors.

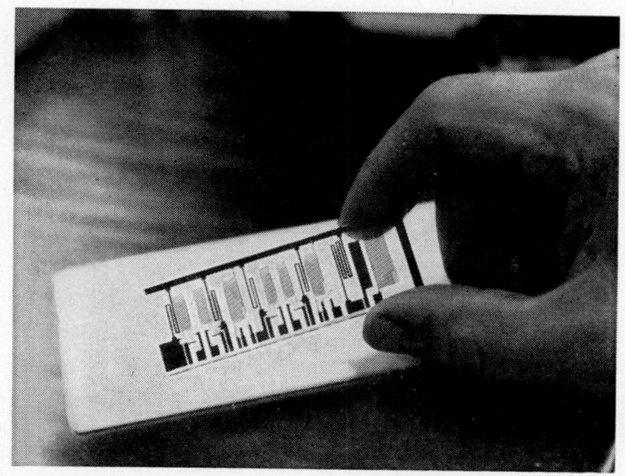

FIG. 4-14. Glass substrate containing tantalum nitride resistors made by sputtering in a partial nitrogen atmosphere. (*Bell Telephone Laboratories.*)

is similar to the process of gold doping used by Maissel. On load tests for 1,000 hr, tantalum-nitride resistors vary less than one tenth of one per cent. Figure 4-14 shows a tantalum-nitride resistor pattern.

2. Film Capacitors

The fabrication of film capacitors requires at least three deposition steps: the lower-plate area, the dielectric film, and the upper-plate area. Equation (4-6) yields the capacitance value:

$$C = \frac{0.225K(N-1)A}{t} \tag{4-6}$$

where C = capacitance, pf
K = dielectric constant
N = number of plates
A = area, in.^{2}
t = dielectric thickness, in.

Multiplate film capacitors are usually avoided because of cost and yield problems. A wide variety of materials and processes have been used to fabricate film capacitors, but the choice is narrowed considerably in microelectronic applications. Here, the dielectric must withstand temperatures of a few hundred degrees centigrade to satisfy military storage requirements, and to be compatible with the temperature profiles used in the fabrication of resistors and the attachment of other components. These requirements make it necessary to use inorganic films which are either evaporated in situ or obtained by the oxidation of metal films.

To evaporate a dielectric, the material should meet the following criteria:

1. It should not decompose during the evaporation-deposition process.

2. It should reach a vapor pressure of about 10 μ between 1000 and 1800°C (a very low evaporation temperature indicates an undesirably high atomic mobility; a very high evaporation temperature generates source problems).

3. It should not be hygroscopic or water-soluble.

4. It must adhere well to dielectric substrates and metals, should be hard, and should not crack when temperature-cycled.

Only few dielectric materials meet these requirements. Examples are SiO, CaF_2, MgF_2, ZnS, and, under certain deposition conditions, Al_2O_3 and a number of rare-earth oxides and fluorides. A more complete list of materials and deposition conditions is given by Holland.[12] Maddocks and Thun[47] as well as Siddall[48] have investigated the use of various dielectric materials in vacuum-deposited capacitors.

The yield of short-circuit-free capacitors is improved by an SiO undercoat and a proper choice of the metal used for the capacitor plates. Metals of high evaporation temperature such as chromium, nickel, or iron result in a higher percentage of failures, probably due to penetration of the dielectric by the highly energized metal atoms. Gold, an obvious choice because of its low evaporation temperature and high conductivity, gives poor results with certain dielectrics such as SiO. Its high mobility may cause in these cases short circuits due to grain boundary diffusion. Of all metals investigated in combination with all-deposited dielectrics, aluminum yields the fewest short circuits. Copper is preferable, however, because of its better resistance against electrocorrosion. To improve the adhesion of copper deposits, they are provided with an underflash of chromium.

Silicon monoxide is the most commonly used film-capacitor dielectric. Its dielectric constant of 6.0 is strongly dependent on the deposition conditions and can thus only be reproduced to about $\pm 10\%$. Its breakdown strength of 2×10^6 volts/cm is excellent, even exceeding that of mica. SiO is easily oxidized in the residual atmosphere of the vacuum system and becomes then hygroscopic.

Zinc sulfide yields a comparatively high dielectric constant of 8.2, good reproducibility ($\pm 5\%$), and small low-frequency losses when deposited from an aperture source. Its d-c breakdown strength of 2×10^5 volts/cm is one order of magnitude lower than that of SiO.

Magnesium fluoride, with a dielectric constant of 6.5, good reproducibility, excellent breakdown strength (2×10^6 volts/cm), and very low d-c leakage current (less than 1 μa at 2×10^6 volts/cm), exhibits the best dielectric properties of the materials studied to date. The films are hard and durable, even in thicknesses up to 2 μ. However, the control and cleanliness of the deposition conditions is more demanding then for SiO.

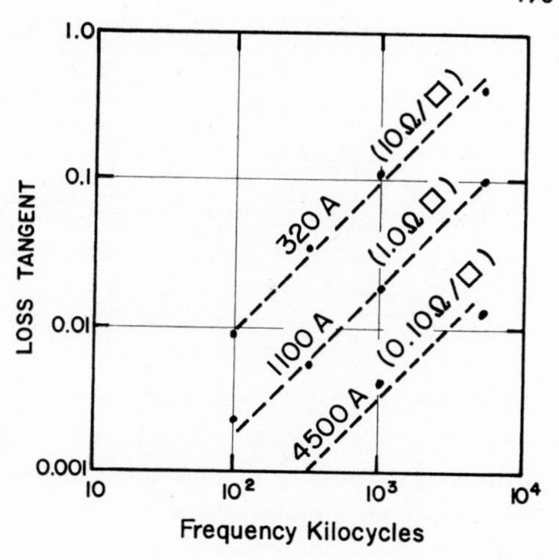

FIG. 4-15. Loss tangent versus frequency for an evaporated magnesium fluoride 1,000-pf capacitor with plates and leads of various thicknesses.

All these materials show extremely low dielectric losses at very high frequencies. The Q factor of film capacitors, therefore, is limited by the plate impedances. Figure 4-15 shows the observed loss tangent as a function of plate thickness. This graph shows the importance of optimizing the plate configuration for the shortest possible current flow and of providing plates of maximum thickness.

Besides stoichiometrically evaporated capacitor dielectrics, a second important class of dielectrics consists of oxide layers of silicon, aluminum, titanium, or tantalum, grown from deposits of the base metal either by thermal or by electrochemical oxidation.

Rudenberg et al.[49] have compared SiO_2 thermally grown on a polished Si wafer with TiO_2 films obtained by the oxidation of Ti films evaporated on glass. TiO_2 capacitors exhibited a specific capacitance of 1 μf/(cm^2), dissipation factors as low as 0.004, low leakage current, and good high-temperature performance. The SiO_2 capacitors showed a similar capacitance. It is interesting that the thermally grown SiO_2 layers, in contrast to evaporated SiO_2, are not noticeably hygroscopic. Smith and Ayling[50] sputtered reactively the oxides of silicon, tantalum, niobium, zirconium, and titanium, using a sputtering atmosphere of oxygen and a cathode of the pure base metal. Some of their results are listed in Table 4-2. Sputtered titanium oxide resulted in ceramic characteristics, while sputtered silicon dioxide gave substantially the characteristics of bulk silica. Both publications contain no detailed information on yield and life-test data.

Further advanced is the development of tantalum capacitors obtained by the thermal or electrolytic oxidation of sputtered tantalum films.[51,52] Typical deposition conditions are: an argon glow discharge of 1 ma/cm^2 at 3,000 volts, a deposition rate of 10 A/sec, and a deposition time of

TABLE 4-2

Cathode	Power factor, %	Resistance, ohms/farad	Dielectric constant
Tantalum........	3.0	200	14
Niobium.........	7.0	95	39
Zirconium.......	4.5	450	25
Titanium	6.0	190	62

about 15 min. This results in tantalum films with a sheet resistance of about 5 ohms/square. Anodizing of the films is usually performed in aqueous solutions such as 1% Na_2SO_4 and at voltages of about 100 volts. Because of the insulating properties of the oxide, the maximum thickness of the formed oxide layer is a function of the anodizing voltage only, which results in a relatively easy control of the anodizing process, as compared with thermal oxidation.

Because of the different thermal-expansion coefficients of tantalum oxide, tantalum metal, and substrate, the formation of microfissures in the oxide cannot be avoided entirely.

The search for high specific capacitances recently has led to a study of ferroelectric films such as barium titanate and barium–strontium titanate with regard to their usability as capacitor dielectrics.[53–55] These ferroelectric materials, when driven to saturation, exhibit a hysteresis loop with two permanent states, and can thus be utilized not only in linear capacitors but also in bistable storage elements. In evaporated ferroelectric films, dielectric constants up to 1,000 have been observed, but research in this area has not yet advanced sufficiently to permit a judgment regarding the uniformity, reproducibility, and yield achievable.

3. Integrated Passive Networks

The term *integrated passive networks* refers to individual or multiple networks of thin-film passive component parts and conductors, arranged on a single substrate to form specific circuit functions. These networks range from individual circuits to complex assemblies containing both circuits and their interconnections. The film-circuit elements may be arranged in a single plane for ease of manufacture or in multilayered form to maximize the packaging density. Until film diodes and transistors become available, microminiature active devices are attached to the film networks after the deposition process is completed.[56]

Circuit Design. The first step in the design of integrated passive networks is the selection of a basic circuit which not only provides the desired function but also is readily fabricated in film form. Circuits requiring large values of inductance or capacitance are generally easier to obtain by using attached discrete component parts for these elements. In addition, the selected circuit should minimize the use of precision component parts to obtain increased yield and reduced cost. Volume reduction of the package is facilitated by minimizing the power dissipation of the circuitry chosen. Finally, the circuit design should minimize the number and types of active elements which have to be attached, in order to simplify production of the circuit and to reduce its cost. Component-part tolerances of $\pm 5\%$ can be readily obtained with good yields, and a power dissipation of less than 100 mw/circuit can be handled with simple cooling techniques. As will be shown later, mask design and fabrication can be simplified if repetitive circuit patterns are used.

Circuit Topological Layout. The geometric arrangement of the film circuit is detailed in a topological layout of the film component parts. This layout must include the size and shape of all components, and take into consideration the limitations of the fabrication process and the practical tolerances which can be met in mask fabrication.[57] As an example, the topological layout of a typical digital-computer NOR circuit[58] will be described (Fig. 4-16). This layout places two circuits in an area of 0.435 by 0.310 in., and includes six 4,000-ohm resistors, two 1,000-ohm resistors, two 40-pf distributed *RC* input capacitors, suitable input and output circuit lands, lands for attaching transistors, and the power- and ground-distribution lines. Cermet resistors of 250 ohms/square were used in this design.

The 4,000-ohm resistors cover a 15- by 240-mil² rectangle, whereas the 1,000-ohm resistors with their higher power dissipation require an area

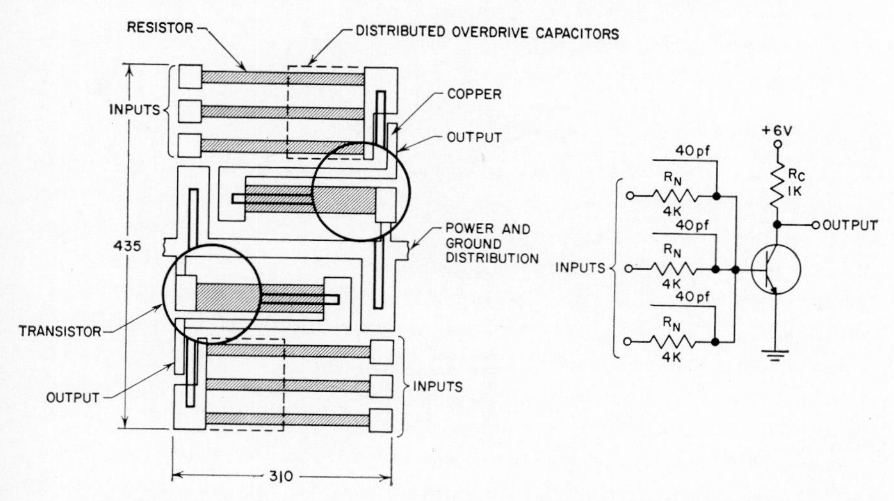

FIG. 4-16. Topological layout of digital-computer NOR circuits.

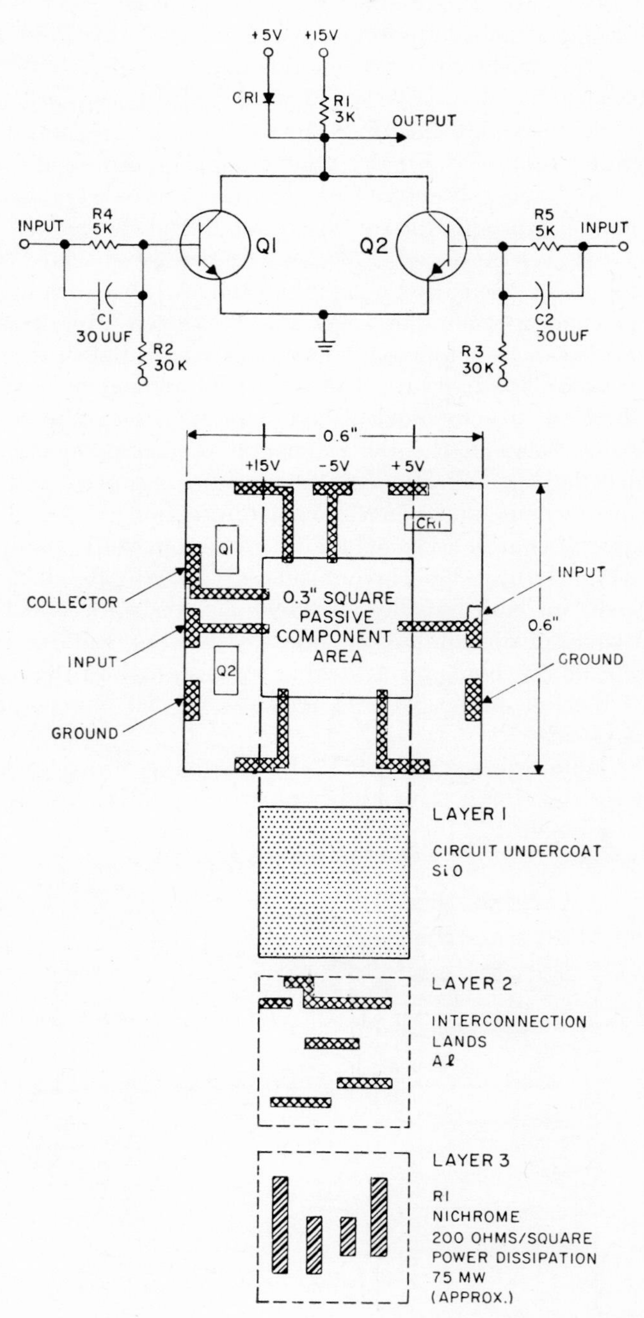

FIG. 4-17. Topological layout of

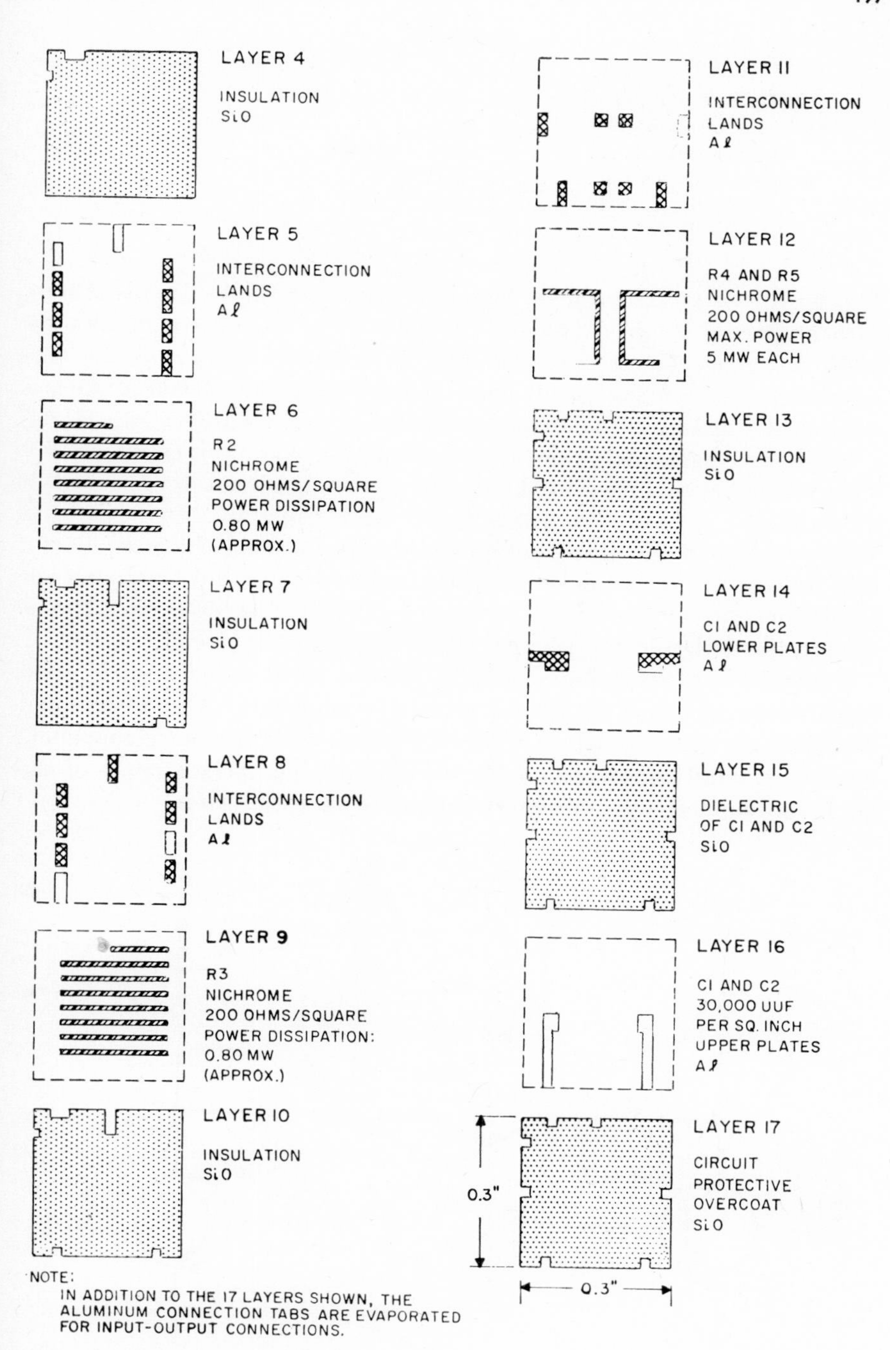

17-layer OR NOT switching circuit.

TABLE 4-3

Circuit type	*Number of evaporation layers*
OR NOT switching circuit	17
TRL switching circuit	11
Multivibrator	11
Audio amplifier	13
I-f amplifier	12
R-f amplifier	12

46.5 by 185 mils. The resistor land area was chosen to be 25 mils square, with a spacing between lands of 10 mils. The copper conductors are 5,000 A thick. Power is distributed by a deposited transmission line positioned between the circuit pair. Silicon monoxide 30,000 A thick is used as the dielectric for crossover insulation and overdrive capacitors.

For denser packaging, multilayer techniques have been employed.[59,60] This technique was used in the design of the six circuit types listed in Table 4-3.

To prevent open circuits in such a multilayered circuit, the crossing of a thin resistor film over thicker deposits (such as is used for conductors or insulation) should be avoided. Figure 4-17 shows the topological layout used in the design of the 17-layer OR NOT switching circuit.

Extensive multilayering generally results in a reduced yield, and requires complex masking and evaporation equipment. Also, distributed electrical effects caused by the close proximity of the film components must be carefully analyzed during the design. The present state of the art favors simpler layouts requiring fewer evaporation steps.

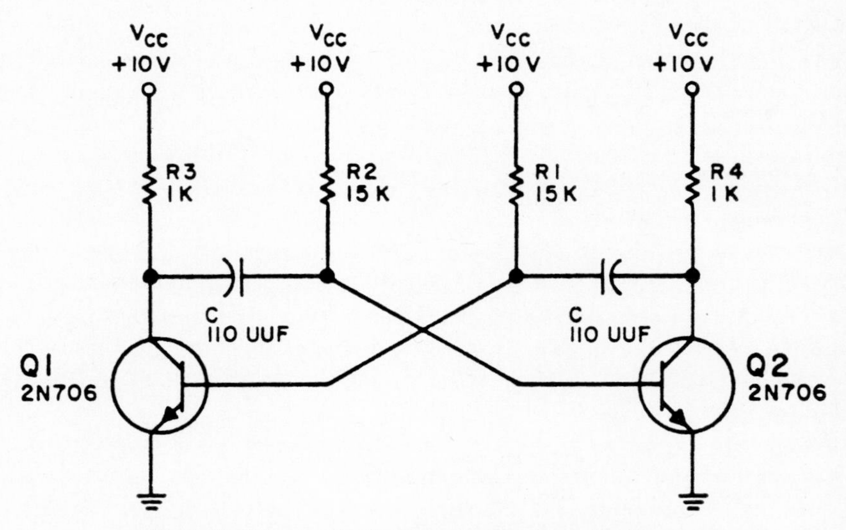

FIG. 4-18. Film multivibrator-circuit schematic diagram.

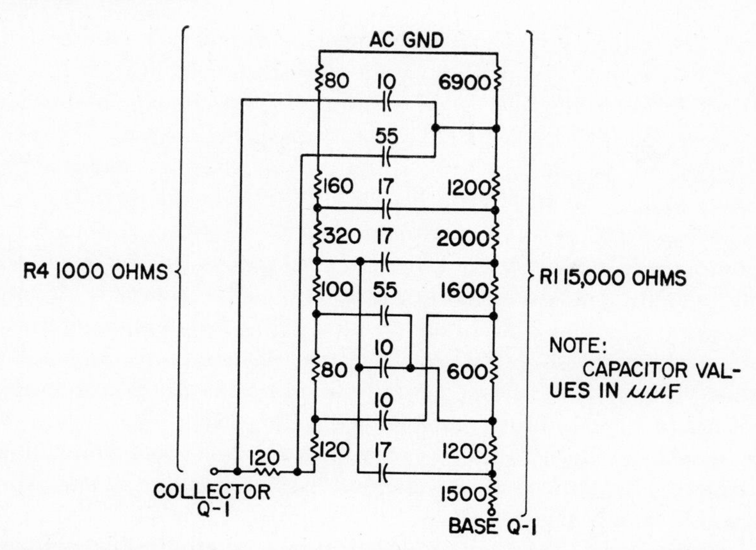

FIG. 4-19. Circuitry required to simultate distributed effects in multilayered film-multivibrator circuit.

Distributed Resistor-Capacitor Networks. By combining resistive, dielectric, and conducting layers, distributed *RC* networks can be formed which are not available with bulk components.[61,62] Such distributed parameter characteristics have been studied particularly on low-pass filters, high-pass filters, and phase-shifting networks. Reductions in number of component parts required for particular functions may be obtained. By using distributed networks for notched-filter designs, it is possible to eliminate the need for inductances in some tuned amplifiers.

Unwanted distributed effects, caused by the very close proximity of film components,[63,64] must also be considered. Figure 4-18 shows the schematic diagram of a simple film-multivibrator circuit which was fabricated using extensive multilayering. Initially, the performance of this circuit (Fig. 4-20*a*) was markedly different from the standard bulk-component version (Fig. 4-20*d*). By simulating the intercomponent capacities of the film circuit, it was possible to duplicate the initial film-circuit waveshape (Fig. 4-20*b*). Figure 4-19 shows the circuitry required for this simulation. Working backward from this circuitry, it was possible to redesign the layout of the multivibrator circuit to obtain the desired performance. Figure 4-20*c* shows the waveforms of the multivibrator after redesign.

Masking. Film-circuit patterns are produced by masking techniques or a combination of etching and masking. In the first method, the circuit logic is broken down into multilayers of conducting and insulating films vacuum-deposited through appropriate masks. In the second method, the majority of the interconnecting circuitry is etched from a previously deposited film, with only the line crossings and their insulation deposited through masks. This method is particularly advantageous when circuit density makes masking and mask registration impractical.

Both methods require the same initial layout and photographic-reduction techniques. A 10× scale layout of the selected circuitry is first drawn as a composite on a coordinate graph. The coordinates between the circuit pairs represent the interconnecting lines (for instance, 0.010-in.-wide conductors on 0.025-in. centers). This complete layout is then broken into the individual mask designs for depositing the segmented horizontal and vertical conductors, resistors, lands, and capacitor elements. Figure 4-21 shows the individual masks which are needed to deposit a 2.5- by 3.5-in. interconnected circuit panel.

The masks are fabricated by any of three methods, depending on the required precision: photoetched glass, arc-eroded metal, and conventional metal machining.

For photoetched glass masks, appropriate photographic artwork is furnished to the Corning Glass Works which etches from Fotoform B glass. The original pattern for this artwork is cut in a studnite ruby-red film at 10× scale on a Haag Streit Coordinatograph. A pair of fixed diamond cutters is used, assuring constant-conductor line width

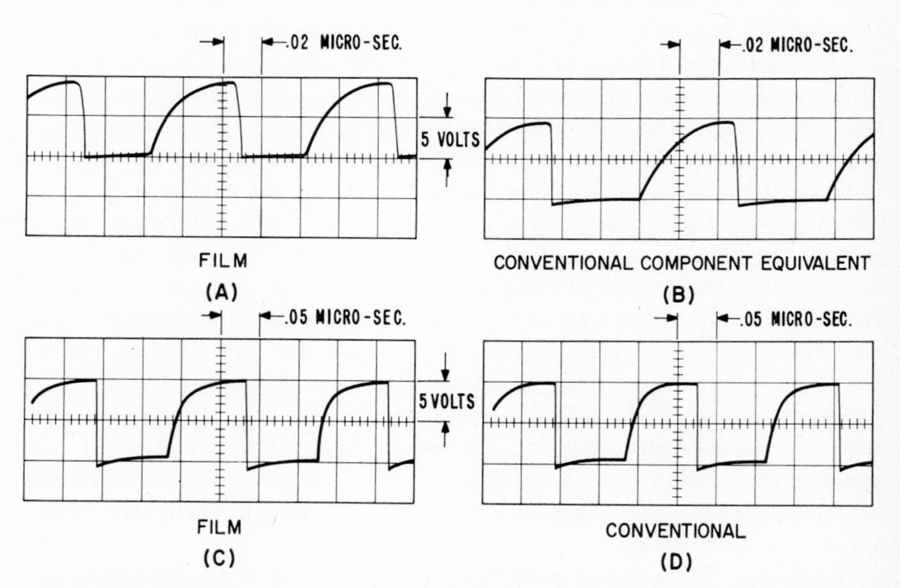

FIG. 4-20. Collector waveforms of film multivibrator: (*a*) Initial film-multivibrator waveform; (*b*) waveform of conventional-component simulation; (*c*) redesigned film-multivibrator collector waveform; (*d*) conventional-component multivibrator collector waveform.

FIG. 4-21. Various fabrication levels of 2.5- by 3.5-in. film-circuit panels with the corresponding evaporation masks. (*International Business Machines Corporation.*)

throughout the pattern. The coordinate plotter is capable of locating the fixed diamond-cutting tool to within ±0.0015 in. over a table area of 45 by 45 in.

The artwork cut in the film is reduced to nominal size within ±0.0015 in. on a Saltzman camera. The resolving power of the lens used in the reduction is 150 lines/mm.

In the event that the deposited circuit pattern is repetitive, the reduced artwork is stepped off and repeated on a Rutherford composer. This machine is capable of handling a photographic film or glass plate 42 by 32 in. An image can be reproduced and located over this area to within ±0.001 in. from a given reference line.

The expansion effects of temperature and humidity for photographic film and plates are given in Table 4-4.

Day-by-day environmental variations in the preparation of the photographic artwork for masking can severely limit the registration accuracy of consecutive depositions. A 20% change in humidity alone would vary the line-to-line accuracy over a 3-in. length to ±0.003 in.

Line openings from 0.005 to 0.015 in. are etched in the glass by Corning to a tolerance of ±0.001 in. The line-to-line tolerance over a 3-in.

TABLE 4-4. EXPANSION EFFECTS OF TEMPERATURE AND HUMIDITY FOR PHOTOGRAPHIC FILM AND PLATES

Type change*	Variation in./linear in. of film	
	Glass plate	Cronar film†
20% humidity increase	0.000	+0.0009–0.0011
20°F temperature increase	+0.00023–0.00033	+0.00026–0.00033

* Du Pont film data.
† Du Pont trademark.

length is ±0.002 in. The etching tolerances are dependent on the pattern density and the mask thickness. The reproduction of identical line widths for resistor masks may vary to ±0.0015 in. when repeated 200 times on the same mask.

Arc-eroded masks[65] are more readily adaptable to a complex repetitive pattern, such as resistors, requiring precise dimensional control over a large number of lines or patterns. In this machine process, the metal mask blank is submerged in an electrolyte, and an eroding arc bridges from the toolhead to the workpiece, reproducing the tool shape in the mask. A minimum of tool wear and the over-all machine-tool precision allow the tool pattern to be accurately reproduced many times over the mask blank. Resistor areas of 0.015 by 0.100 in. can be held to ±0.0003 in. over 200 lines on a 2.5- by 3-in. surface. The relative location of each group or pattern on the same surface may vary by ±0.001 in., which is superior to that of glass masks. Greater accuracy is also realized for a multiple number of identical masks. Figure 4-22 shows an arc-eroded metal mask used in the deposition of 63 identical resistor networks.

The use of conventional machined metal masks is restricted to less critical patterns. Although the line-to-line location tolerance can be machined to within ±0.001 in., the tolerance on a typical 0.010-in.-wide line opening may easily vary as much as 25%.

Since both mask and substrate must be located from common points to maintain proper registration of the multilayered depositions, the linear coefficient of expansion of the materials used is critical. The expansion difference between Fotoform B masks and the Pyrex* glass substrate (mounted on an Invar holder) at 300°C causes a registration error of 0.0002 in. over 3 in. Invar metal masks under similar conditions cause a registration error of 0.0035 in. over 3 in. No appreciable misregistration due to differences of expansion are experienced at 100°C.

Chemical etching techniques[66] can be used to obtain the desired film patterns without masks. The substrate is coated in the vacuum chamber

* Trademark of Corning Glass Works.

FIG. 4-22. Arc-eroded metal mask for the deposition of 63 identical resistor networks.

with a continuous sheet of the appropriate film material. After removal of the substrate from the coating chamber, the pattern is etched by using a Kodak photo-etching process. This approach is well suited for high-density copper interconnection patterns, since it eliminates a number of evaporation steps. Figure 4-23 shows a high-density etched interconnection pattern.

With increasing pattern densities, conventional forms of masking become impractical. New techniques utilizing xerography and electron or ion beams are investigated.[67,68] The beam techniques start with the deposition of the material to be selectively removed. This film is covered by a resist-producing material, which is exposed to an electron beam in the desired pattern. The first layer is then etched with a molecular beam of an element that makes a volatile compound of the material being etched. This technique has produced resolutions in excess of 100 A with very short electron-exposure times at low current densities.

Vacuum Equipment. Although there are a wide variety of methods for evaporating thin films, they generally fall into two classes: the batch process, where each deposition on one or more substrates requires a separate pumping cycle, and the mask-changer method, where all evaporations are completed within one pumping cycle. The latter mode

FIG. 4-23. 63-circuit computer panel containing etched high-density interconnections.

requires special mechanisms for changing masks, substrates, and sources in the vacuum chamber.[69] Figure 4-24 shows one type of mask-source changer. This mask changer holds 24 one-inch-square masks, and can deposit 24 different layers of evaporant upon 8 separate substrates from any of four sources. The processor consists of three rotatable circular plates whose position is controlled by vacuum feedthroughs. The lower plate serves as a shutter, the second plate is the mask carrier, and the top plate is the substrate carrier.

Each of the fabrication techniques has its own particular advantages. Table 4-5 lists a comparison of the two methods.

TABLE 4-5. COMPARISON OF BATCH AND MASK-CHANGER METHODS

Batch	*Mask changer*
No cross contamination of evaporants when using a separate system for each material	Eliminates exposure of each deposition to atmospheric contamination
More adaptable to in-process testing	Minimizes heating, cooling, and pumping cycles
Less complicated tooling	More adaptable to automation and reproducibility
Minimum source capacity requirements	Significant reduction in device completion time

Another approach to semiautomated production equipment, shown in Fig. 4-25, consists of a continuous-flow process capable of the controlled evaporation of four different materials.[70] Exit and entrance magazines with vacuum-lock features are incorporated, and the entire system is of modular design to enable easy additions or subtractions of individual coating stations.

The transport mechanism consists of three major parts: substrate magazine, substrate transport, and mask changer.

There are two substrate magazines in the system. One serves as a load station, the other as an unload station. A high-vacuum valve is placed between the magazine and its adjacent chamber, for loading or unloading the magazine without breaking the vacuum of the main chambers. The magazine has space for 24 substrate holders, 4 by 5¼ in., stacked one above the other. The load magazine is equipped with a preheater, stabilizing the substrate temperature before the substrate reaches the evaporation station. Each magazine is equipped with a pumping port, tied to the roughing line in the adjacent pumping system.

The substrate-transport mechanism consists of modular assemblies connected together in the main vacuum chambers. It removes the

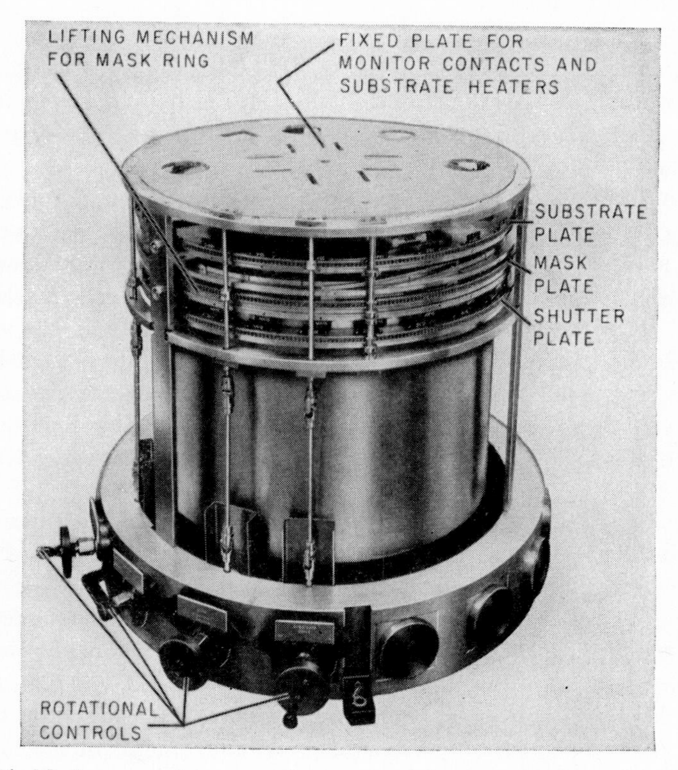

FIG. 4-24. Mask-source changer. (*International Business Machines Corporation.*)

FIG. 4-25. Semiautomatic film-production equipment developed for the Naval Avionics Facility, Indianapolis.

substrate from the entrance magazine, transfers the substrate through the four evaporation stations, and places the finished substrate into the exit magazine. It also provides a mounting for the substrate heater and the mask changer.

One mask changer is used at each evaporation station, a total of four for the system. They are identical and interchangeable. The changer removes a clean mask from the mask cartridge, raises and lowers the mask to register with the substrate holder, and returns the mask to the carriage. It contains a heater, raising the temperature of the mask to that of the substrate. The mask cartridge has space for six masks.

Four source stations are provided, one on the bottom of each main vacuum chamber. The source can be isolated from the chamber by a valve and serviced through an access door without breaking the main vacuum.

The system is capable of many evaporations from the same source at a high rate. A large-volume Drumheller source is used for silicon monoxide, and wire, pellet- or powder-feed mechanisms, providing a continuous supply of material to a heated filament, are used for metal and cermet evaporations. The material reaching the filament is flash-evaporated to completion.

Each chamber is provided with its own controls for pressure, transfer mechanism, and deposition parameters. All movements of the substrate and masks within the equipment are controlled externally, and interlocks are provided to prevent damage if the proper evaporation sequence is not followed.

Film Evaluation. Thin planar films differ from conventional components in form factor, size, and lead configuration. Therefore, new methods have been developed for the rapid measurement of large numbers of integrally deposited thin-film conductors, resistors, and insulators, as well as complex circuit and logic assemblies.

Film resistors are measured for electrical characteristics by contacting the metal lands terminating the resistor. Since the contact resistance of a probe can vary from a few milliohms to several hundred ohms, this method is useful only where very high resistances are to be measured. For more accurate measurements, the four-probe principle[71] is used. In this method, contact resistance becomes essentially unimportant.

Microminiature multiprobe assemblies having many pairs of 0.005-in.-diameter probes designed to contact the resistor lands of circuit patterns have been built and incorporated as part of a semiautomated system for the measurement of film resistors. These systems include punched-card equipment for easy tabulation of the resistor test values.

4. Over-all Packaging Problems

Consideration must be given to the total system design including intercircuit connections, shock and vibration requirements, active elements, and connector designs. Interconnection problems, in particular, have received considerable attention in the literature.[72,73]

Three basic categories of film-circuit packaging schemes have been proposed:[74] individual film circuits, multiple-circuit substrates, and panels with functional interconnections.

FIG. 4-26. Thin-film circuit mounted on transistor header. (*General Electric Company.*)

The individual film-circuit concept uses small discrete substrates that are mounted directly to printed-circuit boards, placed on transistor headers, or stacked and interconnected by means of riser wires similar to those employed with micromodules. Figure 4-26 shows a film circuit deposited on an insulating substrate and mounted in a transistor header. After the semiconductors have been attached, the header is sealed, and

FIG. 4-27. Thin-film circuit assemblies utilizing stacked and interconnected film circuit panels. (*General Electric Company.*)

an entire circuit function is available within one semiconductor package. Figure 4-27 is an example of film-circuit units utilizing a number of individual circuit substrates stacked and interconnected to form one complex plug-in unit. In both approaches, all input and output connections, power and ground are brought into the film networks through land areas around the periphery of the substrates. Soldering or welding techniques are used to attach the individual circuit substrates to a second-level panel. Sputtering techniques have also been used to deposit single-circuit substrates. Figure 4-28 shows a General Instrument Corporation sputtered tantalum tuner circuit deposited on a glass substrate. The various resistor patterns are etched into the tantalum and the part of the remaining tantalum oxidized to form the capacitor dielectric. A thin film of gold is deposited to form the conductors, capacitor plates, and coils.

Multiple-circuit substrates consisting of several individual circuits fabricated on a single substrate and using a common voltage and ground system offer higher component densities and lower fabrication costs. Edge lands are placed around the substrate edges, and methods similar to those for the individual circuit wafers are used for second-level pack-

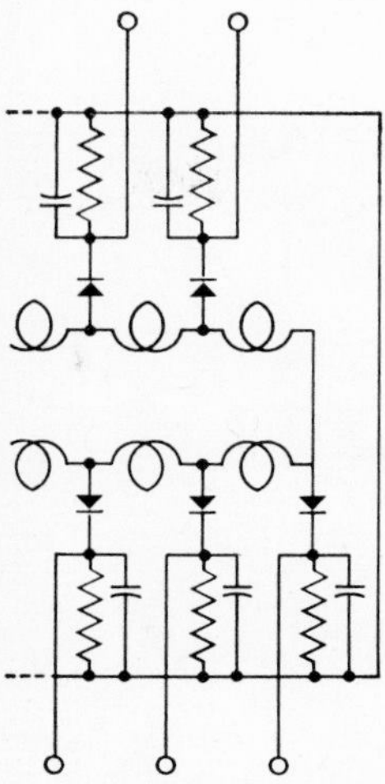

FIG. 4-28. Sputtered tantalum tuner circuit. (*Fortune Magazine.*)

aging. Figure 4-29 shows two typical multiple-circuit assemblies mounted on a printed-circuit panel. Each panel contains 13 individual circuits on a 2.50- by 0.745-in. substrate. Since all the interconnections are contained on the printed panel, the number of unique first-level packages can be minimized. For greater flexibility in production and maintenance at the expense of higher cost, pin connectors as shown in Fig. 4-29 can be used for direct connection to a third-level laminated printed board.

Figure 4-30 shows an interconnected film panel containing 56 circuits with film interconnections.[75] The panel provides six horizontal and four vertical film-interconnection paths which can be connected or insulated at any point of the conductor pattern. The lower edge of the

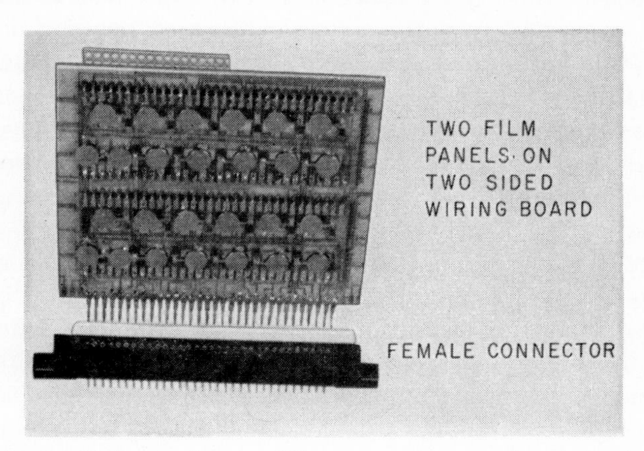

FIG. 4-29. Multiple circuit assemblies mounted on a printed circuit panel.

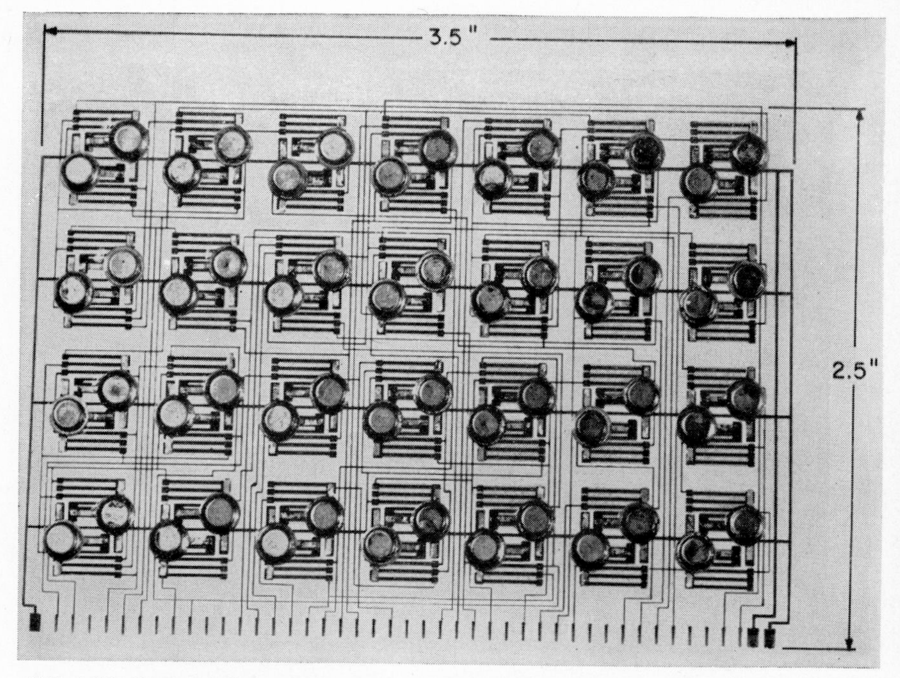

FIG. 4-30. Integral film panel containing 56 NOR circuits with their interconnections.

panel contains 44 connection tabs for interpanel connections and power. Panels such as these contain a substantial amount of interconnections, and represent functional subassemblies of considerable logic power.

Integrated-circuit connection techniques for system assembly have been achieved through multilayered film networks and by multilaminated printed boards. Both techniques can be used simultaneously at different packaging levels and in various combinations, depending on the type of basic circuit package used and the desired flexibility between packaging levels.

In cases where hermetically sealed packages are not provided, the extreme thinness of films makes it mandatory to prevent corrosion and electrocorrosion due to high ambient temperatures and humidities, by applying protective overcoats.

As a first coating, evaporated SiO proves to be relatively effective as long as care is taken to obtain a dense, nonhygroscopic layer by utilizing appropriate deposition conditions. Exposed conductor lines may be protected by reinforcing them with a solder layer. In addition, however, organic overcoats are applied as the final encapsulation.

Where a replacement of elements attached to the film panel is required, thin conformal coatings are used. Good results have been achieved with such materials as Dri Film 88, made by General Electric, or Sylkyd 1400, a Dow-Corning product. Double coatings consisting of a water-repellent material and a material resisting water penetration may yield additional protection. For throwaway packages, the potting of film circuits and panels in plastics such as epoxies is customary. To avoid excessive stresses due to the setting of the material and different expansion coefficients, materials and fillers should be selected carefully.

Glass overcoats are superior to plastics, but little work has been done up to the present in this area.

Active Devices. The practical application of thin-film active devices is still a few years hence. As an interim measure, microminiature active devices in cans or as chip semiconductors can be attached to substrates containing film passive component parts. Surface-passivated and glass-encapsulated chips in either single or multiple units provide low-cost, high-performance devices which can be automatically attached to film-conductor lands through soldering techniques.

The simultaneous attachment of many encapsulated transistors has been accomplished by a solder-reflow process.[76] Film panels with pre-tinned lands are registered with a transistor-positioning fixture shown in Fig. 4-31, and the entire assembly is heated to the solder-reflow temperature. Devices attached in this manner have proved very reliable under wide environmental and shock conditions. The joints had resistance of less than 0.01 ohm, and a minimum force of 50 psi was required for joint separation.

Active devices have also been packaged into small multidevice blocks,[77] which are attached to the film panels in a manner similar to that described above. Figure 4-32 shows a block containing seven diodes and one transistor. Higher active device-packaging densities and increased circuit-layout flexibility are achieved by this method. Inductors and large-valued capacitors, which are not available in film form, can be similarly packaged.

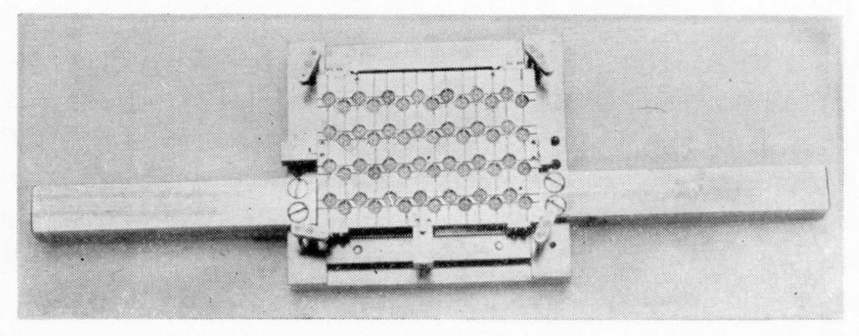

FIG. 4-31. Jig for positioning 56 transistors to be soldered to a 2.5- by 3.5-in. film-circuit network.

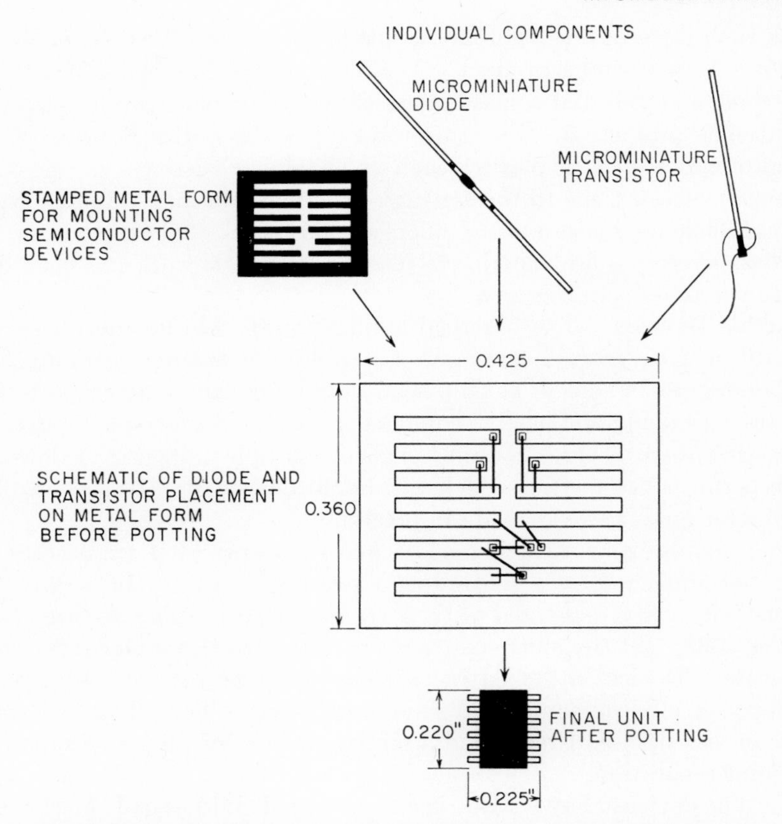

FIG. 4-32. Active element block containing seven diodes and one transistor.

Thermal Evaluation. The high packing densities achieved with film circuits requires the careful consideration of the associated thermal problems. Circuit size reductions have progressed at a faster rate than power levels, resulting in an increased power density. Film circuitry depends primarily on the substrate material for diffusing the heat and minimizing hot spots within the films. Analytical and empirical techniques have been proposed for determining the thermal limits of the packaging. Although an analysis of film-circuit temperature gradients can be obtained through the use of standard heat-flow equations, the geometry of most film networks is much too complex to be handled with ease. Kammerer[78] has described a less time-consuming method using a numerical-relaxation technique for determining the thermal gradients (see Chap. 2).

For most applications, conventional cooling techniques can be adapted to thin-film packaging. For unusual requirements, techniques such as Peltier junction cooling have been proposed.[79]

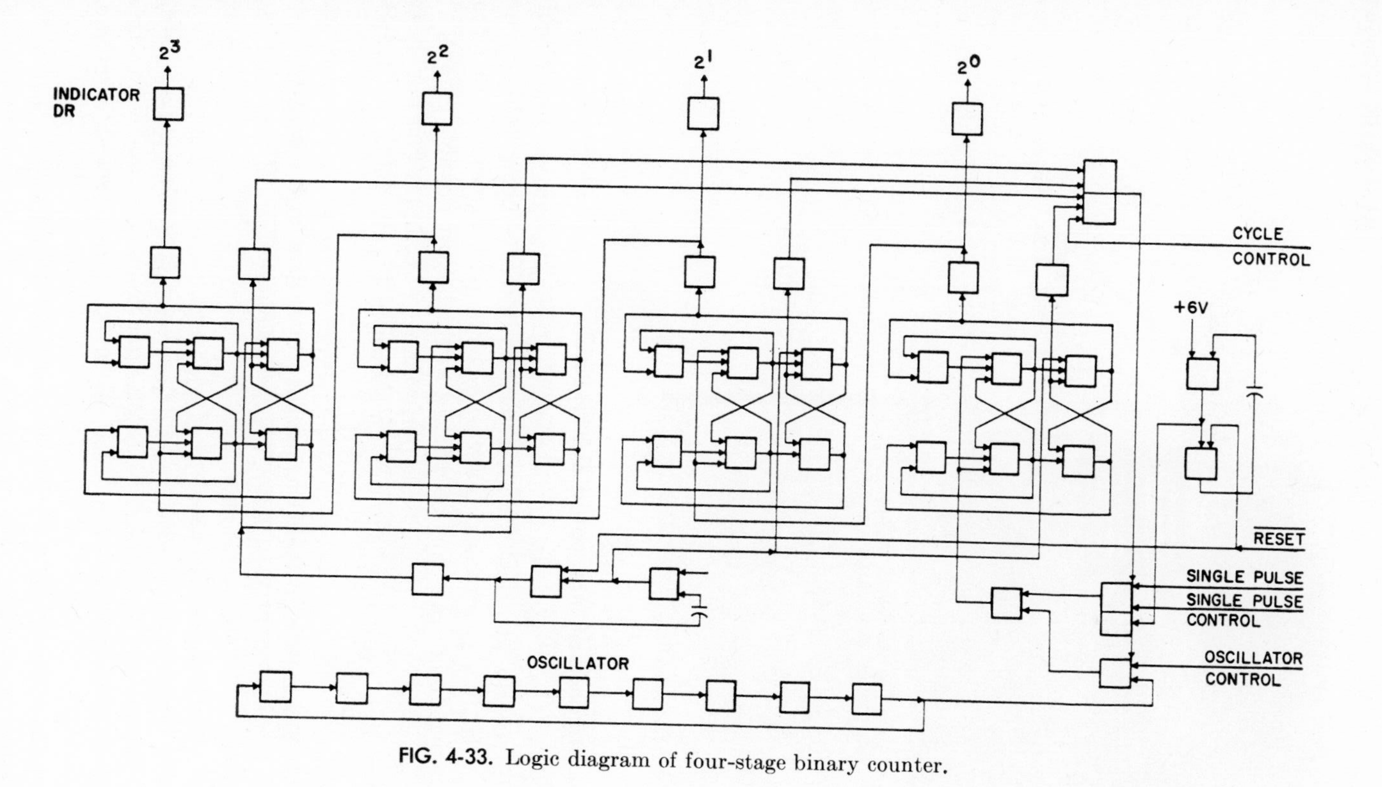

FIG. 4-33. Logic diagram of four-stage binary counter.

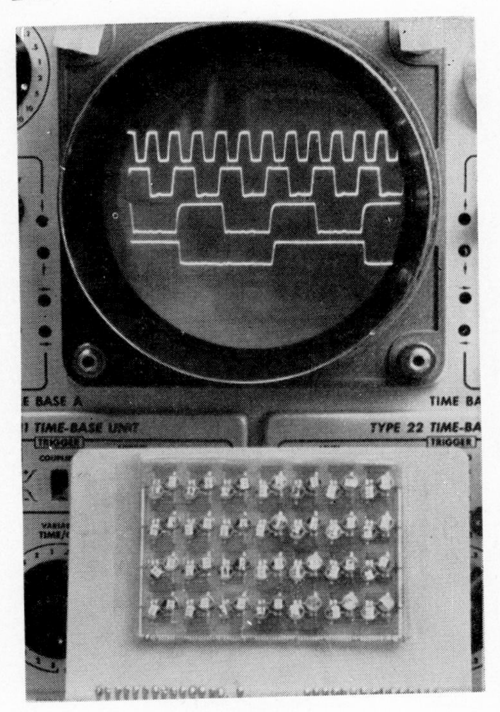

FIG. 4-34. Waveforms of four-stage binary computer operating at 2.5 Mc.

5. Performance of Integrated Subassembly

With a layout minimizing distributed effects, integrated film circuits perform nearly identically with their counterparts using bulk components. In film electronics, the vast pool of past experience in circuit and device design can thus be utilized to the largest possible extent.

An example of the predictable performance of complex film panels was demonstrated with the logic section shown in Figure 4-33. It includes

TABLE 4-6. DYNAMIC CHARACTERISTICS OF FILM NOR CIRCUITS

Fan-in	Fan-out	Average delay (ns)		Average transition (ns)	
		On	Off	On	Off
1	1	30	40	40	80
1	3	25	40	30	80
2	1	30	30	40	40
2	2	25	40	35	50
3	3	40	25	50	40

a four-stage binary counter, a 2.5-Mc oscillator, indicator drive circuits and control circuits. This logic was implemented on a 56-circuit film panel of the type shown in Fig. 4-30, using the basic circuit of Fig. 4-16. The output waveforms (Fig. 4-34) of this panel and the performance of the individual circuits (Table 4-6) are practically identical with those obtainable with conventional components.

The future development potential of film electronics with regard to high-speed performance looks extremely attractive since the use of deposited low-impedance transmission lines and close-circuit spacings should reduce wiring delays to a point where high-circuit repetition rates can be fully utilized.

4-3. SEMICONDUCTOR FILMS

For complete thin-film circuit fabrication, active and passive devices have to be integrally deposited in a compatible manner. The active device functions are generally limited to switching, amplification, rectification, or storage. These functions can be performed by bipolar- or unipolar-type devices. The bipolar devices such as the junction transistor utilize majority- and minority-carrier principles, and require single-crystalline semiconductors with high mobilities and long lifetimes for their fabrication. Thin-film bipolar-device fabrication thus requires substrates that can nucleate single-crystalline semiconductor layers and also fulfill other substrate requirements, such as electrical insulation between adjacent circuits, mechanical and temperature stability, compatibility with the passive components, and, for practical considerations, low cost. The number of substrate materials fulfilling all these requirements is extremely limited. The direct nucleation of a monocrystalline semiconductor film on a polycrystalline or amorphous substrate has, at present, not been accomplished, although some progress has been made in producing large oriented grain growth through recrystallization and strain-annealing techniques. Unipolar-type devices, utilizing tunneling, space-charge, and field-effect principles, can potentially be fabricated from polycrystalline films, and allow compatible deposition onto insulating amorphous or polycrystalline substrates.

Thus, two major approaches for the fabrication of film active devices exist: (1) the deposition of semiconductors onto substrates nucleating single-crystalline films for bipolar-device fabrication, or (2) the development of useful unipolar-type devices in polycrystalline semiconductor films. With significant advances in the area of semiconductor thin-film metallurgy and surface phenomena, a new generation of transistorlike devices will evolve which are fully compatible with electronic film circuits.

1. Film Devices

Several film devices fabricated from polycrystalline semiconductor and dielectric films have already been reported in the literature. Dresner and Shallcross[80] have constructed film diodes by evaporating CdS (1 to 5 μ thick) onto glass substrates coated with Au or an Au-In alloy which makes an ohmic contact to the CdS film. The usual blocking contact consisted of an evaporated Te dot. These units have shown rectification ratios of 10^6 with reverse breakdown voltages ranging from 5 to 15 volts. In high-resistivity CdS diodes, continuous forward currents of 0.1 amp/cm^2 could be passed, whereas in CdS films doped with indium, continuous currents of 10 amp/cm^2 could be withstood. The diodes showed barrier capacitances of the order of 8×10^{-8} farad/cm^2.

A new class of thin-film transistors fabricated entirely by evaporation of all components upon an insulating substrate has been described by Weimer.[81] The device, shown schematically in Figure 4-35, utilizes an insulated control gate, which permits a mode of operation quite different from that of the conventional field-effect transistor.[82] By biasing the gate positively with respect to the source, the drain current can be enhanced by several orders of magnitude without drawing appreciable gate current. The transverse electric field, applied from the gate electrode through an insulator to the semiconductor, induces a charge in the semiconductor surface. The surface conductivity can thus be enhanced or diminished by the applied gate potential and by the nature of the semiconductor-insulator contact. The effective drift mobility μ_d of the electrons is strongly dependent on the density of surface states and traps existing in the semiconductor, and can be expressed by the following relationship:

$$\mu_d = \mu_D \frac{N_F}{N_F + N_T} \qquad (4\text{-}7)$$

where μ_D is the true drift mobility, N_F the free-carrier density, and N_T the density of trapped carriers. Microcrystalline films of CdS have been used by Weimer, but other semiconductors, having a relatively high

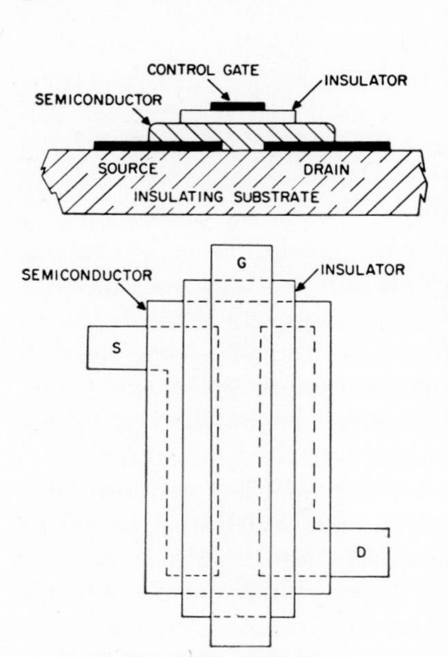

FIG. 4-35. Cross-sectional and plan view of an insulated-gate thin-film transistor. (*P. K. Weimer.*[81])

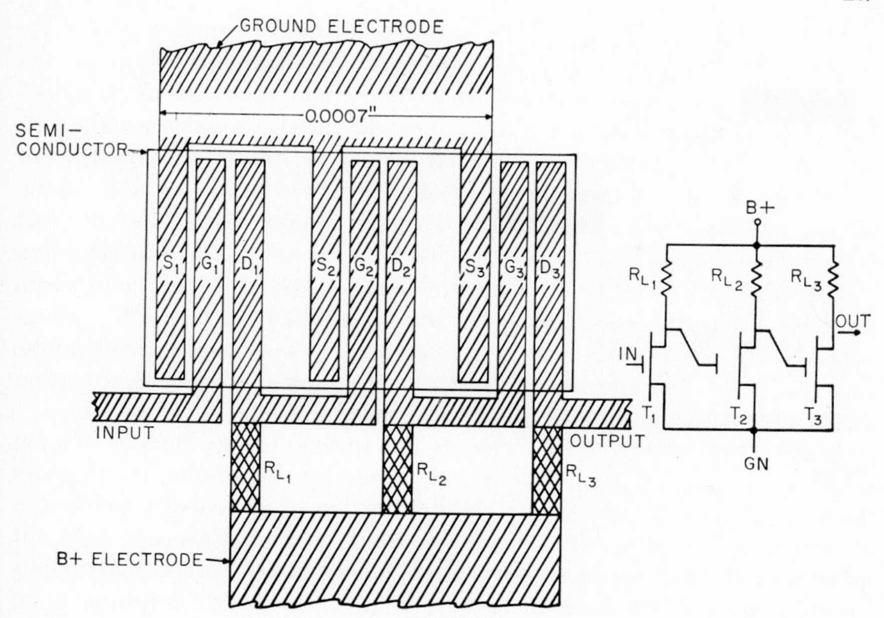

FIG. 4-36. Three-stage amplifier based on the thin-film transistor. (*P. K. Weimer.*[81])

effective drift mobility, can be used. The source and drain electrodes have been formed from metals like gold which make an ohmic contact to the film semiconductor. In Weimer's experimental models, the thickness of the semiconductor was less than 1 μ with a source to drain spacing L from 5 to 50 μ. This small spacing is necessary to achieve a high transconductance g_m and gain-bandwidth product

$$\text{G.B.} = \frac{g_m}{2\pi C_g} = \frac{\mu_d V_D}{2\pi L^2} \tag{4-8}$$

Here, C_g is the gate capacitance, V_D is the source-drain potential, and μ_d is the effective drift mobility of the electrons. The thickness of the dielectric gate spacer and the width of the gate appear implicitly in the gate capacitance.

The required small-device dimensions and the precision evaporation steps needed to fabricate the film transistor, coupled with the necessary control of surface states and trap density in the semiconductor film, make reproducible fabrication of the film transistor more difficult than ordinary bulk devices.

Using masking techniques, active film devices can be fabricated to perform complex circuit functions as shown in Fig. 4-36.

Another new class of film devices employing the principle of tunnel

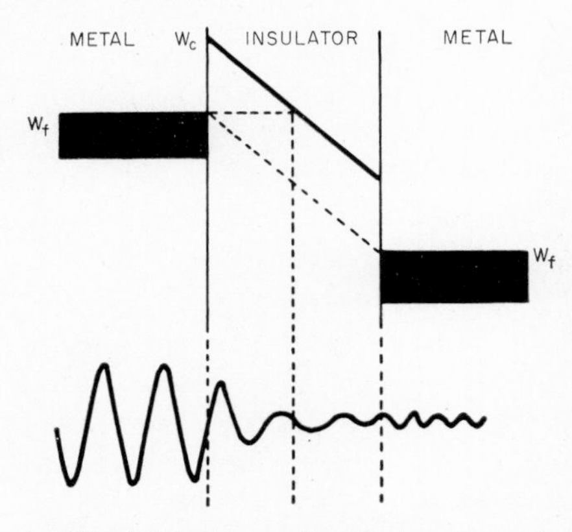

FIG. 4-37. Energy-band structure of metal insulator–metal diode with applied electric field showing wave functions of tunneling electron. (*C. A. Mead.*[83])

emission has been proposed by Mead.[83] Tunnel emission is the phenomenon occurring at a metal-insulator interface when a high electric field is present within the insulator. Electrons in the metal impinge upon the interface and "tunnel" through the insulator forbidden region into the conduction band, as shown in Fig. 4-37. Propagating wave solutions are possible, and the electrons gain energy from the electric field.

Insulating layers thicker than a few tens of angstroms are impenetrable, as the probability of an electron getting through falls off exponentially with the thickness. For applied voltages less than the metal-insulator work function ϕ, the tunneling current density J is proportional to the applied voltage V, demonstrating that the low-voltage tunneling resistance is ohmic.

$$J = \frac{qV}{h^2 s} (2m^* \phi)^{1/2} \exp\left[-\frac{4\pi s}{h} (2m^* \phi)^{1/2} \right] \tag{4-9}$$

At high applied voltages ($qV > \phi$), the current increases very rapidly:

$$J = \frac{q^2 V^2}{8\pi h \phi_s^2} \exp\left[-\frac{8\pi s}{3hqV} (2m^*)^{1/2} \phi^{3/2} \right] \tag{4-10}$$

In these equations, s is the insulator thickness, m^* the electronic effective mass, and q the charge of the electron. Equation (4-10) is similar to the Fowler-Nordheim[84] field-emission relationship. Figure 4-38 shows the tunneling current-voltage characteristics of an Al-Al_2O_3-Al structure[85] and an approximate proportionality of the tunneling current to the effective film area.

An excellent survey of the tunneling phenomenon with a list of references has been given by Chynoweth.[86] A significant feature of the tunnel-emission process is that it constitutes a controlled source of majority carriers.

Diode and triode structures, shown schematically in Fig. 4-39, were constructed by Mead, using metals such as aluminum and tantalum, and insulators such as Al_2O_3, SiO, and Ta_2O_5. At current levels of a few microamperes, the triodes showed current-transfer ratios α up to approximately 0.1. Electrons tunneling through the thin-base region find themselves in the conduction band of the collector insulator, and will be accelerated and collected when the collector is positively biased. The tunnel-emission triode can be characterized in much the same way as the transistor, and the same considerations concerning gain and impedance transformation between input and output apply. Limitations on the current gain stem chiefly from traps in the insulator regions, the surface states at the metal-insulator interfaces, and electron collisions in the base region and emitter insulating layer. High-frequency operation necessitates relatively high current densities through the device. One limitation on the current density is that of space charge in the base-collector insulator. Another serious limitation of the effective current density is given by the self-bias effect which is due to the base-current voltage drop, similar to that found in conventional transistors.

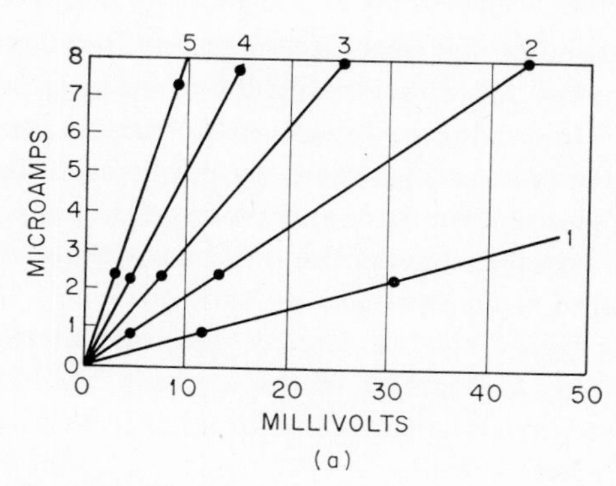

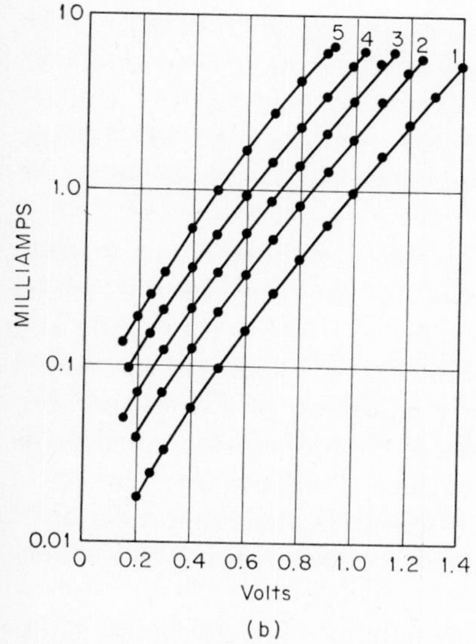

FILM NO.	RELATIVE AREA, UNITS OF 4.7×10^{-3} CM^2	FILM THICKNESS, ANGSTROMS
5	5	48
4	4	45
3	3	48
2	2	47
1	1	48

FIG. 4-38. Al-Al_2O_3-Al tunneling characteristics. (*J. C. Fisher and I. Giaever.*[85])

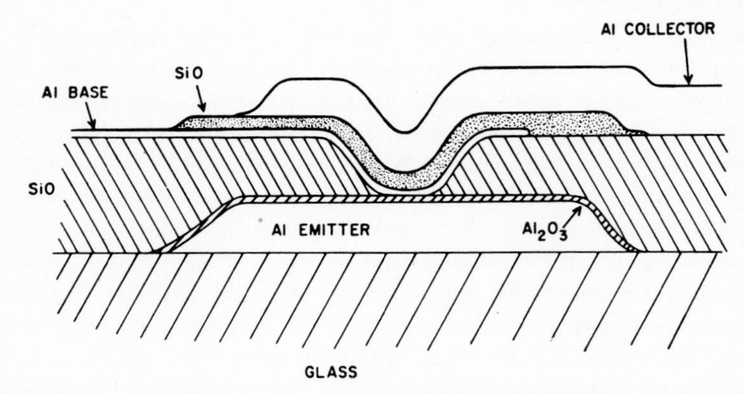

FIG. 4-39. Cross section of experimental tunnel-emission triode. (*C. A. Mead.*[83])

Film devices utilizing, in part, single-crystalline layers of semiconductors hold potential for film circuits when techniques for depositing single-crystalline semiconductor layers on polycrystalline or amorphous substrates are developed. The metal-interface-amplifier, reported by Spratt[87] et al., is such a structure, and its operation supposedly depends on hot electron injection and collection. The source of injected electrons is a metal, separated from the base metal into which the electrons are injected through a thin dielectric film by quantum mechanical tunneling as shown in Fig. 4-40. A fraction of the injected hot electrons possesses sufficient energy to surmount the collector surface barrier and contribute to the collector-circuit current.

Amplifying characteristics similar to those of an *npn* transistor, with current gains up to 0.95 and with a β cutoff at 100 kc, have been observed. The experimental devices consisted of layers of gold for the emitter electrode, 20-A-thick Al_2O_3 for the tunneling insulator, and 100-A-thick aluminum deposited on an *n*-type germanium single-crystal wafer to form a surface-barrier collector.

Hall[88] has proposed that the metal-interface-amplifier might operate as a depletion-layer transistor,[89] instead of the hot-electron model described by Spratt et al, if pinholes are present in the thin oxide and base metal films. Pinholes frequently exist in thin evaporated layers, and since the Al_2O_3 insulator was formed by oxidation of the evaporated aluminum base, their presence would result in the emitter layer being in direct contact with the collector barrier. Furthermore, the mean free path of hot electrons in aluminum is less than the 1,000 A required to obtain current-transfer ratios of the order of 0.9 in the 100-A aluminum-base films.

Electron scattering at the metal-semiconductor interface and backscattering within the germanium collector region will also tend to decrease the current-transfer ratio. The work of Lavine and Iannini[90] shows that the method of device fabrication and the manner in which current and voltage are impressed upon the device during measurement, have an important bearing upon the device characteristics. They conclude that the depletion-layer action is responsible for the observed high current-transfer ratios and power gains, but do not exclude the possibility that hot electrons can be transmitted through thin aluminum films or that part of the current is transferred by such a mechanism.

Recent work by Spitzer et al.[91] has indicated a mean free path of 740 A for one-volt hot electrons in gold. In view of these results, Kahng[92] has made a structure utilizing a 100-A-thick gold film on an *n*-type silicon substrate. A 40-A-thick SiO insulating film separated the gold film from a small mercury ball which was used as the emitter electrode. The use of mercury as an emitter contact minimized film stresses, and permitted the detection of large pinholes in the SiO insulator. The observed values for the current gain were close to the expected values, assuming unity injection and collection efficiencies. These results lend support to the hot-electron transport mechanism for the metal-interface-amplifier.

Heterojunctions between semiconductor compounds were investigated by Aven and Cook.[93] Polycrystalline *n*-type films of CdS were deposited onto *p*-type ZnTe single crystals. The two compounds are similar in lattice constants and band gaps (2.2 ev for ZnTe and 2.4 ev for CdS). Evaporated gold and fused indium formed the external contacts to the ZnTe and CdS, respectively. Rectification ratios up to 5×10^4, and forward-current densities of 5 amp/cm^2 at about 3 volts were achieved. In these diodes, the observed forward-current density is much greater than that calculated according to the diffusion theory, and appears to be strongly dependent on the charge recombinaton through recombination centers in the space-charge region. Similar results have been observed in other wideband-gap semiconductors such as SiC[94] and GaP.[95]

Titanium oxide has also been investigated for film-device fabrication.

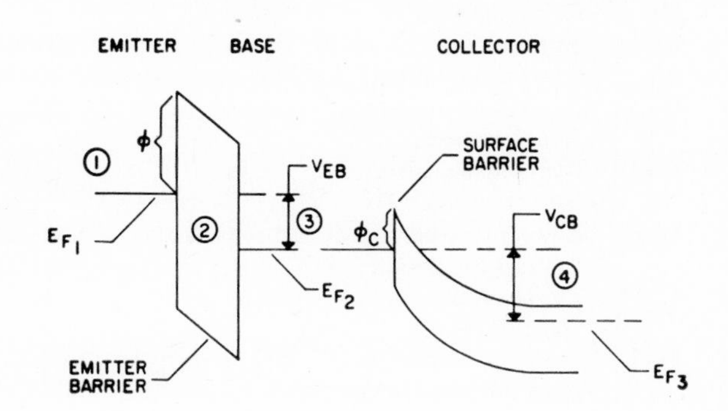

FIG. 4-40. Electron-potential diagram of the metal-interface-amplifier structure. (*J. P. Spratt, R. F. Schwarz, and W. M. Kane.*[87])

Diodes made by evaporating counterelectrodes of gold, bismuth, or silver onto oxidized titanium films have shown rectification ratios up to 10^5 with reverse breakdown voltages ranging from 2 to 20 volts. Since the phase diagram of titanium-oxygen is rather complex, the mechanism of rectification is not well understood, and much basic study is needed to assess the type of devices that can be made.

2. Semiconductor Film Deposition

Semiconductor films can be produced by several techniques such as evaporation in a vacuum, growth from the gas phase through a chemical reaction, or sputtering. The choice of the technique employed is governed by such factors as the class of semiconductors to be deposited, the nature of the substrates used, and the required physical and electrical characteristics of the deposited semiconductor films.

Some of the vapor-growth processes that have been successfully used for the elemental and compound semiconductors are listed in Table 4-7.

TABLE 4-7. VAPOR-GROWTH PROCESSES FOR SEMICONDUCTOR ELEMENTS AND COMPOUNDS

Semiconductor	Type of process	Source material	Ambient or carrier gas	Deposition temperature, °C	References
Si	H_2 reduction	$SiHCl_3$	H_2	1150–1300	115, 116
Si	H_2 reduction	$SiCl_4$	H_2	1200–1300	117, 118, 116
Si	H_2 reduction	$SiBr_4$	H_2	1100–1300	119
Si	H_2 reduction	SiI_4	H_2	1000–1200	120, 121
Si	Pyrolytic decomposition	SiI_4	Vacuum	~1000	122
Si	Disproportionation of SiI_2	Si	I_2	~950	123, 124
Ge	Disproportionation of GeI_2	Ge	I_2	~400	123
Ge	H_2 reduction	$GeCl_4$	H_2	~830	118
SiC	Sublimation	SiC	Vacuum	>2000	125
SiC	Gaseous cracking	$SiCl_4$-toluene	H_2	~2000	126
GaAs	Evaporation–disproportionation	GaAs	I_2	620–690	127, 128
GaAs	Evaporation–disproportionation	GaAs	HCl	900	129
GaP	Reduction and compound formation	P-Ga-Ga_2O_3	Vacuum	~1000	130
GaP	Evaporation–disproportionation	GaP	I_2	....	128
InSb	Complex formation and decomposition	$In(CH_3)_3$-SbH_3	Vacuum	>150	131

Although the less common compounds like sulfides or selenides have been omitted, this list of epitaxial processes for semiconductors is already impressive. Various types of reactions have been employed, using halides, elements, suboxides, or intermetallic compounds as source materials. The vapor-growth process has been most widely used for obtaining isoepitaxial deposits on single-crystalline semiconductor substrates. Heteroepitaxial films such as Ge on GeAs[96] and Ge on GaP[97] have also been investigated. Wafers of single-crystalline semiconductors or materials such as CaF_2,[98,99] which can provide conditions for epitaxial growth, however, are not practical substrates for fabricating large-area thin-film circuits. The chemically corrosive atmosphere used in vapor growth, the difficulty of achieving form-controlled growth through masking, and the poor thickness control of very thin films makes the vapor-growth technology not so attractive for the deposition of film active elements as the vacuum-evaporation technique.

Epitaxial films made by vacuum evaporation have been reported for germanium-on-germanium substrates.[98–102] Kurov[101] has prepared single-crystal films at substrate temperatures of 600°C and pressures around 10^{-6} torr. These films were *p*-type, regardless of the source type evaporated. This is attributed to the large number of vacancy-acceptor levels. Weinreich,[102] however, has found that when highly doped *n*-type germanium of about 0.001 ohm-cm is evaporated, the acceptor centers are overcompensated and the film is *n*-type. Using ultrahigh vacuum conditions ($p < 10^{-8}$ torr), Nakhodkin and Nemtsev[103] have shown that *n*-type germanium films can be obtained from a 26-ohm-cm *n*-type evaporation source.

The threshold of crystallization for germanium films ranges from 350 to 450°C, depending on the evaporation rate and ambient pressure. Single-crystalline growth appears for substrate temperatures above 550°C. Germanium heteroepitaxial film growth has been observed on single-crystal CaF_2 substrates.[98,99] Such growth occurred only for low evaporation rates (10A/sec) and for substrate temperatures above 550°C.

Polycrystalline germanium films, deposited on glass and quartz substrates[98,104,105] showed strong texturization, the degree of which is dependent on the evaporation rate and the deposition temperature.

Little has been reported on the deposition of silicon films by vacuum evaporation. Hass[106] has studied polycrystalline silicon deposits on NaCl and Al_2O_3 substrates that were heated over 600°C. Hale[107] et al. have evaporated single-crystal silicon films onto a silicon single-crystal substrate in a vacuum of about 10^{-6} torr. Electron-beam heating, as well as crucible resistive heating were used to heat the sources. Substrate temperatures above 1125°C and deposition rates of 1 μ/min gave the best films.

p-type film doping was accomplished from a boron-silicon alloy evaporation source, whereas n-type films were obtained by using a two-source system. The n-type doping is very difficult to control because of the large difference in vapor pressures between silicon and the various group V dopants such as phosphorus, arsenic, and antimony.

Polycrystalline films of the group II-VI compound semiconductors such as CdS, CdSe, and CdTe have also been grown by vacuum evaporation.[108–110] The close proximity of the vapor pressures of the compound constituents and the ease of recombinations at the substrate surface permits stoichiometric film growth. From the art of fabricating photoconductive devices, a considerable amount of information concerning the impurity doping and resistivity control of these compounds, in particular CdS, has been obtained. The wideband gap of these semiconductors is well suited for the study of space-charge-limited currents in insulators,[111] a phenomenon which offers considerable potential for new film-device development. Sparse data on electron mobilities in polycrystalline films of CdS and CdSe indicate values ranging from 10 to 200 cm^2/volt-sec.

The group III-V semiconductor compounds, such as InSb and InAs, have been evaporated onto glass and NaCl,[112] but the wide vapor-pressure difference of the constituents caused off-stoichiometric compositions. The Hall electron mobility increased markedly in these films with the thickness of the film and with improved crystallinity.

The electrical film properties of a large number of the semiconducting materials have been measured only on polycrystalline films, since often crystal film-growing techniques have not been sufficiently developed to permit the growth of single-crystalline films onto electrically insulating substrates. The semiconducting properties of the polycrystalline films cannot be fully understood unless their properties can be compared in detail with those of single-crystal films. As an example, Eckart and Jungk[113] have calculated the carrier density of polycrystalline germanium films from the Hall coefficient, and have plotted the resistivity, hole concentration, and hole mobility as functions of the film thickness (Fig. 4-41). Using n- and p-type single crystals with mobilities between 2,000 and 3,000 cm^2/volt-sec as the source material, the carrier concentrations of the thin evaporated germanium films are about two to three orders of magnitude higher than those of the evaporation source material, and are probably caused by surface levels of an acceptor nature. With increasing film thickness, the resistivity and hole concentration decrease, whereas the hole mobility increases. The mobilities are between 20 and 200 cm^2/volt-sec, and are one to two orders of magnitude smaller than in the single-crystalline source material. In contrast to these results, Riesz[114] et al. have made resistivity measurements on polycrystalline germanium films deposited on a synthetic spinel ($3MgO{\cdot}5Al_2O_3$), and

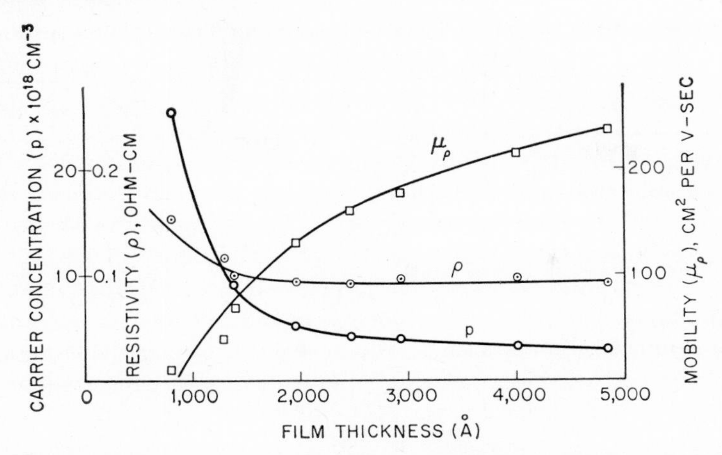

FIG. 4-41. Resistivity, hole concentration, and Hall hole-mobility variation with thickness of polycrystalline germanium films. The p-type single crystals used as sources had measured mobilities around 2,000 cm²/volt-sec. (*F. Eckart and G. Jungk.*[113])

found that the electrical properties of these films were essentially the same as those of single-crystal films deposited on Ge substrates, in spite of the structural differences.

It is clear that much fundamental work is still needed to characterize semiconductor films and to properly exploit their properties in new film-device designs.

4-4. MAGNETIC FILMS

Magnetic-film elements exhibit two stable states, corresponding to positive and negative remanence. A particular element or "bit" in an array of such elements can be selected and magnetized into either one of these states, under which circumstance information is stored in the element. The state of magnetization, corresponding to this information, can be recognized upon interrogation and the information retrieved. An array of such elements or bits can thus be operated as a random-access memory. Thin magnetic films of about 100 A thickness are studied to verify the validity of various two- and three-dimensional models of ferromagnetic phenomena and other physical film properties. Of more practical interest in the field of microelectronics are films with a thickness about 1,000 A because of their potential use as memory elements in a digital computer.

Thin films of 80:20 (by weight) nickel-iron were prepared by Blois[132] who found that the presence of an orienting magnetic field induces a uniaxial anisotropy. Its easy axis of magnetization is aligned parallel

TABLE 4-8. COMPARISON OF FERRITE CORES AND THEIR MAGNETIC FILMS FOR USE IN A COMPUTER MEMORY

Characteristic	Ferrite cores	Thin magnetic film
1. Bistable	Yes	Yes
2. Speed of coincident current switching	1 μsec	3–10 nanosec
3. Drive power	800 ma into 50 ohms	400 ma into 5 ohms
4. Repetition rate before adverse heating	500 kc	At least 5 Mc, probably higher
5. Physical size	1/16 in. diameter	1/16 in. diameter
6. Temperature range of operation	Up to 50°C	Should be higher
7. Reliability	100%	Should be comparable
8. Number in a system	2½ million	Unknown
9. Economy of fabrication	Expensive (~5 cents/bit)	Potentially inexpensive

to the direction of the field. Such anisotropic films which exhibit at a thickness of a few thousand angstroms a single-domain structure have been the subject of extensive investigation over the last few years. Table 4-8, published in 1958, gives a comparison between films and ferrite cores[133] with regard to their potential as computer components. It has since been found that the implementation of thin-magnetic-film memories is a substantial problem, which has been only partially solved, in terms of either the memory itself or the anticipated economic advantage to be realized. However, a thin-film memory of 256 words, 72 bits per word, operating with a cycle time of 0.7 μsec, has been announced commercially.*

1. Technology

The following criteria can be advanced with regard to the desirable properties of a magnetic computer element:

1. Low crystalline anisotropy
2. Low magnetostriction
3. Simplicity of composition
4. Chemical stability
5. Reasonably high flux density
6. High Curie temperature

A material that nearly satisfies these requirements is the binary alloy of nickel-iron with a composition of 81:19 by weight. This composition has zero magnetostriction, changing to positive on the iron side and negative on the nickel side. Above approximately 65% nickel, the binary

* Remington Rand Univac.

alloy exists in the face-centered cubic form. The crystalline anisotropy is at a minimum when the magnetization follows the direction of the cube diagonals. The anisotropy energy can be described by two coefficients: K_1 and K_2. (This anisotropy energy results from the interaction between the electron spins and the magnetic field of the lattice ions. Phenomenologically, the two coefficients are obtained by a Taylor expansion in the direction cosines, making use of the lattice symmetries of the material under consideration.) K_1 is positive on the iron-rich side and goes to zero at about 74% nickel, and K_2 is positive on the nickel-rich side and goes to zero at about 78% nickel. At the 81:19 composition, both K_1 and K_2 are small, with K_2 assuming positive and K_1 negative values. The flux density is 10^4 gauss and the Curie temperature 550°C.[134]

Films of nickel-iron (and magnetic films, in general) have been prepared by the following methods:

1. Vacuum deposition[132,133,135]
2. Electroplating[136–138]
3. Cathode sputtering[139,140]
4. Chemical reduction (electroless plating)[141–143]
5. Pyrolytic methods[144]

The first three of these methods are more commonly used, and yield magnetic films of essentially comparable quality. The chemical-reduction method is a solution method similar to that of Brenner,[142] using hypophosphite to reduce dissolved nickel ions to nickel on a catalytically active surface. The pyrolytic method thermally decomposes an appropriate mixture of metal-organic compounds. A typical process uses a mixture of nickel and iron carbonyls. These carbonyls can be volatilized, mixed, and decomposed at a temperature of a few hundred degrees.

All methods require for good reproducibility and uniformity of orientation a smooth substrate surface. Adequate surface properties have been found on fire-polished glass, mica, and well-polished metal surfaces.[145] To improve the surface uniformity further, it is common practice to coat the surface with silicon monoxide immediately before the deposition of the ferromagnetic film.

There is sufficient difference among the first three methods to necessitate their separate discussion.

Vacuum Evaporation. *Evaporation Sources.* Just about all types of sources conventional to vacuum evaporation have been used. The most common types are the filament and the crucible source (Al_2O_3), the latter heated by either electron bombardment or induction. A critical source problem is the occurrence of fractionation due to the fact that the vapor pressure of iron is greater than that of nickel. It has been found that an 83:17 evaporant composition results in an 81:19 film. A second parameter influencing the magnetic-film properties is the angle of vapor incidence. A uniaxial anisotropy perpendicular to the incident beam can be produced simply by inclining the substrate to the incident beam of evaporating metal.[146] Such an anisotropy causes a deviation or skew of the easy axis. It can be avoided by using a large source-to-substrate distance and by limiting the size of the substrate. Raffel,[147] for example, recommends a maximum substrate size of 1.6 by 1.6 in. for a crucible source with a 12-in. distance from the substrate.

Pressure. Satisfactory films are obtained at pressures of 10^{-5} torr. The importance of the residual gas composition is not known. The influence of oxygen in the magnetic annealing of permalloy[148,149] has raised the question whether a similar or related effect occurs in the induced anisotropy of thin films. Films deposited in ultrahigh vacuum, however, show no superiority of magnetic properties.

Substrate Temperature. According to various references, satisfactory films are obtained at substrate temperatures ranging from 250 to 400°C. The optimum temperature range in reality may be smaller, since accurate substrate-temperature measurements are difficult, and different source geometries can vary considerably the amount of radiation heating supplied to the substrate by the source. The film anisotropy is influenced by the substrate temperature.

Evaporation Rate. High evaporation rates are preferred, and 1,000 A/min is considered a good value. The improvement of film properties with higher rates seems to result more from the decrease in grain size than from the lesser absorption or occlusion of residual gases.

Film Thickness. Measurement of film thickness is based on time of evaporation, or a resistance monitor, or the use of a crystal oscillator.[150]

Cathode Sputtering. This method has been described previously. It differs from the evaporation process in such parameters as a higher pressure (about 10^{-1} torr), the presence of an electric field, and a different material-transport mechanism not requiring a high evaporant temperature. Since the pressure is quite high, collisions occur between the metal atoms and the residual gas molecules, resulting in a diffusion of the metal-vapor stream and an elimination of the angle-of-incidence effects encountered with vacuum evaporation. A further consequence is an improved thickness uniformity over sizable areas.

Films of 81:19 composition have been prepared by starting with an 80:20 source. After a cleanup consisting of a pump down to 5×10^{-6} torr, argon at a pressure of 100 μ was leaked into the system. Using a potential of 3,500 volts and a current of 150 ma with an anode size of 6 in. square, sputtering proceeds at a rate of 15 A/sec for a cathode-to-anode distance of 1 in. Over a 3-in.-square substrate, the resulting thickness uniformity is about 2%. Controlled uniaxial anisotropy is readily obtained with a field of about 20 oersteds, either a-c or d-c. The film properties are comparable in quality to those obtained by vacuum evaporation, and are characterized by a high degree of magnetic uniformity over areas of several square inches. It is further claimed that there is no skew of the easy axis. The references on sputtering report no attempts to control the substrate temperature to the degree common in vacuum evaporation. The substrate is placed on the anode which usually is water-cooled. The receiving surface is exposed to the glow-discharge region. Modified results have been obtained by altering the sputtering conditions.[141]

Electroplating. As with sputtering, no angle-of-incidence effect is observed. Electroplating requires a conductive substrate or conductive-substrate coating. A metal layer, usually gold, is applied by sputtering, or evaporation in the latter case. The conductive coating is generally kept thin (in the vicinity of 100 A), to avoid a deterioration of the original surface smoothness and structure. It has been shown, for instance, that angle-of-incidence effects, obtained by deliberately evaporating gold under high incidence, influence the subsequent magnetic properties obtained after electroplating nickel-iron.[151] As a consequence of the thinness of the conductive layer, a potential drop occurs across the substrate during the electroplating process, accompanied by a nonuniform current density.

The electroplating baths for the deposition of nickel-iron are most commonly sulfate or mixed sulfate-chloride solutions. They contain the corresponding nickel and ferrous salts, boric acid, saccharin (a stress-relieving agent), and a wetting agent. The pH of the bath is generally kept below 3.5 to minimize the oxidation of the ferrous ions and their precipitation. Besides bath composition, the parameters influencing the nickel-iron film properties are bath temperature, current density, and agitation. Since the deposition rates of the components are not the same, one species may be removed more rapidly from the cathode layer, and its replenishment depends upon the diffusion and the degree and type of agitation.

A particularly interesting bath is the sulfamate bath used by Long[138] to prepare 1-μ films of 81:19 Ni-Fe on wires. This bath contains the nickel and ferrous salts of sulfamic acid and boric acid, saccharin, and a wetting agent. Whereas the previously mentioned baths are normally used at low current densities of the order of 3 ma/cm^2 (about 10 A/sec), the sulfamate bath can be used at current densities over 100 times greater. The bath itself is capable of giving a stress-free film of nickel-iron. Long has also used high rates of agitation to introduce turbulence and maintain a cathode layer which has the same composition as the bulk-plating solution.[152] Since the composition dependency on the current density becomes smaller under these conditions, a fairly wide current density can be tolerated for only a 2% variation of the film composition.

The magnetic properties of the films obtained by electroplating are again quite comparable to those produced by vacuum evaporation. The films made by Long, although more than 10 times thicker than those of usual interest, compare extremely well with those made by any other method.

2. Magnetic Characteristics of Films

Probably the most important single feature responsible for the current interest in thin magnetic films is the high speed with which the state of magnetization can be reversed. Dependent on film properties and applied fields, three modes of magnetization reversal can be observed on films: by a coherent rotation of the magnetization, (a process which can take place in nanoseconds), by wall motion (a slow process comparable in speed to that found in ferrites), and by an incoherent rotation.

The theory of magnetostatic behavior of a material with uniaxial anisotropy and with changing state of magnetization by a reversal process has been worked out by Stoner and Wohlfarth.[153] For a thin film, the magnetization is restrained to the plane of the film because of the influence of demagnetizing effects. The stable states of magnetization result from the minimization of the free energy. In a simplified model, not considering the crystalline or shape anisotropies, the free energy consists of two terms, the uniaxial anisotropy energy defined by $K \sin^2 \theta$ and the interaction between the magnetization and the applied field ($-\mathbf{H} \cdot \mathbf{M}$):

$$E = K \sin^2 \theta - H_L M \cos \theta - H_T M \sin \theta \tag{4-11}$$

where θ is the angle between magnetization $\mathbf{M}$ and the easy axis. H_L and H_T are the magnetic-field components parallel and perpendicular to the easy axis. To minimize the free energy,

$$\frac{\partial E}{\partial \theta} = H_K \sin \theta \cos \theta + H_L \sin \theta - H_T \cos \theta = 0 \tag{4-12}$$

$$H_K = \frac{2K}{M} \tag{4-13}$$

If only a traverse field is applied ($H_L = 0$), the hard-direction hysteresis loop is found from

$$\sin \theta = \frac{H_T}{H_K} \tag{4-14}$$

and

$$M_T = M \frac{H_T}{H_K} \qquad \frac{H_T}{H_K} \leq 1 \tag{4-15}$$

with the result that the hard-direction hysteresis loop is a straight line through the origin for $\left|\dfrac{H_T}{H_K}\right| \leq 1$. H_K (the value of H_T at which saturation occurs) is the anisotropy field. For $|H_T| > H_K$, M_T is the saturation value.

To obtain the longitudinal (easy-axis) hysteresis loop, H_T is set equal to zero, and $\sin\theta\,(H_K \cos\theta + H_L) = 0$.

To find a minimum, the second derivative is examined:

$$\frac{\partial^2 E}{\partial \theta^2} = 2K(\cos^2\theta - \sin^2\theta) + H_L M \cos\theta \tag{4-16}$$

For $\theta = 0$:
$$\frac{\partial^2 E}{\partial \theta^2} = 2K + H_L M \qquad H_L > -\frac{2K}{M} \tag{4-17}$$

For $\theta = \pi$:
$$\frac{\partial^2 E}{\partial \theta^2} = 2K - H_L M \qquad H_L < \frac{2K}{M} \tag{4-18}$$

The longitudinal component of magnetization, therefore, is always either $+M$ or $-M$. A sharp change from $-M$ to $+M$ occurs at values of $H_L = \pm H_K$. These loops are shown in Fig. 4-42. (In soft magnetic materials, $B = H + 4\pi M$ is approximately equal to $4\pi M$.) For certain values of H_L and H_T, the second derivative passes through zero and the energy minima become maxima. The locus of such field values comprises a critical curve. This is obtained by setting both the first and second derivatives equal to zero and by solving the resulting equation:

$$\frac{H_L}{H_K} = \cos^3\theta \qquad \frac{H_T}{H_K} = \sin^3\theta \tag{4-19}$$

or

$$H_L^{2/3} + H_T^{2/3} = H_K^{2/3} \tag{4-20}$$

The corresponding curve is shown in Fig. 4-43. When the field vector **H** lies inside the critical curve, two stable directions of **M** exist. When it passes through the critical curve, one of the two states become unstable, resulting in a switching of that state to a stable state. The direction of magnetization for a particular **H** is obtained by drawing the tangent from the end of **H** to the critical curve.

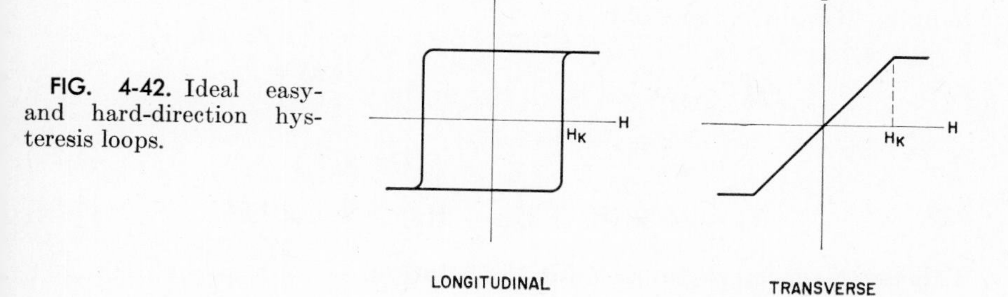

FIG. 4-42. Ideal easy- and hard-direction hysteresis loops.

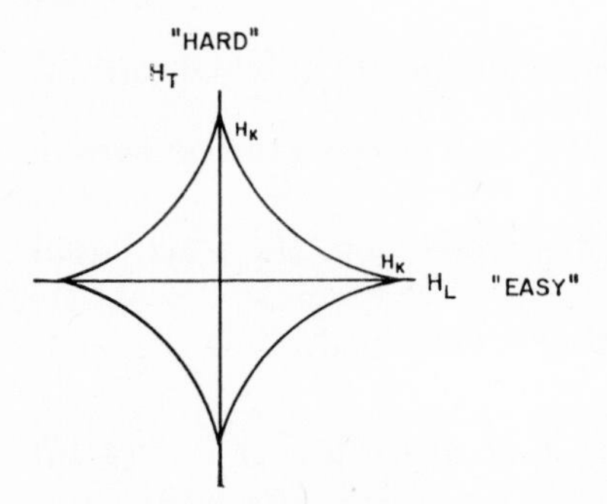

FIG. 4-43. The critical curve $H_L^{2/3} + H_T^{2/3} = H_K^{2/3}$. The line describing the direction of magnetization is tangent to this curve.

The static model described above defines the important operational film parameter H_K, as well as the conditions whereby switching from one stable state to another occurs, namely, by exceeding the field values enclosed by the critical curve. Furthermore, a comparison of the experimental hard-direction loop of an actual film with the theoretical curve (Fig. 4-43) gives a measure of the film quality, indicating how well it approaches the rotational model.

The uniform rotation of a single domain can be described by the following equations:

$$\frac{d\mathbf{p}}{dt} = \mathbf{T} \tag{4-21}$$

$$\mathbf{M} = \gamma\mathbf{p} \tag{4-22}$$

$$\mathbf{T} = \mathbf{M} \times \mathbf{H} \tag{4-23}$$

which combine to

$$\dot{\mathbf{M}} = \frac{d\mathbf{M}}{dt} = \gamma\mathbf{T} \tag{4-24}$$

$\mathbf{p}$ is the angular momentum, $\mathbf{T}$ the torque acting on the magnetization, $\mathbf{H}$ the torque producing the fields, and γ the gyromagnetic ratio. $\mathbf{H}$ contains all the fields which act on the magnetization, including the applied field $\mathbf{H}_a$, anisotropy field $\mathbf{H}_{an}$, demagnetizing field $\mathbf{H}_D$ and a damping field $\mathbf{H}_f$ proportional to the rate at which the magnetization is changing.

$$\mathbf{H}_f = -\frac{\lambda}{\gamma}\frac{\dot{\mathbf{M}}}{M} \tag{4-25}$$

$$\mathbf{H}_D = -4\pi M \sin\psi\, \mathbf{e}_Z \tag{4-26}$$

where ψ is the angle between the magnetization and the xy plane, and $\mathbf{e}_Z$ a unit vector normal to the plane of the film. Introducing the fields defining the torque, one obtains

$$\dot{\mathbf{M}} = \gamma\mathbf{M} \times \left(\mathbf{H}_a + \mathbf{H}_{an} + \mathbf{H}_D - \frac{\lambda\dot{\mathbf{M}}}{\gamma M}\right)$$

$$= \frac{\gamma T_1}{1 + \lambda^2} - \frac{\gamma\lambda}{(1 + \lambda^2)M}(\mathbf{M} \times \mathbf{T}_1) \tag{4-27}$$

$$\mathbf{T}_1 = \mathbf{M} \times (\mathbf{H}_a + \mathbf{H}_{an} + \mathbf{H}_D) = -\mathbf{r} \times \nabla\mathbf{E} \tag{4-28}$$

This is the torque responsible for the rotation.

In a spherical coordinate system and with $\psi = \pi/2 - \theta$, the equations for the ψ and ϕ components are (dropping the λ^2 in the denominator):

$$\dot{\psi} = -\frac{\gamma\lambda}{M}\frac{\partial E}{\partial \psi} - \frac{\gamma}{M}\frac{1}{\cos\psi}\frac{\partial E}{\partial \phi} \tag{4-29}$$

$$\dot{\phi} = \frac{\gamma}{M\cos\psi}\frac{\partial E}{\partial \psi} - \frac{\gamma\lambda}{M\cos^2\psi}\frac{\partial E}{\partial \phi} \tag{4-30}$$

The ψ motion is small because of the large demagnetizing fields, and the z component of the anisotropy is negligible compared to the demagnetizing field. Consequently: $\cos\psi \approx 1$, $\sin\psi \approx 0$, and $\partial E/\partial\psi \approx 4\pi M^2\psi$. The above equations become thus

$$\dot{\psi} = -4\pi M\gamma\lambda\psi - \frac{\gamma}{M}\frac{\partial E}{\partial \phi} \tag{4-31}$$

$$\dot{\phi} = 4\pi M\gamma\psi - \frac{\gamma\lambda}{M}\frac{\partial E}{\partial \phi} \tag{4-32}$$

λ has a value between 0.02 to 0.07, $\gamma = 1.76 \times 10^7$ per oersted-sec. To get an idea of the nature of the motion of the magnetization, we assume the torque $\partial E/\partial\phi$ as constant and equal to 1. The solution of the ψ equation is then

$$\psi = 2.2 \times 10^{-3}(1 - e^{-t/10^{-9}}) \tag{4-33}$$

which represents a shallow lifting of the magnetization out of the plane of the film, with a time constant of one nanosecond.

The magnetization thus precesses about the demagnetizing field and undergoes a change in direction of 180° in several nanoseconds. Numerical solutions of the above equations have been obtained by Smith[133] and by Olson and Pohm.[135] Two aspects of the model are particularly interesting, namely, the single-domain character of the film with uniform properties, and the important role of the demagnetizing field.

The switching behavior of a film is complicated by the fact that wall-motion processes occur also. To approach the case of the idealized rotational model, fields with amplitudes much greater than that predicted from the simple model and with fast rise times are needed. The real film is characterized by magnetic inhomogeneities arising from localized compositional variations,[154] crystalline anisotropy, stress anisotropy, structural imperfections, and its finite size. A dispersion of both H_k and the easy-axis direction thus results in a real film.[146,155] Because of the high demagnetizing fields located at the film edges, domains of reverse magnetization exist in the vicinity of these edges. When the magnetic fields are less than critical, and no switching should occur, these walls can be made to traverse the film.[156]

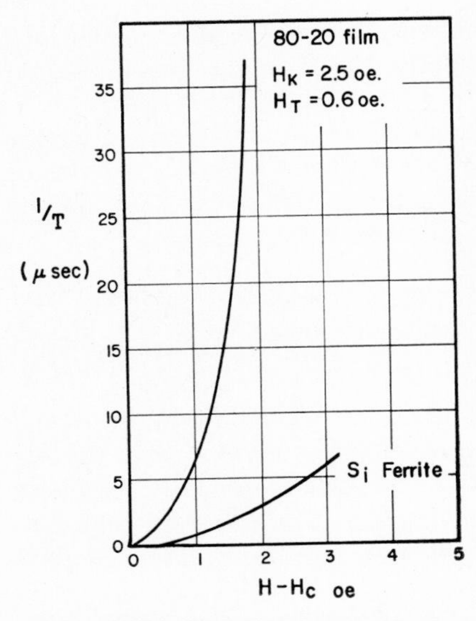

FIG. 4-44. Rotational switching of thin film. T is the switching time. (*Olson and Pohm.*[135])

The net effect is that, dependent on the values of rise time, orientation, and amplitude of the driving field, three modes of magnetization reversal can occur. They are characterized by both waveshape and time of reversal. These modes, as mentioned previously, are wall motion, incoherent rotation, and uniform rotation.[135] Wall-motion switching is expected when the driving fields are smaller than the critical values. It is characterized by switching speeds commonly found with ferrites. Intermediate between the nanosecond speeds characteristic of uniform rotation and the microsecond speeds of wall motion is the switching speed of incoherent rotation. This process has been associated with a nucleation mechanism, rather than with a wall movement. Smith[155] has proposed a theory whereby reversal occurs by labyrinth* propagation, activated by a dispersion in H_k and easy-axis orientation. Initially, rotating low-H_k regions interact magnetostatically with regions of high H_k and thus are raising the threshold for rotation. Labyrinth formation and propagation then occurs. If the amplitude of the drive field is raised to a value exceeding the highest rotational threshold in the film, the incoherent rotation goes over into a coherent rotation.

Figure 4-44 shows some switching curves obtained by Olson and Pohm,[135] and Fig. 4-45 indicates the relationship between the various switching modes and the magnitude and orientation of the drive fields.

During the film preparation, a magnetic field is used to induce uniaxial anisotropy with a well-defined easy axis. The mechanism responsible for the formation of this anisotropy is not known and constitutes an interesting physical problem. A somewhat related problem has been found in the case of bulk nickel-iron where anisotropic changes occur under the influence of a magnetic anneal. The magnetic anneal is observed mainly during passage through the Curie temperature. It thus differs from the manner in which uniaxial anisotropy is introduced in thin films since they are deposited at a substantially lower substrate temperature. In fact, films formed by electroplating at room tempera-

* Labyrinths are regions of reversed magnetization threaded through the film.

ture exhibit anisotropy. This anisotropy of thin magnetic films has generally been associated with the directional ordering of ferromagnetic atom pairs and with the occurrence of anisotropic stresses in the film. The possibility that such anisotropy results from the alignment of imperfections along the annealing field has received consideration because of the work of Nesbitt and Heidenreich.[148,149] Prutton and Bradley[158] show the influence of composition and substrate temperature on the uniaxial anisotropy.

Films with well-defined uniaxial anisotropy can be magnetically annealed and H_k changed. If the magnetization is aligned in the hard direction, a very fast annealing occurs and H_k decreases in value. Segmüller[159] describes two annealing processes: one extremely fast, occurring in seconds, and the other slower, corresponding in times and temperatures with the results of other workers. In the slower process, magnetic annealing occurs in about one hour at 300°C. The anneal mechanism may limit the temperature range over which a uniaxial film can be used as a memory device.

Methods of measuring various magnetic quantities are described in papers by Raffel[147] and Smith.[155]

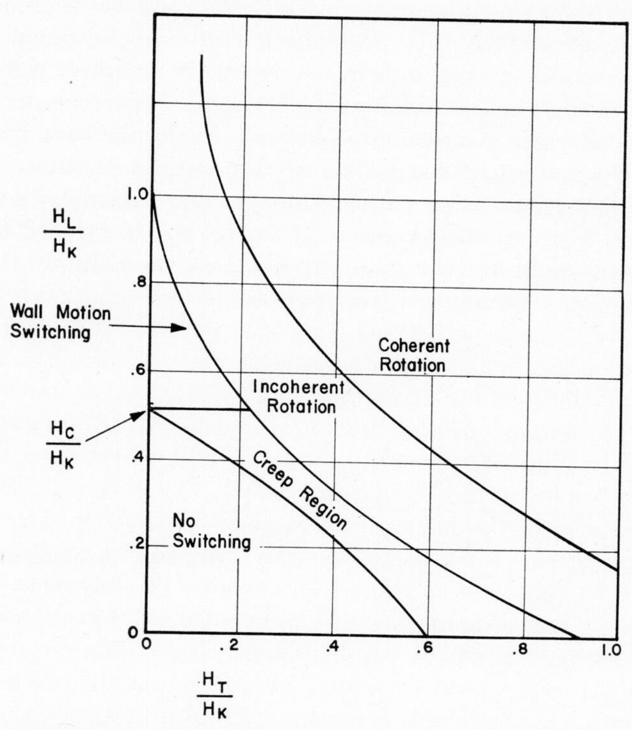

FIG. 4-45. Types of flux reversal. (*Pohm and Mitchell.*[157])

3. Information Storage

A considerable effort has been expended during recent years to develop a practical thin-film storage device for computers. Thin films promise superiority over ferrite cores with regard to speed and fabrication cost. A cost reduction can be expected because films can be produced in a batch process and they simplify the wiring and permit the use of etched-wiring strip-line techniques. In addition, they offer better thermal properties and new modes of usage.

The individual memory bits of a plane are geometrically defined either in terms of individual circular or rectangular spots or in terms of the location of the crossing of the drive lines using a continuous film of nickel-iron.[160] The magnetically active size of the bits varies from 10 mils to greater than 100 mils. The size selected depends on the packing density and signal output desired. The minimum spacing between bits is defined by the permissible bit and drive-line interactions. Circuit considerations such as the use of low-impedance transmission lines, the desirability of well-defined field patterns, and the requirement of minimum inductive and capacitive coupling noise have led to the use of metal or very thin glass substrates.[147,161] The open-flux structure imposes a limitation on the size-thickness ratio, with a resulting limitation in signal strength. A bit 1,000 A thick and 60 mils in diameter, for instance, has a demagnetizing field at its center of about 0.5 oersted. A decrease in size without a simultaneous decrease in thickness would increase the demagnetizing field and cause the bit to contain sizable domains. A bit of the above dimensions has a voltage time product (assuming a triangular signal pulse) of about 0.3 mv-μsec. If the bit size is reduced to 10 mils, the thickness would likewise have to be reduced to maintain the demagnetizing effects. Assuming a thickness reduction by a factor of only 2

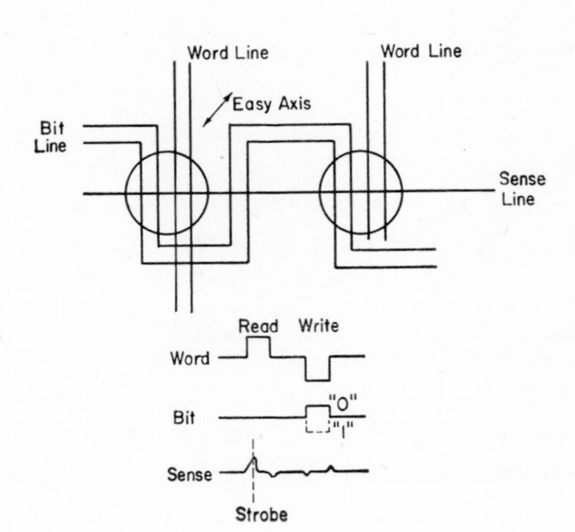

FIG. 4-46. Thin-film memory operated in parallel mode.

the voltage-time product is reduced to 0.025 mv-μsec. Switching in 20 nanosec, and assuming perfect coupling so that all the switched flux is sensed, results in a 1-mv signal. One is therefore, in a high-density thin-film memory, restricted to extremely fast-rising drive fields and can tolerate only extremely low noise level. Thus far, these restrictions have limited the use of such memories to the high-speed low-capacity region.

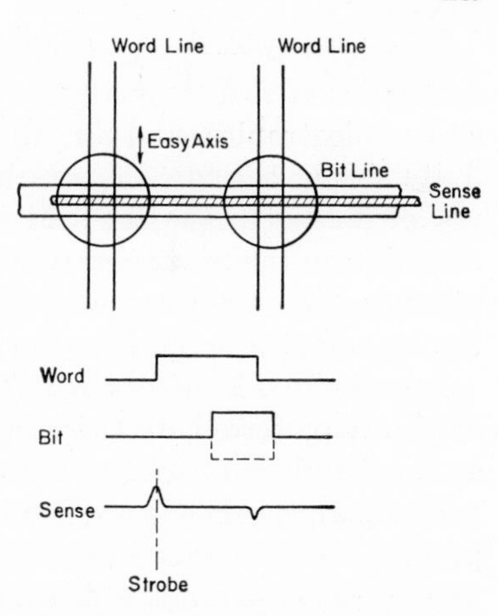

FIG. 4-47. Thin-film memory operated in orthogonal mode.

To attain rotational switching, the film must be driven with a field inclined toward the easy axis. For memory application, this mode of usage[162] could be exploited by a parallel drive system shown in Fig. 4-46. Both the word and bit fields are inclined to the easy axis of the bit, and the memory is operated as a two-dimensional array. A word is written by the coincidence of the word and bit fields with the information dependent on the polarity of the bit pulse. Information is sensed during the rise time of the word pulse. At any given bit location, the word field appears but once during the writing operation, and once during reading. The bit field must be lower than the wall-motion threshold. For fields inclined to the easy axis, irreversible flux changes can occur, usually referred to as *creep*. Dietrich and Proebster[163] have shown that creep sets in at fields substantially below the wall-motion coercive force for fields applied at 45° to the easy axis.

This is shown in Fig. 4-47. The word line which produces the word field is parallel to the easy axis and the bit line is parallel to the hard direction. The sense line runs parallel to the bit line and senses changes of flux along the easy axis. As the word field is turned on, the magnetization rotates into the direction of this field, resulting in a decrease of flux in the easy direction. This decrease can be of opposite sense, depending on the original state of the bit, and consequently produces a positive or negative voltage signal on the sense line. Ideally, the magnetization would rotate back into the easy direction at the end of the word pulse with half the magnetization rotating in one sense and half in the other. If during this time a bit field is introduced, the rotation follows entirely the direction of the bit field, resulting in a defined state of magnetization. This constitutes the writing process. Theoretically, writing can be accomplished by a word field of a magnitude greater than H_k, and a bit field sufficiently large to overcome the skew in the film, but smaller than the wall-motion threshold.

The increased drive requirements are caused by the nonideal characteristics of real films, due to the presence of additional crystalline and stray anisotropies, and due to demagnetizing effects associated with the finite size of the film. Variations of the orthogonal method include the use of a bias in the bit-field direction. (The bias gives a preferential direction to the magnetization in the absence of the bit field.) A single-polarity bit driver can then be used, and the bit field must overcome the bias field to write into the bit.

Another mode of operation[160] slightly rotates the easy direction to obtain a preferred direction after reading. A unipolar bit pulse, with an amplitude sufficient to overcome the skew, is used for writing.

The main problems are fabrication yield, creep, and noise. The yield problem is aggravated by the need of producing simultaneously a large number of bits, all of which must meet the very close tolerance requirements. Additional difficulties arise in the accurate mechanical registration of the wiring to the bit pattern, as well as from magnetostrictive effects encountered when the wiring plane is pressed against the substrate containing the magnetic elements.

In order to obtain a larger signal and avoid the demagnetizing effects associated with thicker films, a multilayered film structure has been suggested. If two bits are placed one over the other with parallel easy axes, the demagnetization effect can be substantially reduced.[147,164] The wiring, including the sense line, would all be placed between the film pair. Such a memory plane has not yet been built, but one would anticipate considerable difficulties in obtaining the required film uniformity and registration.

A simple modification of the rotational switching mode yields a nondestructive readout memory (NDRO). Nondestructive readout results in increased speed since the regeneration cycle can be omitted. The resulting memory is nonvolatile, regardless of how many times it is interrogated. Fast writing, of course, is also possible. Nondestructive readout is achieved by decreasing the word field to a value where the magnetization rotates less than 90°, permitting the bit to return to its original state after reading. The signal output is smaller than that obtained from the DRO type of memory. Since the bit is exposed to varying numbers of pulses in the hard direction, an additional requirement is placed upon the magnetic characteristics of the film: The film must have low dispersion both in angle and H_k, so that repeated hard-direction pulses do not destroy the information. No operating memories of this type have been reported.

Still another modification of the conventional orthogonal mode results in a read-only memory.[165] If a second magnetically hard film is placed over the pulsed bit, the latter will take on a state antiparallel to the hard film. A pulse sufficient to cause the magnetization in the bit to rotate, but insufficient to affect the hard film, will read the information in the bit. At the conclusion of the pulse, the demagnetizing field from the hard film will restore the bit to its original state. In this particular case, the requirements on the interrogating bit are less severe than in the other cases mentioned, since it couples only the information stored in the hard bit into the external circuitry.

4. Magnetic Amplification and Logic

The major shortcoming in the use of magnetic logic is the need to interpose amplification and isolation devices, other than the magnetic element itself, between logic stages. Proebster and Oguey[166] have suggested utilizing the orthogonal switching properties of the thin-film element for this purpose. As previously mentioned, if a field in excess of H_k is applied in the hard direction, a flux change occurs while the magnetization rotates through 90°. When this field is turned off, the initial single-domain film splits into many small domains, half magnetized in one direction and half in the other, parallel to the easy axis. A small field along the easy axis (in principle, only large enough to overcome skew) prevents the splitting and guides the magnetization entirely in one direction or the other, depending on the field direction. A field of 0.3 oersted was found sufficient for this purpose. The source of this field can be derived from another thin film during its millimicrosecond switching. For this purpose, two film elements are connected by a strip transmission line, one film acting as the controlling element, the other as the controlled element (Fig. 4-48). The coupling line lies parallel to the hard direction, and senses the flux change in the easy direction. Depending on the state of the controlling film, a positive or negative voltage is induced on the coupling line, and the magnetization direction in the controlled film is determined by the state of the controlling film.

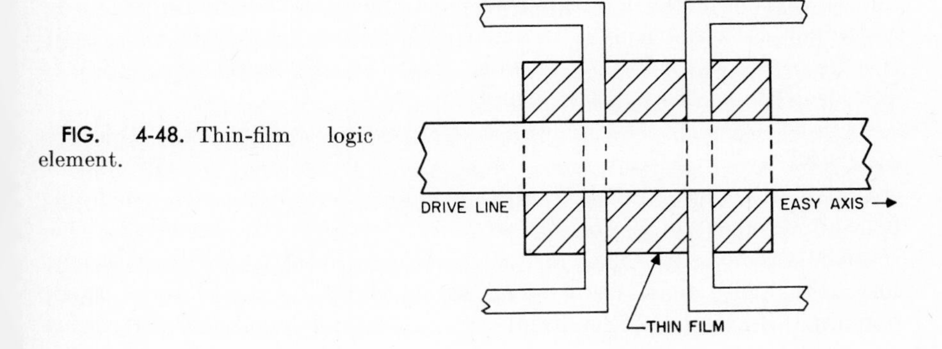

FIG. 4-48. Thin-film logic element.

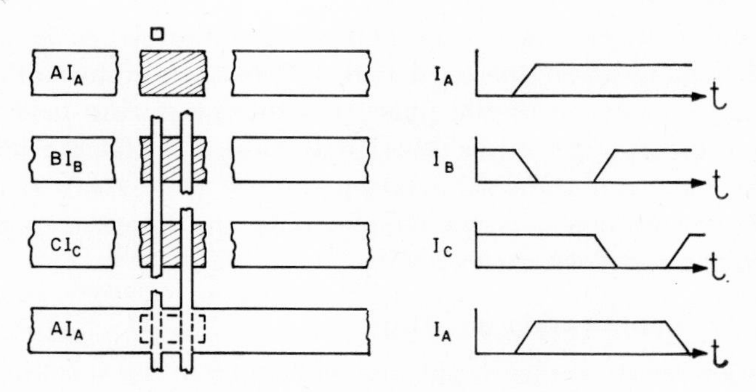

FIG. 4-49. Thin-film shift register.

Since the films are bidirectional, a three-phase pulse system must be used to transfer information in a given direction. To achieve information transfer, the easy-direction field generated by the coupling line must be sufficiently strong to rotate the magnetization. Figure 4-49 shows the timing scheme and the interconnections that can be used to implement a shift register. The magnetization in film B is turned into the hard direction by I_b. When I_a is turned on, the output I_t in the transfer line is either positive or negative, depending on the state of film A. As I_a is turned on, I_b is turned off, and the magnetization of film B is rotated in a clockwise or counterclockwise direction. Film C remains in a hard direction. When I_b is turned on again, the current I_c is turned off, and information is transferred to I_c, with film A maintained in the hard direction.

To attain the control currents necessary for operation, switching voltages must be large. The large current, in turn, requires fast rise-time pulses, and results thus in very rapid switching. The impedance of the coupling lines must be chosen low enough to transmit currents sufficiently large to induce rotation. Logic functions could be generated, in principle, on a majority-logic principle by connecting several inputs to a given element. However, it would be difficult to switch the entire film uniformly in this mode of operation, since the control winding would be restricted to a fraction of the film surface. A definite problem with this device is the achievement of a small bit size, since a reduction in size also reduces the voltage output.

A shift register based on wall-motion effects in thin films has been described by Broadbent and McClung.[167] It utilizes the fact that a region of reverse magnetization in a thin film exerts a demagnetizing field on the parts of the film close to it. Since this demagnetizing field reinforces an external field, a smaller field is needed to reverse the magnetization close to it than would ordinarily be required for a single domain. If a magnetic field aiding the growth of reverse magnetization on one side of the region and opposing it on the other is induced by an appropriate winding, it causes the region of reverse magnetization to move along the film to one edge where it can be sensed by a pickup loop.

One interesting logic application of thin films is its use in a phase-locked oscillator. The idea of storing information in terms of signal phase was first proposed by Von Neumann.[168] A magnetic element, used as a nonlinear inductance and associated with a tank circuit, can be excited by a pumping source of frequency ω to oscillate at frequency $\omega/2$. The phase of this oscillation can assume one of two values. Ferrite cores have been used as the nonlinear element and recently thin-film elements, as shown in Fig. 4-50, have been made by the TDK Company in Japan. The possible advantages to be derived from the use of a thin magnetic film are greater speed and lower power dissipation. The basic element and associated circuitry consist of a magnetic film with a d-c bias field and a pumping field of frequency ω along the easy axis. The signal winding follows the easy direction of the film and is connected to a capacitor of appropriate value. Once subharmonic oscillation starts in a given phase, it will continue until either the biasing d-c field is changed or the pump field is turned off. In the absence of an external influence, the phase of the subharmonic will be random, but in the presence of such an influence, the phase can be controlled. This influence can be the signal field from another thin film, so that information can be readily transferred, and logic performed on a majority basis. Phase reversal from one element to another can be accomplished in terms of the winding direction of the control winding.

Such devices have been operated with a pump frequency of 20 Mc/sec. The attractiveness of the phase-locked oscillator and especially its thin-film version resides in the self-consistency of such devices in which (at least in principle) all the functions of logic and amplification can be performed by the same element. A major drawback, at least from a speed point of view is the relationship between pump frequency and information bit rate, which corresponds to a ratio of 10:100. To operate at 1 Mc, for instance, a pump frequency of about 50 Mc/sec is required.

In summary, it can be stated that the main application of magnetic films appears to be the area of digital-computer memories. With present films and for the high-speed ranges which are of greatest interest, the orthogonal mode of operation permits the largest tolerances of film characteristics and is, therefore, best suited for practical implementation.

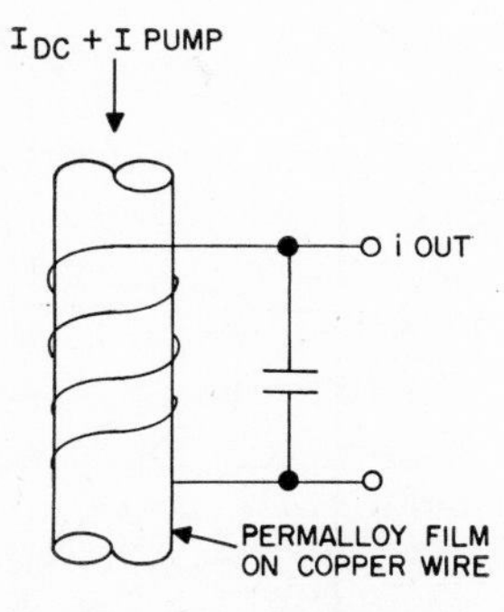

FIG. 4-50. TDK parametron.

4-5. CRYOGENIC FILMS

1. Principles of Superconductivity

The superconducting state—discovered in 1911 by Onnes[169]—represents the total disappearance of the electrical resistance in certain metals below a critical temperature of a few degrees absolute. The critical temperature differs somewhat for the various superconducting metals and alloys, examples of which are mercury, lead, tin, tantalum, and niobium. An external magnetic field or an internal magnetic field generated by a sufficiently large current destroys the superconducting state. This field is defined as a critical field, and varies as a function of temperature. Figure 4-51 shows this variation for several superconducting materials.

According to the accepted BCS theory of superconductivity,[170] a quantum-mechanical force of attraction is induced between pairs of conduction electrons by the potential field of the crystal lattice. The attractive force exceeds the repulsive force below the critical temperature, and the electrons are condensed in the resulting superconducting state into a special low-energy band separated by an energy gap ($\Delta E \approx 4kT_c$) from the conductance band. The value of the gap and the number of superconducting electrons are a maximum at absolute zero, and decrease with increasing temperature until they disappear at the critical temperature. Since a magnetic field reduces and, at sufficient field strength, eliminates the gap also, one can use such a current-induced field to obtain electronic switching.

Most macroscopic superconducting phenomena are adequately

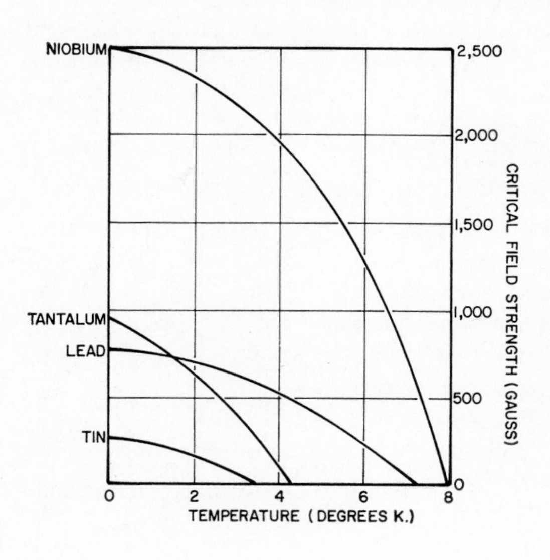

FIG. 4-51. Critical-field-strength curves for several superconductors.

described by the London equations[171]

$$\frac{mc^2}{4\pi Ne^2}[\nabla^2 H] = \mathrm{H} \tag{4-34}$$

and

$$\frac{mc^2}{4\pi Ne^2}[\nabla^2 J] = J \tag{4-35}$$

where m is the electron mass, e the unit charge, N the number of electrons per unit volume, H the magnetic field, and J the superconducting current. These equations include the Meissner effect which demonstrates that the diamagnetism is zero within a superconductor.[172] (This excludes the existence of a magnetic field or magnetic induction from the interior of a bulk superconductor.) The solutions of the London equations, however, are not discontinuous at the specimen surface. Magnetic field and induction decrease rapidly to zero over a thickness

$$\lambda = \left(\frac{mc^2}{4\pi Ne^2}\right)^{1/2} \approx 10^{-5} \text{ cm}$$

called the superconducting penetration depth. A consequence of the finite penetration depth is an increase of the critical field H_c for thin superconducting layers:

$$\frac{H_c\,(\text{film})}{H_c\,(\text{bulk})} = \frac{1}{[1 - (2\lambda/t)\tanh(t/2\lambda)]^{1/2}} \tag{4-36}$$

where t is the layer thickness.

2. Cryotrons

Casimir-Jonken and de Hass[173] built a cryogenic relay or switch in 1935, and Buck[174] proposed the use of this device in computers. Buck's cryogenic switch or "cryotron" consisted of a thin wire of niobium (control coil) wound around a thicker wire of tantalum (gate). A supercurrent flowing through the control coil produces a magnetic field strong enough to destroy the superconductivity of the gate (Fig. 4-52). Since

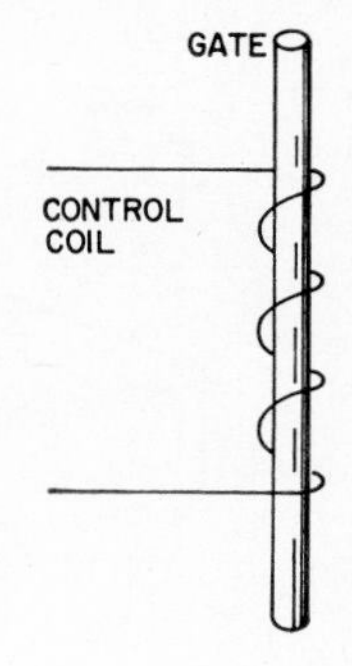

FIG. 4-52. Wirewound cryotron consisting of a thin wire of niobium (control coil) wound around a thicker wire of tantalum (gate). Supercurrent flowing through control coil produces a magnetic field strong enough to destroy the superconductivity of the gate.

a proper geometric layout results in a control current smaller than the gate current, it is possible to design the cryotron with a gain greater than one. Cryotrons can thus drive one another without requiring additional amplifiers.

The transition between the resistive and the superconducting state proceeds in less than a nanosecond. The switching time of a cryotron, therefore, is essentially given by its L/R time constant where L represents the inductance and R the resistance in the resistive state. Wirewound cryotrons have a large L, resulting in a switching time of several hundred microseconds. Because of this slow speed, these cryotrons have not found practical applications in computers.

Major progress with regard to switching speed and compactness was made with the development of thin-film cryotrons. In its simplest form (Fig. 4-53), a film cryotron consists of two crossing strips made of materials with different critical-field curves (for instance, lead and tin) separated by an insulating layer (usually SiO). In the operational range, the lead control stays always superconductive. The magnetic field associated with a current flowing in the control drives the tin gate resistive when it exceeds the critical value. A superconducting base plate, usually of vacuum-deposited lead, compresses the total flux at a given current and thus decreases the self-induction.

To drive one cryotron with another, the required control current must not exceed the value of gate current which drives the gate resistive by its self-induced field. Static gain of a cryotron is thus defined as

$$g = \frac{I_g}{I_c} \tag{4-37}$$

I_g is the critical gate current in the absence of control current, and I_c the critical control current in the absence of a gate current. Figure 4-54 shows the characteristic curve of a film cryotron with a gain greater than unity. In a first approximation, the gain is given by the ratio of gate to control-line width.

Rapid switching is aided in film cryotrons by a number of features.

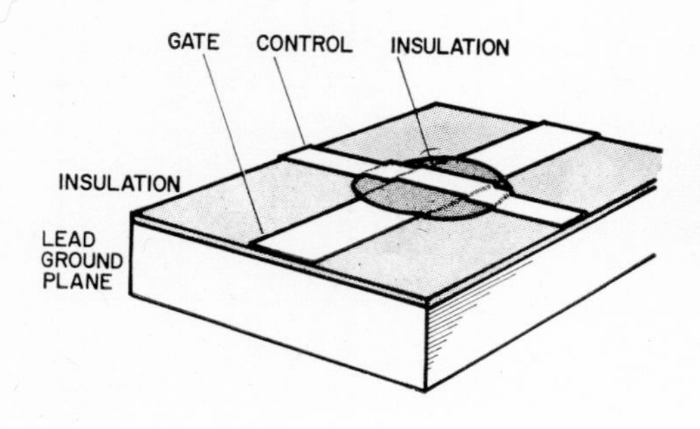

FIG. 4-53. Film cryotron consisting of pair of superconducting strips insulated by SiO. Control film and ground plane are lead; the gate film is tin.

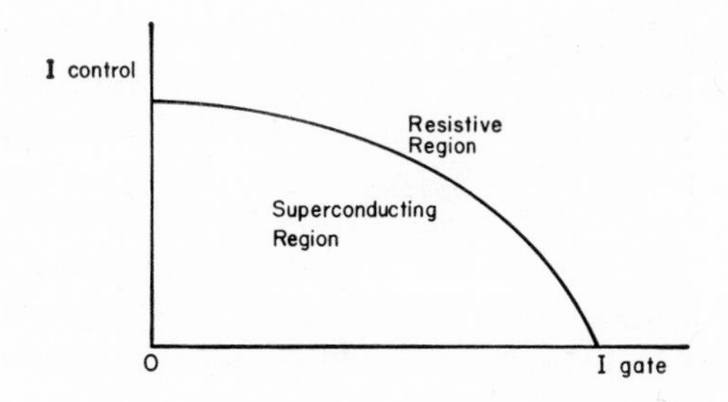

FIG. 4-54. Characteristic curve of a film cryotron with gain greater than one.

The close proximity of control and gate yields a sufficient field strength, with a simple crossing resulting in a small L. Because of the high sheet resistance and small line width possible with deposited films, the resistance of film cryotrons can be chosen a few hundred times larger than the resistance of bulk wires. Quantitatively, the time constant for a thin-film cryotron in a loop, neglecting the contribution of line segments outside the cryotron structure, is given by

$$\frac{L}{R} = \frac{4\pi 10^{-9} t_i w_g{}^2 t_g}{\rho l_g{}^2} \quad \text{sec} \tag{4-38}$$

when the presence of a superconducting base plane is taken into account. L is the inductance of the control, R is resistance inserted in the gate, t_i is the insulation thickness, w_g the gate width (equal to the control length), t_g the gate thickness, ρ the resistivity of the gate material, and l_g the gate length (equal to the width of the control). With practical dimensions ($t_i = 2 \times 10^{-4}$ cm, $w_g = 10^{-2}$ cm, $t_g = 10^{-4}$ cm, $\rho = 10^{-6}$ ohm-cm, $l_g = 2.5 \times 10^{-3}$ cm, see Fig. 4-55), an L/R time constant of 4×10^{-9} sec seems to be obtainable. However, since cryogenic circuits contain additional inductance, the switching speeds achieved with simple crossline cryotrons are considerably slower.

A further shrinking of the significant dimensions is limited by mask

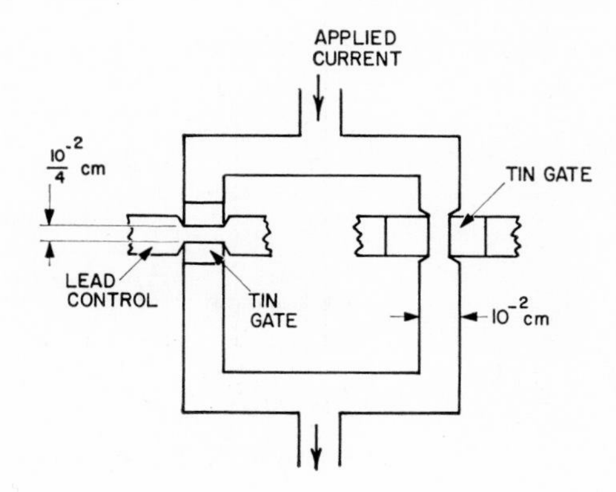

FIG. 4-55. Simple cryotron loop circuit.

tolerances and the fast-decreasing insulation yield with decreasing SiO thickness, but speed can be improved by various other modifications. Gate resistivity can be increased by using as gate material high-resistance alloys or by introducing impurities into the film. The path resistance can be increased further by an in-line geometry of the cryotron,[175] where the control line is placed parallel over the gate line from which it is insulated by an SiO layer. By this technique, the entire gate path can be made resistive. In order to obtain a gain greater than unity, it is necessary to add an additional control line. Its bias current supplies part of the magnetic field needed to switch the gate. Flip-flops of in-line cryotrons have been fabricated with a switching time of 2 nanosec.

The switching speed also is influenced by the film structure. Films evaporated through a mask have ordinarily tapered edges due to the so-called penumbra effect,[176] and vapor-particle reflections from the

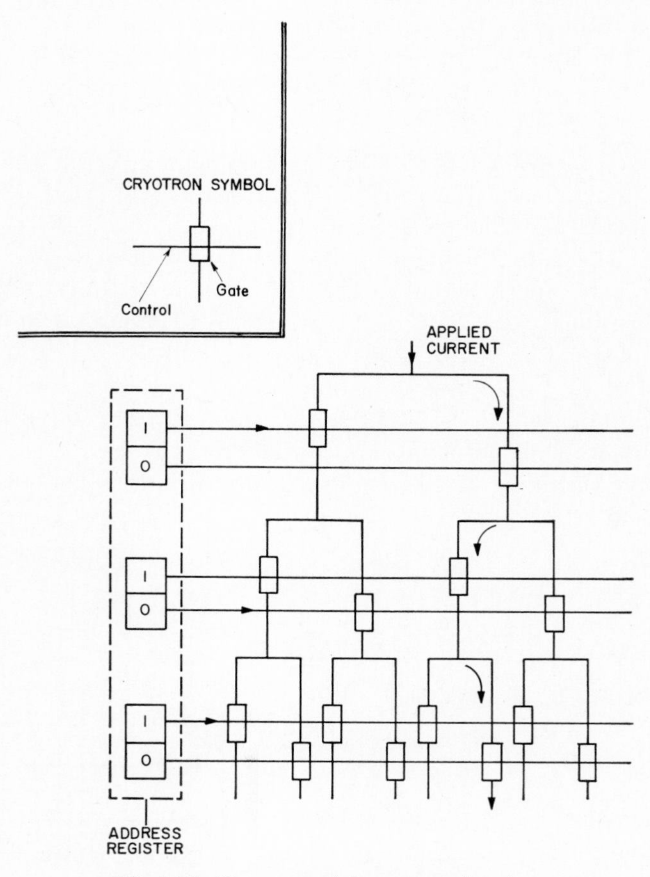

FIG. 4-56. Tree-type eight-output decoder.

mask edges. These tapered film edges result in a broadening of the superconductive transition, with a correspondingly longer switching time. Sharp film edges can be obtained by evaporating tin at a relatively high substrate temperature (100°C) on a substrate seeded with a monolayer of silver or copper atoms serving as nucleation centers. The tin condenses under these conditions in small grains over the seeded surface, and in large coagulated grains at the film edge outside the seeded area. As a result, the film forms a continuous sheet of uniform thickness up to the nucleation edge, and isolated single-grain areas beyond.

Details of cryotron design and performance have been discussed by Rosenberger,[177] Ittner,[178] and Smallman et al.[179]

3. Gating Circuits

The previous paragraph has shown that the cryotron represents a nearly ideal gating element: It switches rapidly, exhibits an infinite impedance ratio between the on and off states, shows gain, is extremely compact, and (as will be discussed later) dissipates very little power. Because of these properties, computer networks such as registers, decoders, and arithmetic units can be built entirely of cryotrons, without the need for auxiliary amplifiers. Figure 4-56, for example, shows a decoder tree which can be expanded to any desired number of outputs. Figure 4-57 demonstrates the common AND and OR functions. In the AND circuit, inputs 1 and 2 must be in the on state to insert resistance in the path of the applied current. In the OR circuit, only one of the inputs need be applied to block the current path.

Figure 4-58 shows a cryotron flip-flop. Current in the control marked "1" (in) inserts resistance in branch 1, thereby switching the gate current to branch 0 (and vice versa). In either side of the loop, the current flows indefinitely until a control current drives the current-carrying

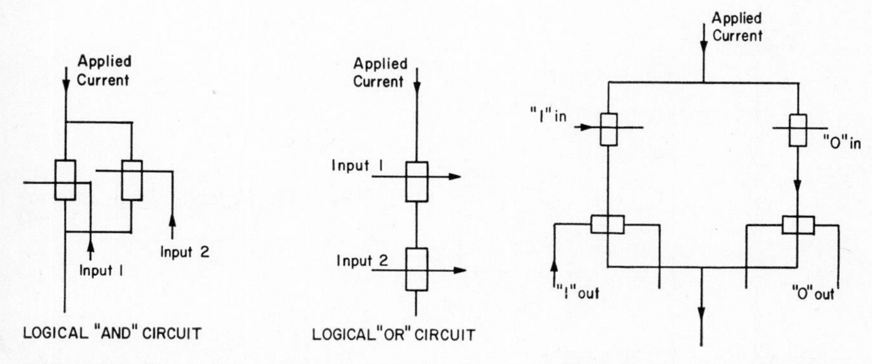

FIG. 4-57. Typical logic circuits formed by cryotrons.

FIG. 4-58. Cryotron flip-flop.

gate resistive. The cryotron flip-flop represents thus a bistable storage element. The "out" cryotrons permit an interrogation of the flip-flop regarding its state in either the 0 or 1 position.

4. Memory Circuits

The flip-flop described in the previous article is a useful bistable storage element, and logic as well as memory could be constructed entirely out of cryotron networks. For storage applications, however, simpler cryogenic devices have been developed.

A memory cell using only three cryotrons for the bit itself has been described by Haynes[180] (Fig. 4-59). Only one leg of the storage loop of this cell is provided with a write cryotron (marked I), and can thus be switched to the resistive state. By driving gate I resistive, the applied current ceases to flow in the resistive leg and diverts to the remaining superconducting path. Thus, a net field links the loop. When the gate returns to the superconducting state and the applied current is removed also, this net field is trapped in the loop, and a circulating current flows in the loop sustaining the stored magnetic field. This may correspond to the storage of a one bit. The absence of the loop current then represents a zero.

Circulating current inserts resistance in the gate of cryotron II. During the read cycle, the read pulse switches cryotron III to the resistive state, thereby inserting resistance in the direct path of the applied read current, while the bypass loop can be either in the resistive or superconducting state. A resistance in the read circuit corresponds, therefore, to a stored current, whereas no resistance indicates the absence of current in the storage loop. In a typical memory system, the read current is diverted to other cryotron circuits which provide a superconducting path to perform the necessary logic functions.

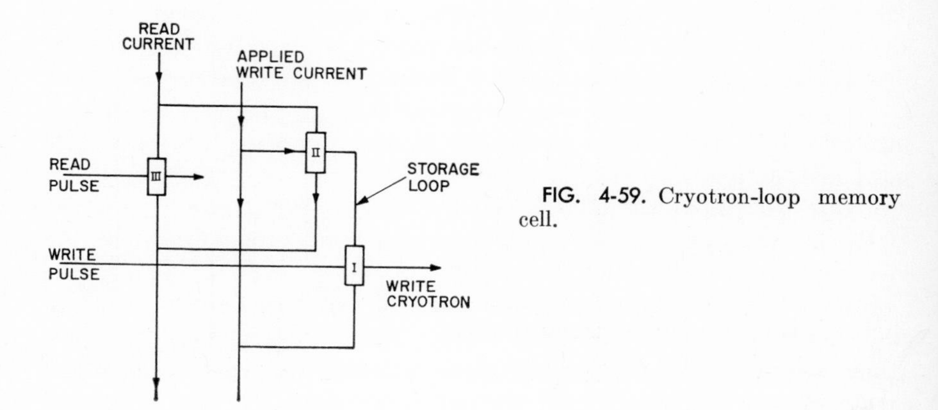

FIG. 4-59. Cryotron-loop memory cell.

FIG. 4-60. Eight-word 40-bit memory plane evaporated in one pump down. (*International Business Machines Corporation.*)

An eight-word, 40-bit memory plane with a cycle time of 4 μsec utilizing this type of cell is shown in Fig. 4-60. The memory organization provides for nondestructive read. Because of the relative complexity of this cell as compared to the sheet-film approach discussed later, packing densities are limited to about 100 cells/in.^{2}

In certain applications, such as information retrieval, pattern recognition, or inventory control, the entire storage content must be compared with specific input information. This can best be accomplished with a memory addressed directly by storage content rather than location. The organization of such an associative or data-addressed memory is shown in Fig. 4-61. These memories require a large number of nondestructive readout elements, and it seems that the cryotron is one of the few electronic elements with the potential of making large associative memories feasible from the viewpoints of cost, component density, and power dissipation.

Figure 4-62 shows the layout, and Fig. 4-63 a photograph of an associative memory panel using cryotrons which has been developed by the General Electric Research Laboratory.

Crowe[181] has proposed a simple memory cell based on a principle differing from the cryotron (Fig. 4-64). The cell consists of two superconducting paths in the ground plane following the edges of the two semicircle cutouts. Both have the bar in common. As long as ground

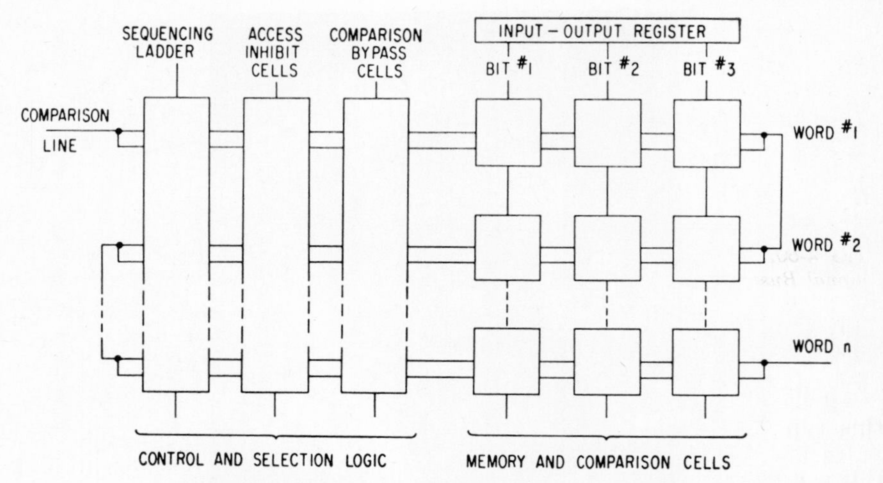

FIG. 4-61. Organization of an associative memory.

plane and bar are superconducting, no magnetic field can loop the bar through the cutouts.

Current flowing through the drive lines induces an equal and opposite current in the bar. If both the x and y currents flow in the same direction, the bar current exceeds the critical value, and the bar becomes

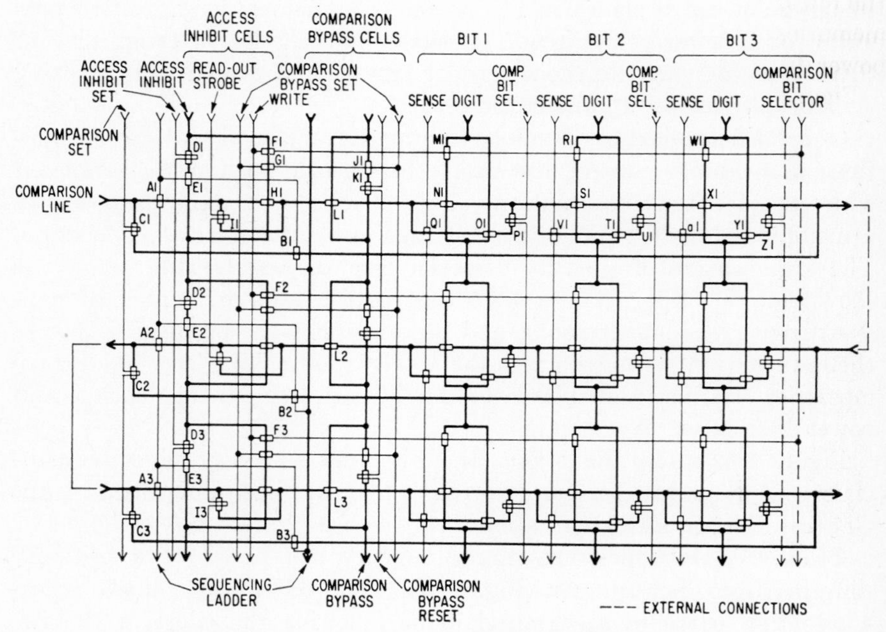

FIG. 4-62. Layout of a 3-bit associative memory panel.

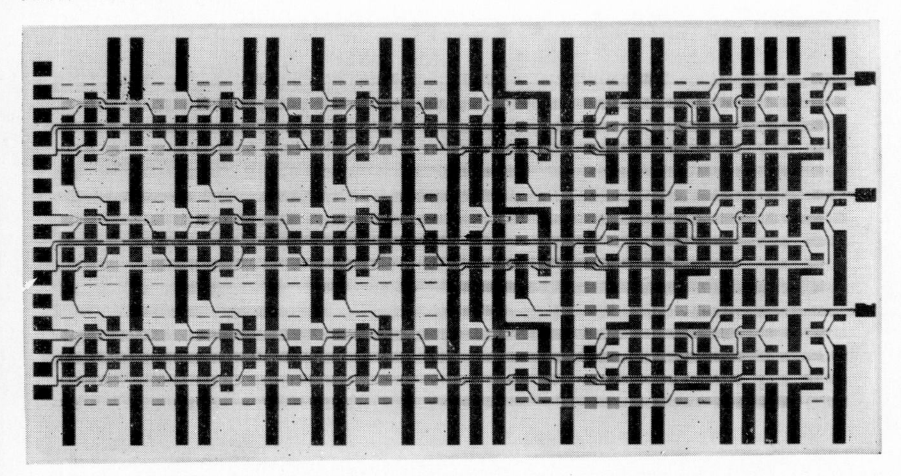

FIG. 4-63. Photograph of the cryogenic memory panel. (*General Electric Company.*)

resistive. Now the flux generated by the drive lines can loop the bar since part of the previously superconducting loops have become resistive. Since the induced current in the bar does not immediately reduce to zero, some heating of the bar takes place. For the memory cycle shown in Fig. 4-65, it is assumed that the heat generated during switching holds the bar normal for a sufficient period so that the induced currents in the cell are able to decay to zero. To operate this cell in a normal read-write cycle, an initial priming pulse is required, which biases the cell to 60% of the critical-current value.

Tests of this type of memory indicate that a memory cycle of less than one microsecond is possible. The measured cell characteristics are as follows:

Cell switching time.........	60 nanosec
Drive-current amplitude....	300 ma
Signal-to-noise ratio........	10

A memory cell with a simpler geometry has been investigated by Burns, Alphonse, and Leck.[182] This cell resembles in its operation the Crowe cell, but consists only of intersecting drive lines over a superconducting ground plane. Coincident currents at the drive-line intersection generate critical

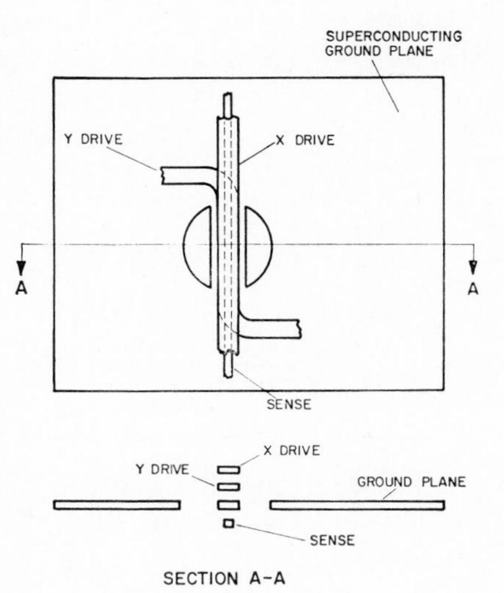

FIG. 4-64. Crowe memory cell. (*International Business Machines Corporation.*)

magnetic fields in two of the quadrants. Instead of providing in these quadrants cutouts as in the Crowe cell, discontinuities of the film are achieved by a coarse grain structure. The stored field can be reversed by reversing the polarity of the drive-line currents. A sense line coupled to all cells of the plane is used to detect the switching of a selected cell.

This type of cell has a major advantage: Packing densities of at least 10,000 cells/in.² should be possible. Tests of this memory device show switching times in the order of 50 nanosec. With drive currents of 100 ma, the output signal is several millivolts, and the signal-to-noise ratio is greater than 10:1. This combination of desirable characteristics make this memory attractive for large-capacity high-speed requirements.

5. Superconducting Tunnel Amplifier

In Art. 1, the presence of a band gap in superconductors in accordance with the BCS theory has been discussed. This band gap can be studied experimentally by employing the electron tunneling effect[183–185] which already has been described in Art. 1 of Sec. 4-3. The required film structure consists of two superimposed films of different superconducting metals, separated by an insulating layer 50 to 100 A thickness. The electron-band structure of this film combination is shown in Fig. 4-66, together with its current-voltage characteristic. Schematic diagram I shows the relation of the two band gaps without voltage. At the small band gap, a certain number of electrons are thermally excited, but thermal excitation at the larger gap is negligible. The Fermi levels, of course, are at the same potential. In schematic diagram II, voltage is applied and elevates an increasing number of excited electrons on the left, over the upper limit of the right-sided band gap. These electrons

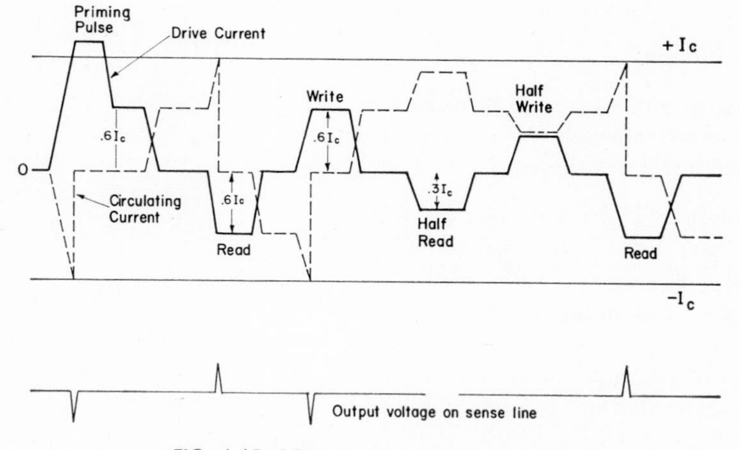

FIG. 4-65. Memory cycle of Crowe cell.

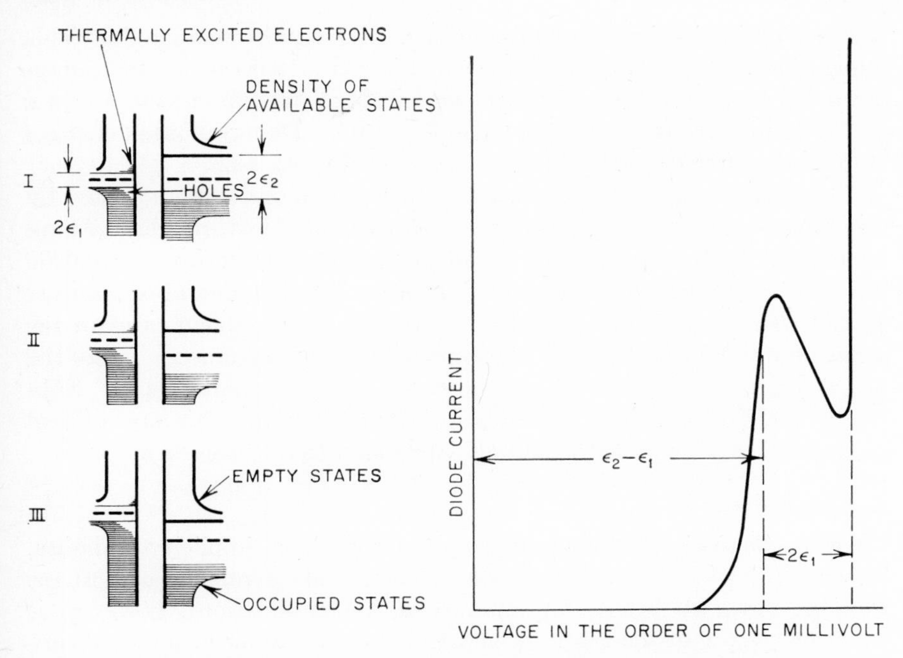

FIG. 6-66. Band structure and characteristic of superconducting tunnel device.

can then tunnel through the insulator. When the applied voltage reaches a value equal to $(\epsilon_2 - \epsilon_1)$, all excited electrons are sufficiently energetic to tunnel. If the voltage is increased further as in schematic diagram III, the number of available electrons stays constant, but the density of available states decreases, corresponding to a decrease of the tunnel current and a section of the current-voltage curve with negative resistance. An increase of the voltage beyond a value of $(\epsilon_2 + \epsilon_1)$ makes electrons below the band gap available for tunneling, and increases the current rapidly.

Because of its negative-resistance region, the structure described above can be used as an amplifying device. By modulating the current with an external magnetic field, a control characteristic similar to a triode vacuum tube can be achieved. Since the device is extremely small and requires little power, it offers attractive possibilities in microminiature systems where liquid helium temperatures are available.

6. Fabrication Process

The prototype fabrication of cryogenic networks up to now was largely restricted to vacuum-deposited memory panels of limited complexity. Even a panel containing in the order of 10^2 cryotrons such as is shown in Fig. 4-60, however, requires about 20 different evaporation steps because of its intricate layout pattern. To obtain reasonable yields under these conditions, the entire panel is completed during one pump-down by use of a mask and source changer. A typical installation in a 36- by 36-in. vacuum tank is shown in Fig. 4-67. Details of mask changer (top) and source changer (bottom) are given in Fig. 4-68.

The mask changer holds 21 masks, which can be randomly selected by mechanical controls with O-ring sealed rotating feedthroughs. Alignment pins register the mask to the substrate with a tolerance of ± 0.0005 in. The rotating source changer contains three evaporation sources loaded with silicon monoxide, lead, and tin. Evaporation rates in the order of 100 A/sec are used. The source-substrate geometry limits the film-thickness variation over the substrate to approximately 4%. Substrate temperature, film thickness, and deposition rate are critical parameters which are automatically controlled to 5% accuracy.

7. Conclusions

Cryogenic devices offer many advantages, such as simplicity of design, high speed, ideal switching characteristics, and extreme compactness. Perhaps most attractive for microelectronic applications, however, is their low power dissipation. A cryotron loop in either of its superconducting equilibrium conditions, for example, dissipates no power at all. The amount of energy dissipation during switching is so small that a

FIG. 4-67. Evaporator for deposition of cryogenic film panels.

FIG. 4-68. Interior of coating chamber with mask and source changers. (*International Business Machines Corporation.*)

network of one million cryotron loops, each switched 10 million times per second, requires only about one watt of power.

On the debit side, the power must be dissipated at extremely low temperatures and thus requires about 1,000 watts of cooling power for each watt generated by the electronics. Since closed-cycle refrigerators can be built relatively cheaper and more compact for large outputs, cryogenic electronics is particularly suited for large and complex systems.

To utilize the performance potential of cryogenic electronics, the simultaneous deposition of a large number of components on a single substrate is required. The associated yield and reliability problems have not yet been fully solved, particularly with regard to the insulation. The present state of development makes it thus difficult to predict the future potential of cryogenic electronics.

REFERENCES

[1] G. F. Mansbridge, British Patent 19, 451, October, 1900.
[2] P. Alexander and E. L. Cranstone, British Patent 551,757, September, 1941.

[3] P. Godley, *Iron Age*, vol. 161, p. 90, 1949.

[4] H. G. Wehe, *Bell Labs. Record*, vol. 27, p. 317, 1949.

[5] D. A. McLean and H. G. Wehe, *Proc. IRE*, vol. 42, p. 1779, 1954.

[6] National Defense Research Committee (U.S.), division 14, Reports 360, 521, and 534.

[7] A. Schulze and H. Eicke, *Deut. Elektrotech.*, vol. 6, p. 616, 1952.

[8] L. Holland, *Vacuum*, vol. 1, p. 23, 1951.

[9] Libbey-Owens-Ford Glass Co. and H. A. McMaster, British Patent 632,256, October, 1942.

[10] R. Gomer, *Rev. Sci. Instr.*, vol. 24, p. 993, 1953.

[11] H. G. Dill, *Semicond. Prod.*, vol. 5, p. 30, April, 1962.

[12] L. Holland, "Vacuum Deposition of Thin Films," John Wiley & Sons, Inc., New York, 1958.

[13] S. Dushman, "Scientific Foundations of Vacuum Technique," 2d ed., John Wiley & Sons, Inc., New York, 1962.

[14] American Vacuum Society, *Vacuum Technol. Trans.*, Pergamon Press, New York, 1954 to 1961.

[15] L. Holland and G. Siddall, *Vacuum*, vol. 3, pp. 245, 375, 1953.

[16] K. Wehner, *Advan. Electron.*, vol. 7, p. 239, 1955.

[17] G. K. Wehner, *J. Appl. Phys.*, vol. 30, p. 1762, 1959.

[18] D. E. Harrison, *J. Chem. Phys.*, vol. 32, p. 1336, 1960.

[19] C. A. Neugebauer et al., "Structure and Properties of Thin Films," John Wiley & Sons, Inc., New York, 1959.

[20] H. Mayer, "Physik dünner Schichten," vols. I and III, Wissenschaftliche Verlagsgesellschaft, m.b.H., Stuttgart, 1950 and 1955.

[21] R. E. Thun, G. F. Caudle, and E. R. Pasciutti, *Rev. Sci. Instr.* vol. 31, p. 446, 1960.

[22] K. H. Behrndt and R. W. Love, *Natl. Symp. Vacuum Technol. Trans.*, vol. 7, 1960; published 1961.

[23] G. R. Giedd and M. H. Perkins, *Rev. Sci. Instr.*, vol. 31, p. 773, 1960.

[24] R. J. Dow, *Proc. Electron. Components Conf.*, Washington, D.C., May, 1962.

[25] R. G. Chambers, *Proc. Roy. Soc. London*, ser. A, vol. 202, p. 378, 1950.

[26] R. B. Dingle, *Proc. Roy. Soc. London*, ser. A, vol. 201, p. 544, 1950.

[27] A. C. B. Lovell, *Proc. Roy. Soc. London*, ser. A, vol. 157, p. 311, 1936.

[28] A. Sommerfeld and H. A. Bethe, "Handbuch der Physik," vol. 24, Springer-Verlag, OHG, Berlin, 1933.

[29] R. B. Belser, *J. Appl. Phys.*, vol. 28, p. 109, 1957.

[30] H. L. Bullis and W. E. Isler, *Proc. Symp. on Microminiaturization of Electron. Assemblies*, Diamond Ordnance Fuze Laboratories, Sept. 30–Oct. 1, 1958, pp. 31–43.

[31] N. J. Doctor and E. M. Davies, *Elec. Mfg.*, vol. 62, p. 94, 1958.

[32] D. Hoffman and J. Riseman, Theoretical and Experimental Investigation on Thin Film Resistance Elements, Scientific Reports, Contract AF33(616)-3443, ASTIA AD 130996, June 15, 1956.

[33] J. T. Doherty, Study of Physical Characteristics of Thin Film Resistance Elements, ASTIA Scientific Report, Contract AF33(616)-3453, Mar. 1, 1956.

[34] J. Riseman, *Trans. N.Y. Acad. Sci.*, vol. 19, p. 503, 1957.

[35] D. Hoffman and J. Riseman, *Natl. Symp. Vacuum Technol. Trans.*, vol. 4, Boston, Mass., October, 1957.

[36] J. J. Bohrer, W. E. Hanth, and S. J. Stein, *Proc. Electron. Components Symp.*, Washington, D.C., May, 1956, pp. 166–170.

[37] S. J. Stein and J. Riseman, *Proc. Electron. Components Symp.*, Washington, D.C., May, 1956, pp. 171–174.

[38] J. G. Simmons and L. I. Maissel, *Rev. Sci. Instr.*, vol. 32, p. 642, 1961.

[39] L. Maissel, J. Simmons, and M. Casey, *IRE Trans. Component Pts.*, vol. CP-8, p. 70, 1961.

[40] L. I. Maissel, Electrical Properties of Sputtered Tantalum Films, *Vacuum Technol. Trans.*, Pergamon Press, New York, 1962.

[41] P. M. Schaible and L. I. Maissel, Grain Boundary Diffusion in Sputtered Tantalum Films, *Vacuum Technol. Trans.*, Pergamon Press, New York, 1962.

[42] E. H. Layer, C. M. Chapman, and E. R. Olson, *Proc. 2d Natl. Conf. Military Electron.*, 1958, p. 96.

[43] E. H. Layer, 1959 *Vacuum Technol. Trans.*, Pergamon Press, New York, 1959, p. 210; published 1960.

[44] M. Beckerman and R. E. Thun, *Vacuum Technol. Trans.*, Pergamon Press, New York, 1961.

[45] M. Beckerman and R. L. Bullard, *Proc. Electron. Components Conf.*, Washington, D.C., May, 1962, p. 53.

[46] D. Gerstenberg and E. H. Mayer, *Proc. Electron. Components Conf.*, Washington, D.C., May, 1962, p. 57.

[47] F. S. Maddocks and R. E. Thun, *J. Electrochem. Soc.*, vol. 109, p. 99, 1962.

[48] G. Siddall, *Vacuum*, vol. 9, p. 274, 1960.

[49] H. G. Rudenberg, J. R. Johnson, and L. C. White, *Proc. Electron. Components Conf.*, Washington D.C., May, 1962, p. 90.

[50] E. E. Smith and S. G. Ayling, *Proc. Electron. Components Conf.*, Washington, D.C., May, 1962, p. 82.

[51] R. W. Berry and D. J. Sloan, *Proc. IRE*, vol. 47, p. 1070, 1959.

[52] N. W. Silcox and L. I. Maissel, Some Factors Controlling Gross Leakage Currents in Sputtered Tantalum-film Capacitors, *J. Electrochem. Soc.*, in press.

[53] R. C. DeVries, *J. Am. Ceram. Soc.*, vol. 45, p. 225, 1962.

[54] A. Moll, *Z. Angen. Phys.*, vol. 10, p. 410, 1958.

[55] C. Feldman, *J. Appl. Phys.*, vol. 27, p. 870, 1956.

[56] H. B. Martin, State-of-the-Art Survey of Electronic Microminiaturization, Report NADC-EL-6079, Dec. 29, 1960.

[57] J. J. Bohrer, Thin Film Circuit Techniques, Professional Group on Component Parts, p. 37, June, 1960.

[58] F. F. Jenny, Circuit Design and Parameters in Thin-film Technology, International Solid-state Circuits Conference, Philadelphia, Pa., February, 1962.

[59] W. N. Carroll and F. F. Jenny, Microminiaturization Utilizing Thin Film Technology, Northeast Electronics Research and Engineering Meeting, November, 1960.

[60] USASRDL, Thin Film Circuit Functions Final Report, Contract DA36-039 sc-84547, July, 1960.

[61] C. Holm, Distributed Resistor Capacitor Networks for Microminiaturization, Royal Radar Establishment, Malvern, WORCS, England, Technical Note 654, August, 1959.

[62] P. S. Castro and W. W. Happ, Distributed Parameter Circuits and Microsystem Electronics, National Electronics Conference, Chicago, Ill., October, 1960.

[63] W. N. Carroll and F. F. Jenny, *Electronics*, vol. 34, p. 90, 1961.

[64] USASRDL, Thin Film Circuit Fabrication Program, Final Report, Contract DA36-039 sc-87387, 1960.

[65] F. Jaques and J. Schmidt, Breakthrough in Mold-making Electro-erosion, *Mod. Plastics*, December, 1960.

[66] E. D. Olsen, Fine Line Etched Wiring, Microminiaturization of Electronic Assemblies, Diamond Ordnance Fuze Laboratories, March, 1958.

[67] K. R. Shoulders, Research in Microelectronics Using Electron-beam-activated Machining Techniques, Stanford Research Institute, September, 1960.

[68] D. A. Buck and K. R. Shoulders, An Approach to Microminiature Printed Systems, *Proc. Eastern Joint Computer Conf.*, Philadelphia, Pa., December, 1958.

[69] Evaporator Makes Deposited Microcircuits, *Electronics*, January, 1962, p. 68.

[70] Thin Film Production Techniques, NAFI, Contract N163-9142(x).

[71] L. B. Valdes, *Proc. IRE*, vol. 42, p. 420, 1954.

[72] N. J. Doctor and E. L. Hebb, *Proc. Symp. on Microminiaturization of Electron. Assemblies*, Washington, D.C., September, 1958.

[73] R. G. Counihan, National Electronics Conference, Chicago, Ill., October, 1960.

[74] M. M. Perugini and N. Lindgren, Microminiaturization, *Electronics*, Nov. 25, 1960.

[75] USASRDL, Planar Integration of Thin Film Circuit Units, 1st Quarterly Report, Contract DA36-039 sc-87246, 1961.

[76] USASRDL, Planar Integration of Thin Film Circuit Units, 3d Quarterly Report, Contract DA36-039 sc-87246, June, 1961.

[77] A. E. Lessor, J. W. Skerritt, R. E. Thun, and D. S. Weed, *Proc. Electron. Components Conf.*, Washington, D.C., May, 1962, p. 32.

[78] H. C. Kammerer, Aviation Conference, Los Angeles, Calif., March, 1961.

[79] D. K. Allison, *Proc. Electron. Components Conf.*, Washington, D.C., May, 1962.

[80] J. Dresner and F. V. Shallcross, *Solid-State Electron.*, vol. 5, p. 205, 1962; *IRE Trans. Electron Devices*, vol. 8, p. 422, 1961.

[81] P. K. Weimer, IRE-AIEE Device Research Conference, Stanford, Calif., June, 1961; International Solid-state Circuits Conference, Philadelphia, Pa., February, 1962; *Proc. IRE*, vol. 50, p. 1462, 1962.

[82] G. C. Dacey and I. M. Ross, *Bell System Tech. J.*, vol. 34, p. 1149, 1955.

[83] C. A. Mead, *J. Appl. Phys.*, vol. 32, p. 646, 1961; *Proc. IRE*, vol. 48, pp. 359, 1478, 1960.

[84] R. H. Fowler and L. Nordheim, *Proc. Roy. Soc. London*, ser. A, vol. 119, 173, 1928.

[85] J. C. Fisher and I. Giaever, *J. Appl. Phys.* vol. 32, p. 172, 1961.

[86] A. G. Chynoweth, *Progr. Semicond.*, vol. 4, p. 97, 1959.

[87] J. P. Spratt, R. F. Schwarz, and W. M. Kane, *Phys. Rev. Letters*, vol. 6, no. 7, p. 341, 1961. Solid State Devices Research Conference, Stanford, Calif., June, 1961.

[88] R. N. Hall, *Solid-State Electron.*, vol. 3, p. 320, 1961.

[89] W. W. Gaertner, *Proc. IRE*, vol. 45, p. 1392, 1957.

[90] J. M. Lavine and A. A. Iannini, *Solid-State Electron.*, vol. 5, p. 109, 1962.

[91] W. G. Spitzer, C. R. Crowell, and M. M. Atalla, *Phys. Rev. Letters*, vol. 8, p. 57, 1962.

[92] D. Kahng, *Proc. IRE*, vol. 50, p. 1534, 1962.

[93] M. Aven and D. M. Cook, *J. Appl. Phys.*, vol. 32, p. 960, 1961.

[94] R. N. Hall, *J. Appl. Phys.*, vol. 29, p. 914, 1958.

[95] M. Gershenzon and R. M. Mikulyak, *Electrochem. Soc. Abstr.*, Chicago, Ill., 1960, p. 215.

[96] J. C. Marinace, *IBM J. Res. Develop.*, vol. 4, p. 280, 1960.
R. L. Anderson, *ibid.*, p. 283.

[97] L. J. VanRuyven and W. Dekker, *Physica*, vol. 28, no. 3, p. 307, 1962.

[98] G. Via and R. E. Thun, Vacuum Society Symposium, Washington, D.C., 1961.

[99] J. Marucchi and N. Nifontoff, *Comp. Rend.*, vol. 249, p. 435, 1959.

[100] S. A. Semiletov, *Soviet Phys.-Cryst.* (*English Trans.*), vol. 1, p. 542, 1956.

[101] G. A. Kurov, S. A. Semiletov, and Z. G. Pinsker, *Soviet Phys. Doklady* (*English Trans.*), vol. 1, p. 604, 1956.

[102] O. Weinreich, G. Dermit, and C. Tufts, *J. Appl. Phys.*, vol. 32, p. 1170, 1961.

[103] N. G. Nakhodkin and V. P. Nemtsev, *Radiotekhn. i Elektron.*, vol. 5, p. 1669, 1960.

[104] J. E. Davey, *J. Appl. Phys.*, vol. 32, p. 877, 1961.

[105] K. V. Shalimova, V. I. Vardanyan, and I. P. Ternovikh, *Nauchn. Dokl. Visshei. Shkoly Radiotekhn. i Elektron.*, no. 4, p. 232, 1958.

[106] G. Hass, *Z. Anorg. Chem.*, vol. 257, p. 166, 1948.

[107] A. P. Hale and B. D. James, IRE Electron Devices Meeting, Washington, D.C., October, 1960.

[108] I. P. Kalinkin, L. A. Sergeeva, V. B. Aleskovskii, and L. P. Strakhov, *Soviet Phys. Solid State* (*English Trans.*), vol. 3, p. 1922, 1962.

[109] B. Goldstein and L. Pensak, *J. Appl. Phys.*, vol. 30, p. 155, 1959.

[110] J. M. Gilles, and J. Van Cakenberghe, *Nature*, vol. 182, p. 862, 1958.

[111] A. Rose, *Phys. Rev.*, vol. 97, p. 1538, 1955.
A. Rose and M. Lampert, *Phys. Rev.*, vol. 113, p. 1227, 1959.

[112] K. G. Gunther, *Naturwissenschaften*, vol. 45, no. 17, p. 415, 1958; *Z. Naturforsch.*, vol. 13a, no. 12, p. 1081, 1958.

[113] F. Eckart and G. Jungk, *Ann. Physik*, vol. 7, p. 210, 1961.

[114] R. P. Riesz and L. V. Sharp, Solid State Devices Research Conference, Stanford, Calif., June, 1961.

[115] J. E. Allegretti, D. J. Shombert, E. Schaarschmidt, and J. Waldman, "Vapor Deposited Silicon Single Crystal Layers," Electronic Chemicals Division of Merck & Co., Inc., Rahway, N.J.

[116] Ch. D. LaFond, *Missiles and Rockets*, vol. 7, pp. 24–28, December, 1960.

[117] H. H. Loar, H. Christensen, and J. J. Kleimac, AIEE-IRE Joint Conference on Research of Semiconductor Devices, Pittsburgh, Pa., June, 1960.

[118] H. C. Theurer and H. Christensen, Electrochemical Society, 118th Meeting, Houston, Tex., Oct. 9–13, 1960.

[119] R. C. Sangster, E. F. Maverick, and M. L. Croutch, *J. Electrochem. Soc.*, vol. 104, p. 317, 1957.

[120] G. Szekely, *J. Electrochem. Soc.*, vol. 104, p. 663, 1957.

[121] H. J. Beatty, R. Glang, E. S. Wajda, and W. H. White, U.S. Signal Corps, 1st Quarterly Progress Report, Contract DA36-039 sc-87395, p. 10, 1960.

[122] F. B. Litton and H. C. Anderson, *J. Electrochem. Soc.*, vol. 101, p. 287, 1954.

[123] 11 papers on Vapor Growth of Ge and Si, *IBM J. Res. Develop.*, vol. 4, pp. 248–304, July, 1960.

[124] E. S. Wajda and R. Glang, *Proc. AIME Conf. Semicond. Materials*, Boston, Mass., August, 1960.

[125] K. M. Hergenrother, S. E. Mayer, and A. I. Mlavsky, Silicon Carbide, *Proc. Conf. Silicon Carbide*, Boston, Mass., Apr. 2–3, 1959, p. 60.

[126] Silicon Carbide, 5 papers by various authors, *Proc. Conf. Silicon Carbide*, Boston, Mass., Apr. 2–3, 1959, pp. 67–114.

[127] A. Hagenlocher, Recent News Paper 2, Electrochemical Society 119th Meeting, Indianapolis, Ind., May 1–3, 1961.

[128] G. R. Antell and D. Effer, *J. Electrochem. Soc.*, vol. 106, p. 509, 1959.

[129] R. L. Newman and N. Goldsmith, Recent News Paper 3, Electrochemical Society, 119th Meeting, Indianapolis, Ind., May 1–3, 1961.

[130] M. Gershenzon and R. M. Mikulyak, *J. Electrochem. Soc.*, vol. 108, p. 548, 1961.

[131] J. N. Keith and E. H. Thompkins, Armour Research Foundation, Final Report 3819-4, Project C819.

[132] M. S. Blois, Jr., *J. Appl. Phys.*, vol. 26, p. 975, 1955.

[133] D. O. Smith, *J. Appl. Phys.*, vol. 29, p. 264, 1958.

[134] R. M. Bozorth, "Ferromagnetism," D. Van Nostrand Company, Inc., Princeton, N.J., 1956.

[135] C. D. Olson and A. V. Pohm, *J. Appl. Phys.*, vol. 29, p. 274, 1958.

[136] I. W. Wolf and V. P. McConnell, *Proc. Am. Electroplaters' Soc.*, 1956.

[137] J. C. Lloyd and R. S. Smith, *J. Appl. Phys.*, vol. 30, p. 274S, 1959.

[138] T. R. Long, *J. Appl. Phys.*, vol. 31, p. 123S, 1960.

[139] I. W. Wolf, *J. Appl. Phys.*, vol. 33, p. 1152, 1962.

[140] M. H. Francombe and A. J. Noreika, *J. Appl. Phys.*, vol. 32, p. 99S, 1961.

[141] A. J. Noreika and M. H. Francombe, *J. Appl. Phys.*, vol. 33, p. 1119, 1962.

[142] A. Brenner and G. E. Riddell, *J. Res. Natl. Bur. Std. Res. Paper* 1835, pp. 39, 385, 1947.

[143] R. J. Heritage and M. T. Walker, *J. Electron. Control*, vol. 7, p. 542, 1959.

[144] P. H. Eisenberg, U.S. Patent 2,827,399, 1958.

[145] *U.S. Dept. Comm. Office Tech. Serv. P. B. Rept.* 151525, Remington Rand Univac, November, 1958.

[146] D. O. Smith, M. S. Cohen, and G. P. Weiss, *J. Appl. Phys.*, vol. 30, p. 1755, 1960.

[147] J. I. Raffel, T. S. Crowther, A. H. Anderson, and T. O. Herndon, *Proc. IRE*, vol. 49, no. 1, p. 155, 1961.

[148] R. H. Heidenreich and F. W. Reynolds, "Structure and Properties of Thin Films," John Wiley & Sons, Inc., New York, 1959.

[149] E. A. Nesbitt, R. H. Heidenreich, and A. J. Williams, *J. Appl. Phys.*, vol. 31, p. 2285, 1960.

[150] P. Oberg and J. Lingenjo, *Rev. Sci. Instr.*, vol. 30, p. 1053, 1959.

[151] M. Lauriente and J. J. Bagrowski, *J. Appl. Phys.*, vol. 33, p. 1109, 1962.

[152] T. R. Long, Electrodeposition Symposium, Electrochemical Society Conference, 1961.

[153] E. C. Stoner and E. P. Wohlfarth, *Proc. Roy. Soc., London*, ser. A, vol. 240, p. 74, 1948.

[154] W. W. L. Chu, J. E. Wolfe, and B. C. Wagner, *J. Appl. Phys.*, vol. 30, p. 272, 1959.

[155] D. O. Smith and K. J. Harte, *J. Appl. Phys.*, vol. 33, p. 1399, 1962.

[156] S. Middlehoek, Ferromagnetic Domains in Thin Ni-Fe Films, Thesis, University of Amsterdam, 1961.

[157] A. V. Pohm and E. N. Mitchell, International Solid-state Circuits Conference, Philadelphia, Pa., Feb. 10–12, 1960.

[158] M. Prutton and E. M. Bradley, *Proc. Roy. Soc.*, vol. 75, p. 557, 1960.

[159] A. Segmüller, *J. Appl. Phys.*, vol. 32, p. 895, 1961.

[160] E. M. Bradley, *J. Brit. IRE*, vol. 20, p. 765, 1960.

[161] W. E. Proebster, International Solid-state Circuits Conference, Philadelphia, Pa., February, 1962.

[162] E. E. Bittman, International Solid-state Circuits Conference, 1959.

[163] W. E. Proebster and W. Dietrich, International Solid-state Circuits Conference, 1961.

[164] J. C. Suits and E. W. Pugh, *J. Appl. Phys.*, vol. 33, p. 1057, 1962.

[165] L. J. Oakland and T. D. Rossing, *J. Appl. Phys.*, vol. 30, p. 545, 1959.

[166] W. E. Proebster and H. J. Oguey, International Solid-state Circuits Conference, Philadelphia, Pa., Feb. 10–12, 1960.

[167] K. D. Broadbent and F. J. McClung, International Solid-state Circuits Conference, Philadelphia, Pa., Feb. 10–12, 1960.

[168] J. Von Neumann, U.S. Patent 2,815,488, 1957.

[169] H. K. Onnes, *Commun., Phys. Lab. Univ. Leiden*, no. 119b, 1911.

[170] J. Bardeen, L. N. Cooper, and J. R. Schrieffer, *Phys. Rev.*, vol. 108, p. 1175, 1957.

[171] F. and H. London, *Proc. Roy. Soc. London*, ser. A, vol. 149, p. 71, 1935.

[172] W. Meissner and R. Ochsenfeld, *Naturwissenchaften*, vol. 21, p. 787, 1933.

[173] J. M. Casimir-Jonker and W. J. de Hass, *Physica*, vol. 2, p. 935, 1935.

[174] D. A. Buck, *Proc. IRE*, vol. 44, p. 482, 1956.

[175] D. R. Young, *Brit. J. Appl. Phys.*, vol. 12, August, 1961.

[176] M. E. Behrndt, R. H. Blumberg, and G. R. Giedd, *IBM J. Res. Develop.*, vol. 4, no. 2, April, 1960.

[177] G. B. Rosenberger, *Solid-State Electron.*, vol. 1, no. 4, pp. 388–398, 1960.

[178] W. B. Ittner, III, *Solid State J.*, July-August, 1960.

[179] C. R. Smallman, A. E. Slade, and M. L. Cohen, *Proc. IRE*, September, 1960.

[180] M. K. Haynes, ONR Symposium Report ACR-50, *Proc. Symp. Superconductive Tech. for Computing Systems*, May, 1960.

[181] J. W. Crowe, *IBM J. Res. Develop.*, vol. 1, p. 294, 1957.

[182] L. L. Burns, G. A. Alphonse, and G. W. Leck, *IRE Trans. Electron. Computers*, vol. 10, September, 1961.

[183] I. Giaever, *Phys. Rev. Letters*, vol. 5, pp. 147, 464, 1960.

[184] J. Nicol, S. Shapiro, and P. H. Smith, *Phys. Rev. Letters*, vol. 5, p. 461, 1960.

[185] S. Shapiro, P. H. Smith, J. Nicol, J. L. Miles, and P. F. Strong, *IBM J. Res. Develop.*, vol. 6, p. 34, 1962.

CHAPTER 5

SEMICONDUCTOR INTEGRATED CIRCUITS

By Gordon E. Moore

5-1. INTRODUCTION

If miniaturization were the sole justification for microelectronics, an improvement of the order of a factor of 10^3 in volumetric efficiency could be achieved merely by elimination of the packing voids in conventional electronics. The cost of this type of assembly can be easily justified (see Chap. 3) for special applications where volume and weight are at a premium. The excitement created by microelectronics, however, extends far beyond size and weight alone. The potential achievements in the reduction of cost and in the improvement of reliability point the way to an entirely new realm of allowable system complexity. The actual achievements to date, though considerably more modest than the publicity might lead one to believe, suggest that microelectronics will be employed extensively from now on. An important approach to this exciting field is provided by semiconductor integrated circuitry. This should not be viewed as a completely independent entity from other approaches, for example, thin films, but rather as a complementary approach evolving toward the same general goal but arising from a different technological background. The origin of this evolutionary process sprang naturally from the transistor and diode technology. It is being extended to include more and more complex combinations of operations.

Before we proceed further, some definition of semiconductor integrated circuitry is required, because this term or similar terms have been applied to a wide variety of levels of sophistication. There are two extremes in semiconductor integrated circuitry. The first of these is the *chip approach* wherein individual components, such as transistors, resistors, and diodes, are produced on separate pieces of material; then these separate components are mounted and interconnected in a single package to produce a circuit function by what is, in reality, a microassembly technique.

The second extreme makes the entire electronic function in and upon a single piece of semiconductor material having many components or regions that are isolated or interconnected electrically, as the circuit requires. In general, in this latter approach, all the intraconnections within the functional block itself will be done by batch processing on large numbers of circuits. The only individual assembly operations are associated with mounting the final circuit function in a package, so that it can be connected conveniently to the outside world.

All levels between these two extremes are practiced. The level of integration will, in general, be determined at any time by the degree of development of the technology and such economic considerations as the number of identical circuits needed and the time scale.

The approach employing individual components offers the advantage of complete flexibility, analogous to that obtainable in conventional electronics. Individual chips are interconnected by lead-bonding techniques essentially identical with those employed in transistor manufacture. A circuit made from individual chips mounted on a metallized ceramic substrate is shown in Fig. 5-1. More elaborate schemes wherein the individual components are mounted in precise locations on a substrate and where the separations between individual components are filled with a material suitable for thin-film interconnections have been proposed. None of these schemes have been developed adequately enough so that they can be seriously considered at the present.

This chapter will be concerned primarily with the other extreme of semiconductor integrated circuits—those which are produced entirely

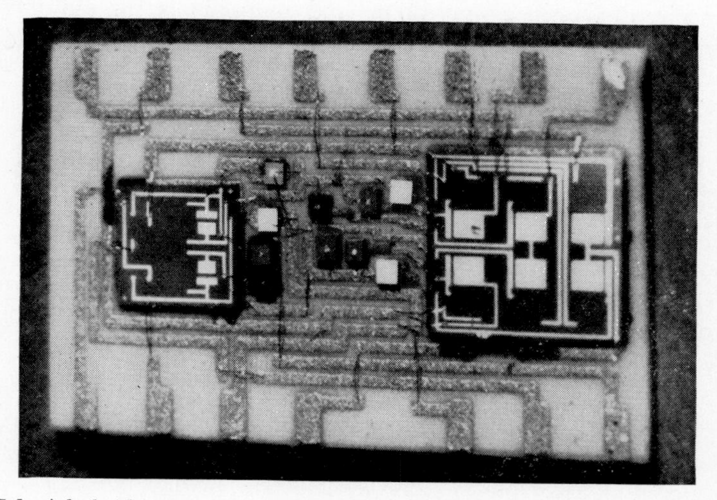

FIG. 5-1. A hybrid integrated circuit made from separate diode and transistor chips and resistor arrays, mounted upon a metallized ceramic substrate. The size of the ceramic substrate is 0.325 by 0.575 in. (*General Instrument Corporation.*)

in a single monolithic block of material. Examples are solid circuits produced by Texas Instruments, Inc., and micrologic circuits produced by Fairchild Camera and Instrument Corporation. Intermediate cases can usually be considered as the single-chip approach applied to smaller monolithic blocks: for example, the molecular electronics used by Westinghouse in the AN/ARC-63 transceiver.[1] The only semiconductor materials which need be considered for their applicability to integrated circuits, at present, are germanium and silicon. Since the trend toward integrated circuits arose after the silicon technology was well developed, and indeed was, to a considerable extent, stimulated by this technology, germanium has not been an important material. Accordingly the discussion will be restricted to silicon. It is unlikely that other semiconductor materials, such as, for example, gallium arsenide, will be of importance in the field of integrated circuitry for other than a few highly specialized applications.

The large-scale manufacture of silicon mesa transistors by batch processing had been accomplished before the rise of semiconductor integrated circuits. Such mesa transistors are made in an array covering a slice of silicon material of the order of one inch in diameter. Typical arrays of double-diffused silicon transistors are shown in Fig. 5-2. Such a wafer may contain as many as 1,000 transistor structures. These mesa transistors are completed electrically while still in wafer form. They are positioned in a precise geometric array. The manufacturing operation beyond this stage consists of separating these structures, mounting them, and selling them to a customer, who proceeds to reconnect them to form circuits. Since this dicing and assembly is the major expense in making small transistors, one is led rather naturally to consider per-

FIG. 5-2. A typical array of double-diffused silicon transistors. Each ring-and-dot pattern is a transistor structure. Only the metallized emitter and base contacts are visible in this photomicrograph. The silicon wafer is approximately one inch in diameter. The flat edge is used for indexing during process.

forming interconnections between the structures without loss of index, in order to decrease total system-manufacturing costs.

Several problems arise when this is considered. First, the surfaces of mesa transistors are extremely sensitive to ambient, resulting in relatively low yields from devices at the wafer stage to good final products. This problem is aggravated by increasing the number of devices or by attempting to run interconnections over the surfaces. Interconnections by bonding of leads can be accomplished successfully. However, this is a unit-by-unit operation, and reinserts the expensive portion of semiconductor-device manufacture while decreasing yields because of the increased number of possibilities for imperfections as the complexity of the interconnected unit increases.

The next important step in making semiconductor integrated circuitry practical was achieved with the development of the planar transistor structure.[2] The schematic cross sections of a single planar transistor structure and a mesa transistor are shown in Fig. 5-3. The advantages of the planar structure are intimately tied to the silicon dioxide layer covering the region where the junctions intersect the surface. The oxide layer is an effective barrier to the deleterious effects of the ambient on the junction surfaces. In addition, it supplies a relatively flat surface, upon which thin metal intraconnections can be applied. Although this oxide layer is actually a micron or less in thickness, it has a dielectric strength ranging to several hundred volts, allowing metal films to pass over the junction regions without effect. Thus, in the planar structure, the interconnection problem for the individual components is solved. In order to make complete circuits, it is necessary to add the capability of making other elements, such as resistors and capacitors, as well as to achieve the required electrical isolation of these components. This chapter will discuss the technology and device structures available for the

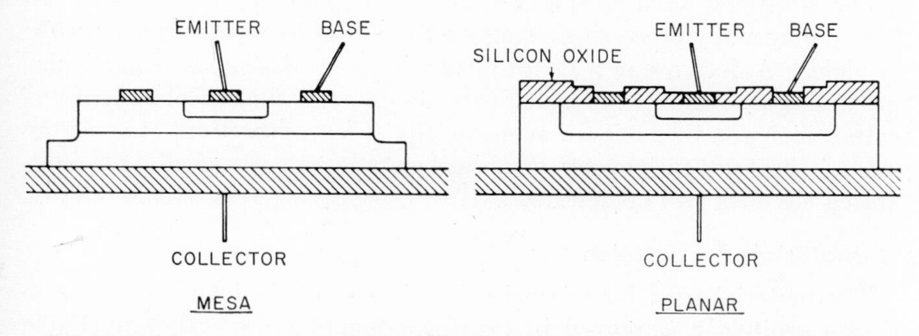

FIG. 5-3. Schematic cross sections of mesa and planar transistors. The metal emitter, base, and collector contacts normally employed are shown. Note the silicon oxide layer covering the intersection of the junctions with the surface in the planar case.

manufacture of semiconductor integrated circuits and will try to define the limitations and potentialities of the technology.

5-2. THE TECHNOLOGY

The critical role played by the available technology in determining the applicability of microcircuitry in a given application cannot be overestimated. It determines what can be done at all, as well as what can be done economically. Using known physical phenomena, integrated solid-state structures can be conceived which would perform almost any useful electronic function presently performed by conventional electronics. Only a small subclass of this class of structures is accessible, because of the limitations imposed by the available technology. Thus, the problem in making functional electronic blocks does not reside merely in the conception of structures which, if made, would perform the function, but rather in the conception of a structure which is economically accessible through the available technology.

The technological capability determines into which class the structure will fall—if it will be a manufacturable item suitable for design into operating equipment; if it will be a structure whose "feasibility" can be established by the construction of a few samples by highly trained laboratory technicians; if it will be a one-of-a-kind structure made through the extensive labors of engineers adjusting and selecting until operation is finally achieved, suitable for description in a technical publication or contract report; or if it will fall into the class of *gedanken* structures, unmakable, but with principles of operation consistent with established physical laws.

Each major addition to the pool of technologies has a tremendous fan-out in that many more circuit functions can be made. As well as major additions, such as epitaxial growth, important contributions can be made by many processing changes which result in appreciable increases in yield. A decrease of a factor of two in the shrinkage at a particular processing operation can result in the doubling of the complexity of circuits which can be achieved economically. In this section, some of the more important technology areas associated with semiconductor integrated circuitry will be described.

1. Materials Preparation

The material used for microcircuitry is essentially identical with the silicon ordinarily employed in transistor manufacture. It is available commercially as pure polycrystalline silicon, as doped single crystals, and as epitaxial wafers.

Since the technology for preparing crystals of silicon suitable for microcircuitry is well established, it will not be discussed in detail here. Suffice it to say that both Czochralski crystals pulled from a crucible of molten material and zoned melted material can be used. Slicing with a diamond impregnated saw and lapping with silicon carbide abrasive is straightforward and widespread throughout the semiconductor industry.

Any differences in material preparation that exist which are significant occur subsequent to this lapping operation. It is, of course, generally agreed that a high degree of cleanliness is required for the high-temperature processing operation, such as diffusion.

There is disagreement as to which method of surface preparation results in wafers most suitable for the various processes. It is likely, in fact, that the most suitable surface for a given process will differ from that for a different step in the operation. For example, an etched surface very desirable for diffusion might have sufficient surface irregularities so that the precision of the masking operations would be influenced, both by affecting the coating operation, and by making poor contact with the mask during exposure. Essentially the choice comes down to that between etched surfaces and mechanically polished surfaces. Etched surfaces are generally considered to be more nearly perfect on a microscopic scale, while on a macroscopic scale a well mechanically polished wafer is superior. As usual with two alternatives of this sort, one can think of combinations of the two processes which he hopes will supply the advantages of each.

For example, electrochemical etching can be employed using an extremely thin film of electrolyte.[3] This results in selective dissolution of the high points on the wafer, resulting in a flat surface comparable with that achieved by mechanically polishing, while the surface itself should have the perfection expected from an etched surface. However, this technique is relatively tedious and involves such undesirable problems as attaching electrodes to the wafer so that electropolishing current can be passed.

At the other extreme, mechanically polished wafers can be etched slightly either directly or by such techniques as oxidation followed by removal of the oxide, a technique which results in an extremely uniform removal of the damaged surface layer while preserving the flat surface.

The simplest surface preparation employed consists merely of etching the wafers after they have been cleaned, subsequent to lapping, in a polishing chemical etch, usually consisting primarily of nitric acid and hydrofluoric acid, occasionally with an addition of a moderator, such as acetic acid. For compositions relatively rich in nitric acid, a nonselective etching occurs, which results in a shiny surface blemished only by a relatively uniform orange-peel effect.

For epitaxial growth, the surface preparation is perhaps even more critical than for the diffusion and masking operations. Here again, one is torn between etching and mechanical polishing. In order to allow the reader some latitude in which to exercise his creativity, no specific preference for either method will be given.

2. Diffusion

This is presently the most convenient, flexible, and well-controlled method for obtaining the impurity-concentration distributions necessary for device fabrication. An extensive literature has developed concerning diffusion in semiconductors.[4] Study of diffusion in semiconductors has lead to a greatly improved understanding of the details of the diffusion process itself.

For device fabrication in silicon, one is primarily concerned with the diffusion of impurity elements from the third and fifth groups of the periodic table. These impurities, in general, enter the lattice substitutionally at the regular silicon lattice sites. The diffusion mechanism consists of the jumping from one lattice site to an adjacent vacancy. Such substitutional diffusion in crystals like silicon requires temperatures approaching the melting point to proceed at reasonable rates.

Some other impurities of interest diffuse by an interstitial mechanism. These, in general, are much more rapid diffusers. Examples of the latter class are gold and nickel in silicon. In order for diffusion to occur, it is necessary that a concentration gradient or, more precisely, a chemical potential gradient be established. Several treatises on the diffusion mechanism itself exist.[5]

In device fabrication, one is usually concerned with the diffusion from a surface of the semiconductor material. Two boundary conditions are of special interest in the diffusions employed for device manufacture. The first of these occurs when the concentration of the impurity on the surface is maintained constant throughout the diffusion process. In this case, a useful approximation to the concentration profile obtained is given by the complementary error-function distribution.

$$C(x) = C_0\left(1 - \frac{2}{\sqrt{\pi}} \int_0^{x/2\sqrt{Dt}} e^{-\lambda^2}\, d\lambda\right) = C_0 \operatorname{erfc} \frac{x}{2\sqrt{Dt}} \tag{5-1}$$

where $C(x)$ is the concentration a depth x below the surface; C_0 is the surface concentration; D, the diffusion constant, is the constant of proportionality between the diffusion flux and the concentration gradient; and t is the time.

The other case of special interest occurs when a fixed amount of impurity Q is deposited on the surface at the beginning of the diffusion and held constant throughout the diffusion. Under these conditions, the useful approximation is the gaussian distribution given by

$$C(x) = Q(\pi Dt)^{-1/2} \exp\left(\frac{-x^2}{4Dt}\right) \tag{5-2}$$

The above relations hold fairly well for diffusion depths greater than several microns or for low concentrations. With shallow diffusions and in high concentrations deviations often appear. Reasons for these deviations can be understood qualitatively, but because they are often dependent on the specific parameters of the diffusion system under consideration, it is desirable to allow for them empirically. The reasons for these deviations include such factors as the dependence of D upon concentration, the time required to obtain C_0 or Q at its final value, the fact that the surface is usually moving during the diffusion process, either by evaporation or oxidation, and interactions between impurity atoms.

Diffusion Systems. Three different types of diffusion systems must be considered. In *closed-tube diffusion*, the semiconductor wafers are sealed in an ampule with the impurity source. In *vacuum diffusion*, a controlled pressure of the diffusant impurity is established in a vacuum chamber, and diffusion proceeds inward while simultaneously the semiconductor material evaporates. In *open-tube diffusion*, an atmosphere is established to produce the desired boundary conditions by flowing gas over the semiconductor wafers.

With silicon in the closed-tube system, one usually encapsulates the wafers and source in a fused silica ampule. The concentration of impurity on the silicon surface can be controlled in a variety of ways. One can limit the total quantity of impurity added, one can use with an alloy of the impurity some neutral diluent such as lead or tin to reduce the equilibrium vapor pressure of the impurity, or one can use a doped silicon source. In the last case, the impurity concentration is determined by the ratio of the relative surface areas of the sample to the total surface area of silicon, assuming that the volume of the ampule is relatively small and that its surface does not act as a sink for the impurities.

The advantage of the closed-tube system is flexibility. A single diffusion furnace can be used for a large number of different diffusions with no problems of cross contamination. The disadvantages of the technique for device manufacture, however, greatly outweigh this advantage of flexibility. It is difficult by this technique to process large batches of material; the sealing process is annoying at best; and it does not allow one to take full advantage of oxide-masking techniques for geometry control.

Vacuum diffusion has not been used extensively for device manufac-

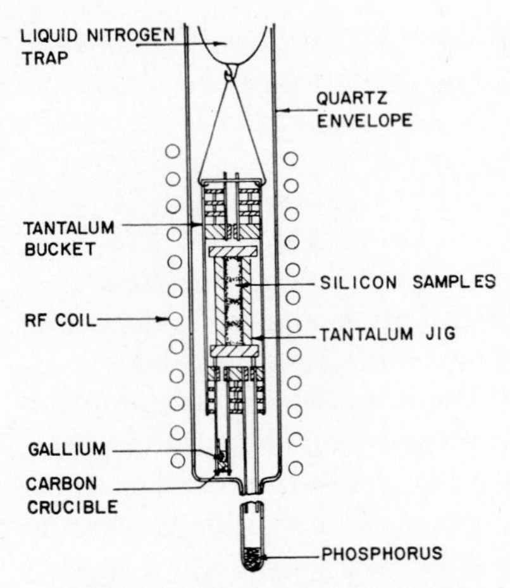

FIG. 5-4. Cross section of a vacuum station used to diffuse phosphorus and gallium simultaneously into evaporating silicon. (*R. L. Batdorf and F. M. Smits.*[6])

ture. Figure 5-4 shows an apparatus employed for the simultaneous vacuum diffusion of phosphorus and gallium into evaporating silicon.[6]

A dynamic vapor pressure of gallium and phosphorus is established by controlling the temperatures of the reservoirs of the elements. These are adjusted to give the desired surface concentration. The entire jig is heated by the r-f field to the diffusion temperature. A vacuum is maintained with the liquid nitrogen trap and pumps beyond the trap. The tantalum jig, as well as supporting samples, contains heat shields to help maintain a uniform diffusion temperature. The vapors of gallium and phosphorus pass from the sources up past the silicon wafers, establishing a concentration dissolved in the surface layer of the silicon. The surface of the silicon evaporates at a constant rate when all the oxide has been eliminated.

An interesting feature of vacuum diffusion is that because of the constant evaporation rate of the semiconductor material, a steady-state condition is approached wherein the impurity profile is exponential with depth and independent of time. This suggests a process wherein control of time and temperature need not be especially precise in order to obtain a reproducible profile. However, the expense of vacuum-diffusion apparatus combined with the disadvantage that oxide masking is not possible has limited it to a laboratory technique, useful in the study of the fundamental diffusion processes.

Open-tube diffusion is the generally employed method for the fabrication of diffused-silicon device structures. The remainder of this section will be concerned with this diffusion technique. The principal advantages of the open-tube technique are its applicability to large-scale batch processing and the use of oxide masking to control the geometry of the diffused regions.

There are many variations of the open-tube process, depending upon the impurity source and the desired impurity concentrations. It is usually convenient to employ a two-step diffusion process, wherein the impurity is deposited on the surface or in a shallow surface layer during the first step, and diffused to the desired depth during a subsequent operation. The reason for the preference of the two-step process is that surface damage to the silicon wafers results at the high diffusion temperatures necessary to get deeper diffusion in a nonoxidizing atmosphere.

On the other hand, if the atmosphere is made oxidizing, the formation of SiO_2 masks at least partially against all common group III and group V impurities, except gallium. The simultaneous oxidation competing with the source deposition results in a relatively uncontrollable process. The lower temperatures ordinarily employed in the first step or *predeposition process* are below those where appreciable surface evaporation takes place and, accordingly, can be carried out in a nonoxidizing atmosphere.

In general, the idea during predeposition is to establish an atmosphere containing the desired impurity surrounding the surface of the semiconductor to be diffused. Ideally one has an equilibrium between the volatile species in the surrounding atmosphere and the dissolved species on the silicon surface. The atmosphere can be established in a variety of ways which will be discussed below for the specific impurities. Ordinarily, as shown in Fig. 5-5, the silicon wafers stand in a silica boat, although there are cases where laying them down flat on a silica plate seems to improve the uniformity of the diffused layer.

Donor Impurities. Phosphorus, arsenic, and antimony are the important donor impurities in silicon. The diffusion constants[7] and solubilities[8] of these impurities as a function of temperature are shown in Figs. 5-6 and 5-7, respectively. Arsenic and antimony are relatively slow diffusers compared with phosphorus. All three impurities are commonly employed, although phosphorus is the most common.

Phosphorus Diffusion. Phosphorus diffusion is performed most conveniently in a two-step process. Red phosphorus, P_2O_5, $POCl_3$, and PCl_3 are the usual sources to consider. Red phosphorus and P_2O_5 are solids at the source temperatures employed. Figure 5-8 shows a schematic diagram of a diffusion system suitable for use with P_2O_5 or red phosphorus.

This is a two-zone system in which the silicon wafers are held at the temperature desired, usually of the order of 1000 to 1200°C, and the source is held at a considerably lower temperature, in the vicinity of 200°C for P_2O_5. The temperature increases monotonically from the source furnace to the silicon wafers, so that material evaporated at the

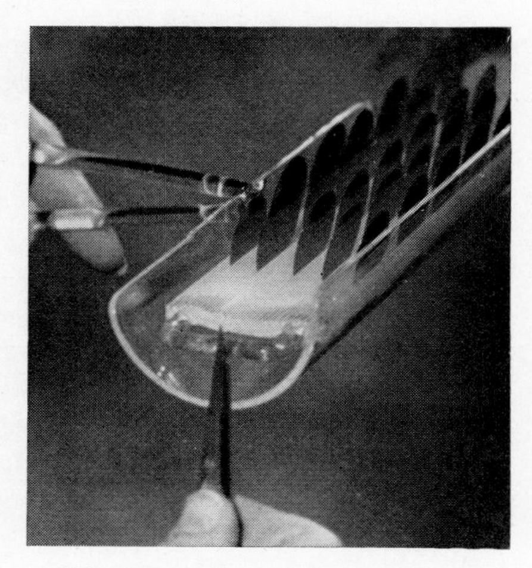

FIG. 5-5. Silicon wafers in a silica boat ready for insertion into a diffusion furnace.

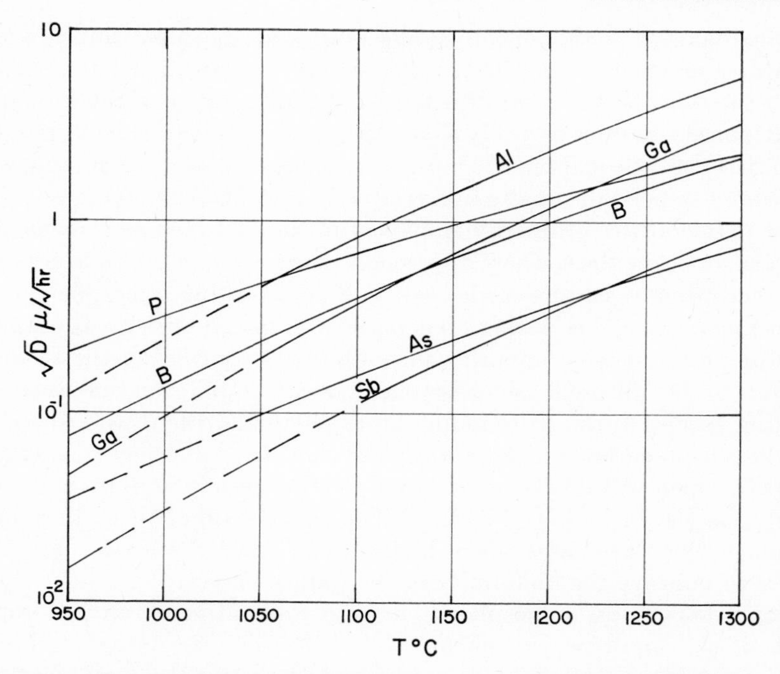

FIG. 5-6. Square root of the diffusion constant versus temperature for the common donor and acceptor impurities in silicon.

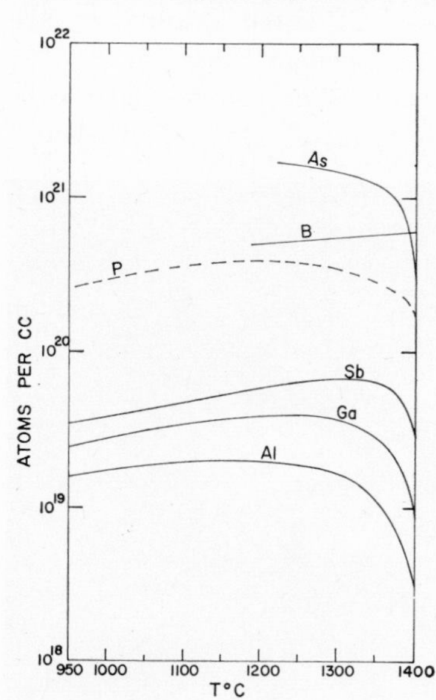

FIG. 5-7. Solid solubilities of common donor and acceptor impurities in silicon versus temperature. (*F. A. Trumbore.*[8])

source will not condense on the furnace walls. A carrier gas such as argon or nitrogen passes over the source, ideally becoming saturated with the vapor of the source at the source temperature; carries through a quartz wool filter to remove any particles picked up by the flowing gas stream; and passes over the silicon wafers and out the exhaust vent.

The quantity of phosphorus deposited on the silicon wafers is roughly an exponential function of source temperature, and depends upon the silicon wafer temperature and the time of deposition. P_2O_5 is a convenient source and works well if kept dry.

The principal problem concerning P_2O_5 is moisture pickup while loading the boat and transferring it to the source furnace. Relatively small quan-

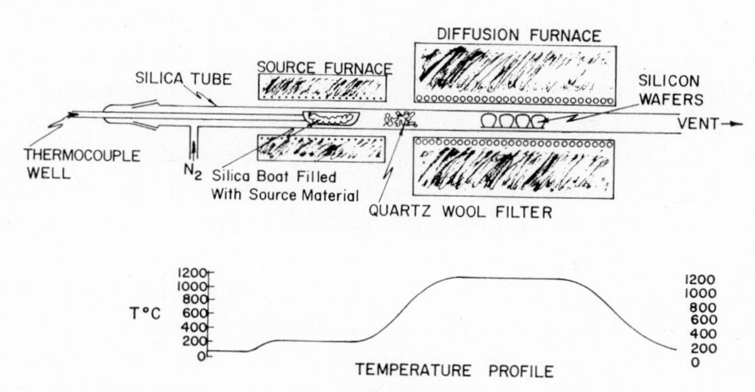

FIG. 5-8. Schematic diagram of a diffusion system suitable for use with P_2O_5 or red phorphorus source material. A typical temperature profile is also shown.

tities of moisture appreciably affect the apparent vapor pressure from the source. By controlling the source temperature with dry P_2O_5 one can perform phosphorus diffusions with surface concentrations from below 10^{16} to above 10^{20} impurity atoms/cm^3. This wide range of available concentrations makes P_2O_5 an extremely flexible dopant source.

Red phosphorus is unreliable as a source, since at the temperatures required to obtain appreciable vapor pressure, it gradually converts to another form of phosphorus, with a corresponding decrease in the equilibrium vapor pressure.

$POCl_3$ and PCl_3 are low-boiling liquids. A suitable system for the use of these source materials is shown diagrammatically in Fig. 5-9. An inert gas is saturated with the vapor or liquid, and introduced into the diffusion furnace. The amount of impurity introduced can be varied by varying either the temperature of the impurity source or the fraction of the gas stream that gets saturated from the source. With the halogens present, there is always the danger of pitting of the silicon surface by chemical reaction to form silicon tetrahalide. This is an especially important problem with PCl_3.

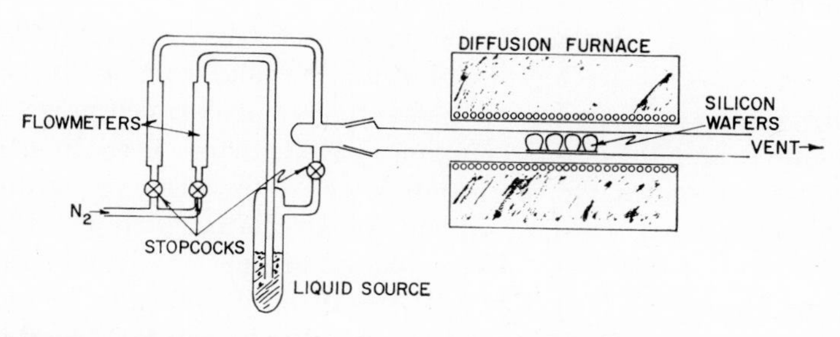

FIG. 5-9. Schematic diagram of a system suitable for the use of liquid source material, such as $POCl_3$.

Many other compounds of phosphorus can be employed but results obtained with P_2O_5 or $POCl_3$ are adequate for device manufacture.

After predeposition, the diffusion to achieve the desired depth can be performed in any oxidizing atmosphere, depending upon the amount of oxide necessary to prevent surface damage by evaporation and upon the thickness of the oxide layer desired for subsequent processing. Dry oxygen, wet oxygen, wet nitrogen, and pure water vapor are common oxidizing atmospheres.

Arsenic Diffusion. As_2O_3 is a convenient source for the diffusion of arsenic. It is used in a manner very similar to P_2O_5.

Antimony Diffusion. For antimony, Sb_2O_3 is a convenient source material, although it has appreciably lower vapor pressure than either P_2O_5 or As_2O_3. Source temperatures using Sb_2O_3 are generally 700 to 1000°C. Where separate source and diffusion furnaces are employed, one must be especially careful to maintain a monotonically increasing temperature.

A more convenient system for antimony employs a single furnace, with separately controlled source. At high concentrations of antimony, there appears to be a new phase formed on the surface that acts as a strong source. This makes it inconvenient to control antimony diffusions above about 10^{19} impurity atoms/cm^3. There is also a tendency for pitting to occur during antimony diffusion, which can be very troublesome. Like arsenic, it is an extremely slow diffuser. This is often the principal reason for choosing arsenic or antimony instead of phosphorus.

Acceptor Impurities. Boron and gallium are the most important acceptor impurities in silicon. The diffusion constants of these and of aluminum are included in Fig. 5-6 as a function of temperature.[7] Occasionally, because of its extremely high diffusion coefficient, aluminum is worth considering for special structures. The solid solubility versus temperature is also shown for these impurities in Fig. 5-7.[8] Boron is the only one with adequate solubility for use as an efficient emitter in transistors, or for other applications requiring extremely high impurity concentrations. Boron diffuses at about the same rate as phosphorus. Gallium diffuses somewhat more rapidly.

Boron Diffusion. Boron is the most extensively employed donor diffusant because of its high solid solubility and because it is oxide-masked. There is, however, no general agreement concerning the best boron source and diffusion technique. B_2O_3, when used in a system similar to that shown for the group V oxides in Fig. 5-8, results in a variety of problems, most of which are associated with the fact that B_2O_3 and SiO_2 are miscible. After a period of time, the fused silica tube gets extremely sticky from the B_2O_3, resulting in difficulty in removing sample boats, and the cracking of furnace tubes if they are cooled to room temperature. The dissolution of the boron in the silica also results in establishing a reservoir of boron in the system, which ends up as the controlling dopant source.

Successful use of B_2O_3 has been achieved in the box technique[9] shown in Fig. 5-10. In this method, the wafers and dopant source, usually a mixture of SiO_2 and B_2O_3, are placed in a box made of platinum or graphite which is closed on top with a loose-fitting cover. The box is then placed in an ordinary open-tube diffusion furnace. This method is somewhat similar to the closed-tube technique, although it avoids the sealing and evacuation operations.

BCl_3 and BBr_3 are employed as sources. BCl_3 is a gas at room temperature, and can be purchased as the liquid under pressure in cylinders. BBr_3 is a low-boiling liquid. BCl_3 can be metered directly into the gas stream, and apparatus similar to that used for phosphorus halides shown in Fig. 5-9 is suitable for boron tribromide.

Here again, both the halides have a tendency to pit the silicon wafers. This is especially pronounced in BCl_3. This effect can be minimized by adding a small quantity of both hydrogen and oxygen (of the order of 1%) to the inert carrier gas. When there is too little hydrogen, severe pitting occurs, whereas with inadequate oxygen, black deposits, often found during boron diffusion, form on the surface of the wafers.

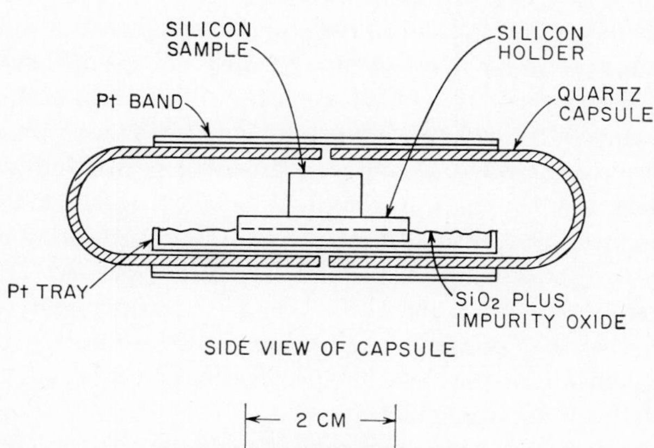

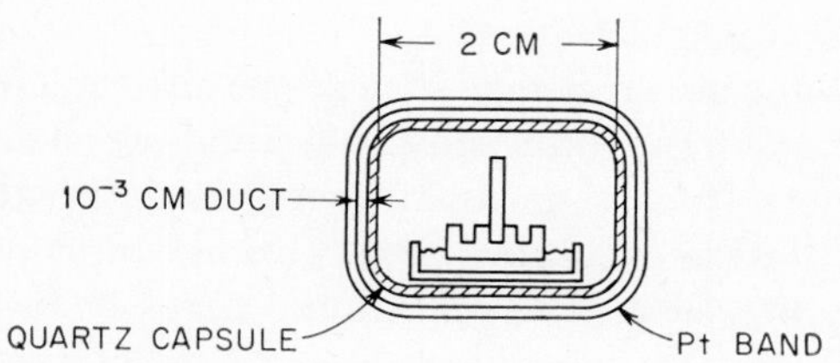

FIG. 5-10. Schematic diagram of apparatus used for boron diffusion by the box technique. (*L. A. D'Asaro.*[9])

The addition of water vapor is not a substitute for adding both hydrogen and oxygen, since the water vapor will hydrolize the boron halide at low temperatures, resulting in B_2O_3 deposits on the furnace tube. The nature of the black deposit which forms with inadequate oxygen or excess BCl_3 is somewhat variable. It contains a large percentage of elemental boron, but also can involve silicon and oxygen. The layer is chemically very inert. When it is necessary to find a material from which to make a container for the universal solvent, it should be a strong contender. Actually, in the boron halide diffusions using hydrogen and oxygen, it may well not be the halide by the time the gas stream arrives at the silicon wafers. Upon looking down a boron furnace operating to achieve relatively high concentrations of boron, one can see solid particles or liquid droplets as a fog in the furnace tube. These are likely to be particles of B_2O_3.

Another method of introducing a boron source is by the controlled oxidation of methyl borate. A small quantity of $(CH_3O)_3B$ is metered into the gas stream with a quantity of oxygen. These react before reaching the wafers to produce a smoke that is probably boron oxide. The other combustion products such as CO_2 do not interfere with this process. Many other boron compounds have been tried, such as diborane, B_2H_4, and pentaborane, B_5H_{10}. To the best of the author's knowledge none are completely satisfactory.

Gallium Diffusion. Gallium diffusion is unique because it is best done as a one-step process.[10] This occurs because at the diffusion temperature, SiO_2 has no measurable effect upon the diffusion of gallium into the underlying silicon. The only important source for gallium diffusion is the refractory oxide Ga_2O_3. Gallium diffusion is performed in a two-zone furnace where the Ga_2O_3 source is held between about 800°C and the diffusion temperature. The gas stream contains a quantity of hydrogen necessary to reduce the Ga_2O_3 to form a volatile species. The reaction at the source is thought to be $H_2 + Ga_2O_3 \rightleftharpoons Ga(\text{or } Ga_2O) + H_2O$. It is relatively easy to approach equilibrium at the source in this system. Hence, the pressure of the volatile species, Ga or Ga_2O, is controlled by the ratio of the pressures of hydrogen to water vapor. This results in an excellent method of source control. Combined with the fact that silicon dioxide does not mask the diffusion, gallium diffusions are extremely uniform and well controlled. The inability to employ oxide masking, however, is a serious drawback in many applications to microelectronics.

Oxide Masking. Frosh and Derrick[11] found that SiO_2 is an effective barrier for preventing many impurities from reaching the underlying silicon. This is an extremely important invention with respect to the construction of complex structures. A fraction of a micron of SiO_2, which is easily grown thermally, prevents phosphorus, arsenic, antimony, and boron from penetrating. In this manner, the extent of diffused areas on the surface of a silicon wafer can be controlled by oxidizing the wafer and exposing only those areas where diffusion is desired.

The oxide for masking can be grown conveniently in a furnace similar to the diffusion furnaces. Either oxygen or water vapor is a suitable oxidizing agent. The properties of the oxides in both cases are very similar, although minor differences have been reported.[12] The growth rate is parabolic in time; that is, dx/dt is proportional to $x^{-\frac{1}{2}}$, where x is the oxide thickness. Figure 5-11 shows oxide thickness versus time, with temperature as a parameter, for pure dry oxygen at one atmosphere pressure, and similar data for pure water vapor at one atmosphere are shown in Fig. 5-12.

The rate of oxidation is essentially independent of the doping in the silicon, except at the very highest impurity concentrations such as those which occur on the silicon surface in the emitter regions of transistor structures. At these extremely high impurity concentrations, the rate of oxidation is somewhat higher. When a multiple oxide masking is

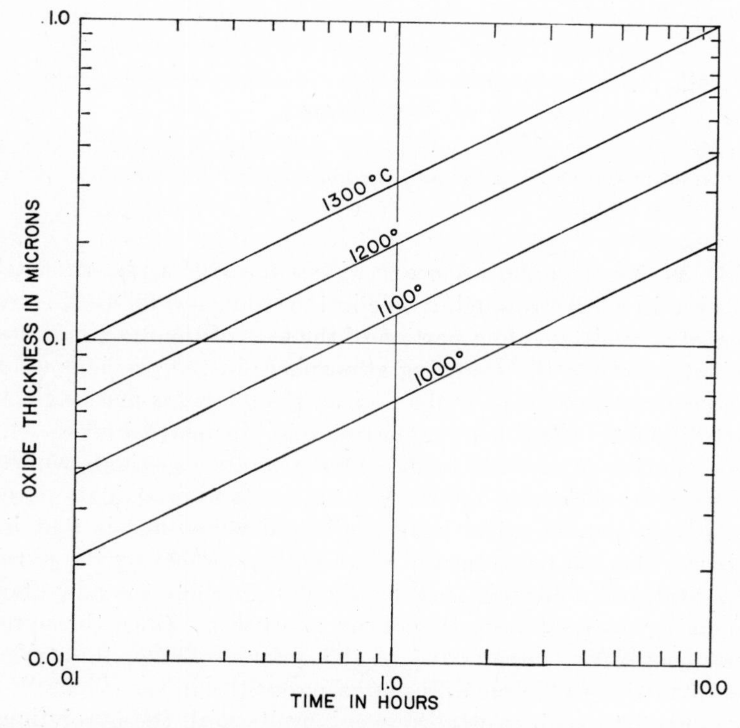

FIG. 5-11. Oxide thickness versus oxidation time in pure dry oxygen with various temperatures for relatively pure silicon. (*Flint, Electrochem. Soc. Electron. Div. Abstr. vol.* 11, *p.* 222, *May,* 1962.)

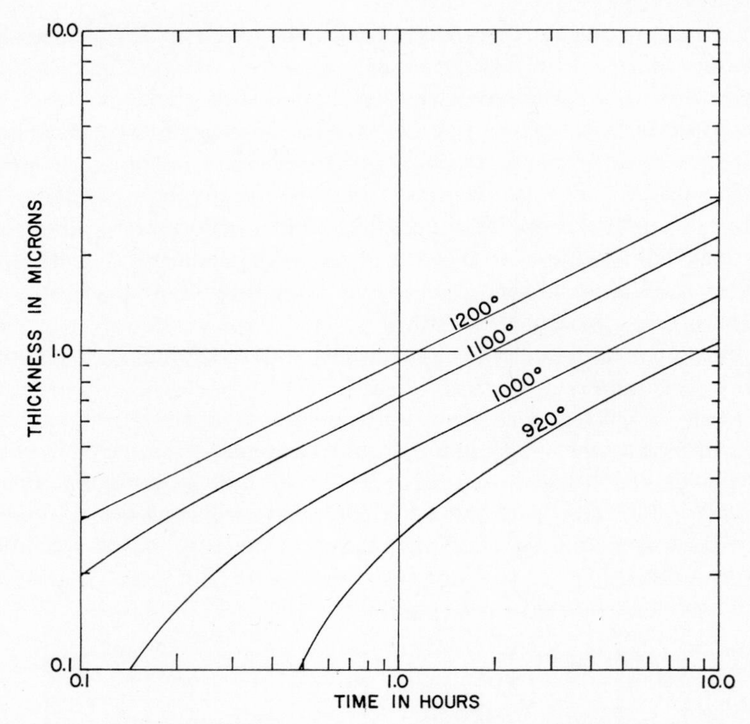

FIG. 5-12. Oxide thickness versus time for various temperatures in pure water vapor at atmospheric pressure. (*Flint, Electrochem. Soc. Electron. Div. Abstr., vol.* 11, *p.* 222, *May,* 1962.)

desirable, as, for example, when one wishes to make a planar transistor by diffusing in a base through a hole in the oxide and follow this by the diffusion of an emitter into a portion of the area of the base, it is usually convenient to adjust the oxidizing atmosphere in the previous diffusion to achieve a regrown oxide of the desired thickness for masking against the next diffusion. This will be more clearly illustrated in Sec. 5-4.

Silicon dioxide layers can be prepared by other methods which, for some special applications, offer advantages. Anodic oxidation results in a relatively porous oxide, whose principal advantage is that it can be prepared at room temperature. Oxide films formed by the pyrolytic decompositions of siloxanes, such as $Si(OC_2H_5)_4$, offer the advantage of not removing material from the silicon substrate. Since the pyrolysis proceeds readily at temperatures of the order of 750°C, the oxide can be deposited with negligible diffusion of impurities in the silicon.[13]

It is difficult to achieve as good a uniformity with the pyrolytic oxide as with the anodic oxide, although its masking properties are similar to those of the thermally grown oxide. The rate of removal of pyrolytic oxide and also of anodic oxide in fluoride-containing etches is considerably greater than for the thermal structure.

Oxidation in high-pressure steam can be made to proceed at relatively low temperatures.[14] It requires an autoclave capable of operation to above the critical pressure of water. Under these conditions, the growth rate of the oxide becomes linear with time and, accordingly, should be good for the production of thick oxides. Since the temperature is relatively low, the problem of relative expansion coefficients of the oxide and silicon which result in cracking of thick oxides prepared in other ways should be minimal.

In general, oxygen and wet inert gas are preferred oxidants for thermal oxidation. One can make adherent layers up to approximately 2 μ thick before cracks and other flaws appear. If an oxidized wafer has part of the oxide removed, and then is reoxidized, the new oxide which grows against the original layer is well bonded, and serves as an effective diffusion mask. The thickness of oxide necessary to mask against the various impurity diffusions is certainly somewhat system-dependent. As an example, Fig. 5-13 illustrates the oxide thickness required to prevent the penetration of phosphorus.[15]

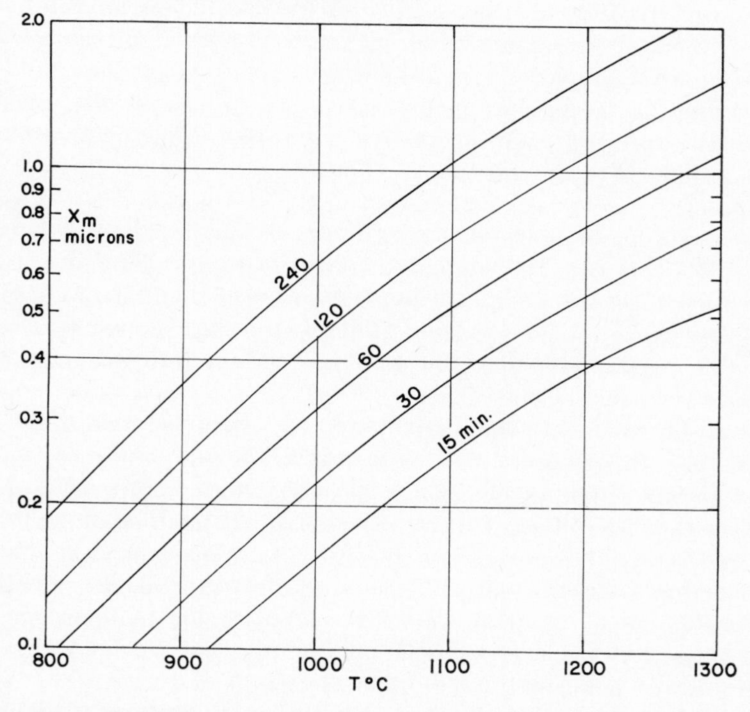

FIG. 5-13. Oxide thickness required to mask against high-concentration phosphorus diffusion for various times. (*C. T. Sah, H. Sello, and D. A. Tremere.*[7])

Evaluation of Diffused Layers for Impurities. For diffusions of donors into p-type or acceptors into n-type obeying one or the other of the distributions given by Eqs. (5-1) and (5-2), one need make only two measurements to evaluate the diffusion length $(Dt)^{1/2}$, and either the surface concentration C_0 or Q, the total quantity of impurity atoms/cm^3. The most convenient measurements to make are those of sheet conductivity of the diffused layer, and the concentration at one depth, usually at a junction where the concentration of the diffused species is equal to the original doping of the substrate. The sheet conductivity is conveniently measured by means of a conventional four-point probe. In diffused layers, the probe spacing, in general, is large compared with the thickness of the layers, leading to the approximation where the measured voltage drop divided by the current, V/I, is independent of probe spacing.[16] This measured resistance is related to the sheet resistivity σ of the diffused layer by $\sigma \cong 4.3V/I$.

Junction-depth measurements are made conveniently by angle lapping through the layer or by grinding a cylindrical groove.[17] In either case, the junctions are delineated by one of a variety of techniques, the most common of which is staining with a solution of hydrofluoric acid containing a fraction of a per cent of nitric acid. This mixture, when applied under strong illumination, preferentially stains p-type material (and very heavily doped n-type). The low-angle bevel or the cylindrical groove gives a mechanical magnification which can be employed to convert the apparent length of the stained areas to junction depth from simple geometric considerations.

However, it is very convenient to employ an interference microscope which superimposes upon the stained pattern an interference pattern, so that the depth from the original surface can be determined by counting interference fringes. Figure 5-14 shows a photomicrograph of a stained-device structure taken through a vertically illuminated microscope employing a sodium lamp as a light source, with an ordinary microscope slide inserted over the specimen.

Fringes obtained with this simple setup are adequate for most diffusion evaluation. In this particular sample, diffusion was performed from an etched surface. The unevenness of such a surface results in an uncertainty of the order of one fringe in the measured junction depths. For more precise measurement, it is necessary to employ more nearly flat original surfaces. By this technique, one fringe corresponds to one-half the wavelength of the light or, with sodium light, to approximately 3,000 A.

Diffusion of Impurities Other than Group III and Group V. Occasionally it is desirable to diffuse other impurities than the common donor and acceptor elements. For example, it is common to diffuse gold into

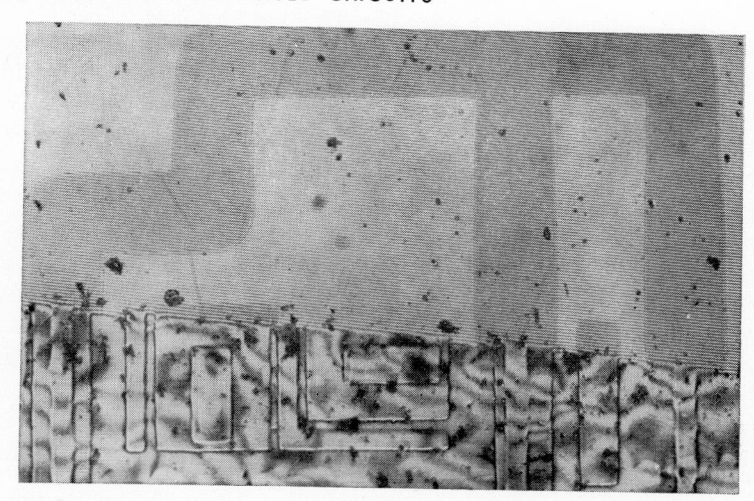

FIG. 5-14. Photomicrograph of a stained-device structure, taken through a vertically illuminated microscope employing a sodium lamp as a light source, with an ordinary microscope slide inserted over the specimen. The originally etched surface shows the outline of the isolation and components. The isolation grid is stained dark on the side of the bevel. The narrow parallel lines along the bevel are interference fringes. Each corresponds to approximately 0.3 μ below the original surface.

silicon device structures, in order to increase the carrier recombination rate. This is important to improve diode recovery times and to decrease the storage of minority carriers in the collector regions of *npn* transistors. Diffusion of gold, which goes by interstitial mechanisms, is rapid, but does not follow one of the simple approximations.[18] It is preferable to establish the desired diffusion schedule for gold by empirical measurement of the parameters one wishes to control.

The usual way of diffusing such impurities is to deposit them on the surface of the wafer, either the front or the back, by evaporation or chemical plating techniques. The wafers are then diffused for the appropriate time-temperature cycle. Since the solubility of these impurities at the diffusion temperatures is usually considerably higher than at lower temperatures, it is necessary to cool relatively rapidly, to prevent precipitation on the surfaces or at defects in the semiconductor material.

Miscellaneous Effects. A variety of phenomena occur when various combinations of oxidations and diffusions are performed. These are often specifically related to the device structures being considered. As a few examples of the effects with which one must contend, we give the following. This list is by no means exhaustive.

When oxide is being grown on a wafer doped with one of the common donor elements, there is evidence that the growing oxide rejects the impurity originally contained within the bulk semiconductor. This

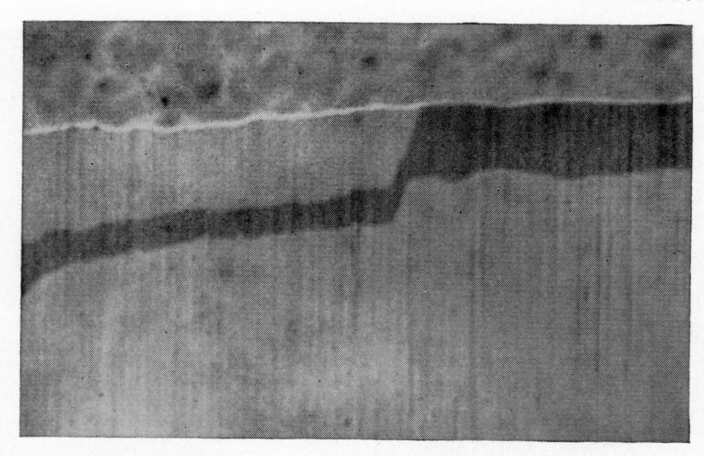

FIG. 5-15. Cross section through an *npn* structure with gallium-base diffusion. The section was prepared by grinding a cylindrical groove in the surface of the etched wafer. The *p*-type material is stained dark. The bumpy area is the original wafer surface. Without the heavy-phosphorus emitter diffusion, the gallium junction depth would have been uniform across the wafer. In this case, where the emitter diffusion is about equally deep as the base diffusion away from the emitter, the enhanced diffusion under the emitter stands out clearly.

results in a concentration of impurity near the surface of the semiconductor. This so-called "snowplow effect" results in the lowering of breakdown voltages on the surface of junctions diffused in under the oxide. It is possible, although usually not worthwhile, to avoid this effect by making guard-ring structures, in which a first diffusion forming a higher breakdown voltage is made in the form of a ring surrounding the diffused area, where the desired junction structure is subsequently prepared.

As an example of the interaction of impurities, Fig. 5-15 shows a cross-section through an *npn* structure made with a gallium base diffusion. One notices that the base-collector junction has penetrated more deeply under the emitter than over the rest of the surface. It can be shown by spreading resistance measurements under the emitter that this additional diffusion is not the result of rejection of impurities by the region of high phosphorus concentration, but rather is the result of an increased diffusion constant for gallium in the region near the collector junction.

This is very similar to the result achieved upon high-energy electron bombardment of a diffused layer.[19] The explanation presented in the case of the electrons was that bombardment increased the vacancy concentration beneath the surface and, hence, increased the diffusion constant. It is possible that strain from the extremely high phosphorus concentration at the surface results in an increased vacancy concentration below the emitter. The same enhanced diffusion under the emitter can be observed in structures with boron as the base impurity. However, because of the lower diffusion constant of boron, it is less pronounced. In *pnp* structures, where heavy-boron emitters are employed, no such enhanced diffusion at the collector junction is apparent.

In the case of boron, in particular, during diffusion in an oxidizing atmosphere, the total quantity of impurity atoms in the silicon is decreased. The reason is that boron is soluble in silicon dioxide, so that the boron on the surface is incorporated into the growing oxide interface. This loss of boron by outdiffusion results in large deviations from the simple diffusion models, especially when one is trying to predict the concentration profiles of deep layers from the depth and sheet conductance of shallow predepositions. The competing effects of oxide growth, outdiffusion and indiffusion are too complex to be taken into account analytically for routine control. Empirical curves again are strongly recommended.

3. Epitaxial Growth

As we shall use it here, the term epitaxial growth refers rather restrictively to the process by which a film of semiconductor material is deposited upon a substrate of the same semiconductor by a gas or vapor process. Though this is not really consistent with the original idea of epitaxy, it can be considered such from the point of view that the doping in the deposited film is usually considerably different from that in the substrate, for structures of interest in microcircuitry applications. Epitaxial growth is a relatively new addition to the semiconductor technology and, as such, has not yet been exploited to the extent of such well-developed technologies as solid-state diffusion.

Microcircuits containing epitaxially grown films are just at the laboratory stage at present. This technique offers what appears to be an extremely flexible method of obtaining impurity-concentration profiles, which are not accessible by any of the other established techniques. As such, it certainly has an important future in microcircuitry. The technology is in a state of very rapid development. In this state of flux, any description of its present standing will be obsolete in a short period of time. Accordingly, we will present a rather brief and general review of the methods and usefulness of the technique.

As with diffusion, several different systems have been employed for epitaxial growth. It has been accomplished by vacuum evaporation of silicon onto a heated silicon substrate;[20] it has been done in a closed-tube system, wherein a halogen is used to transfer semiconductor material from a source maintained at low temperature to the substrate at a higher temperature;[21] and open-tube systems have been employed, wherein silicon is deposited by the reaction of silicon tetrachloride or other chlorinated silanes with hydrogen upon the surface of a heated substrate.[22] The reaction will proceed readily in any of a variety of systems, providing only that the temperature is in the correct range and that the composition of the gas mixture is appropriate.

Single-crystal epitaxial growth of silicon on silicon has been reported over the temperature range from 930 to 1400°C. In general, in order to obtain single-crystal growth at the lower temperatures, it is necessary to employ relatively low growth rates. At temperatures above about 1200°C, growth rates of the order of 1 μ/min up to 5 μ/min result in useful single-crystal films.

Growth at higher rates leads to increasingly imperfect films and finally to polycrystalline deposition. The growth rate is determined primarily by the concentration of chlorinated silane in the gas feed. Usual mole fractions in the gas are of the order of 0.01. The temperature dependence of deposition rate is system-dependent. According to Tung,[23] the temperature dependence of the reaction corresponds to an activation energy of about 25 kcal/mole, while Corrigan[24] claims that in his apparatus no apparent dependence upon temperature is observed. In most systems, there seems to be a broad range of temperature in the vicinity of 1150 to 1300°C, where the effective growth rate is independent of temperature to a good approximation. Since this is also the desired temperature range in which to work to grow films of good quality, the importance of careful control of the substrate temperature is minimized.

Both hot-wall resistance-heated furnaces[25] and cold-wall r-f-heated reactors have been employed.[26] These include a wide variety of geometry and flow configurations. A fairly typical system is diagrammed in Fig. 5-16. It shows an r-f-heated reactor where the silicon substrate wafers are placed upon a silica-coated graphite block to which the radio frequency is coupled. The silicon tetrachloride is contained in a thermostated silica flask through which hydrogen can be bubbled, to

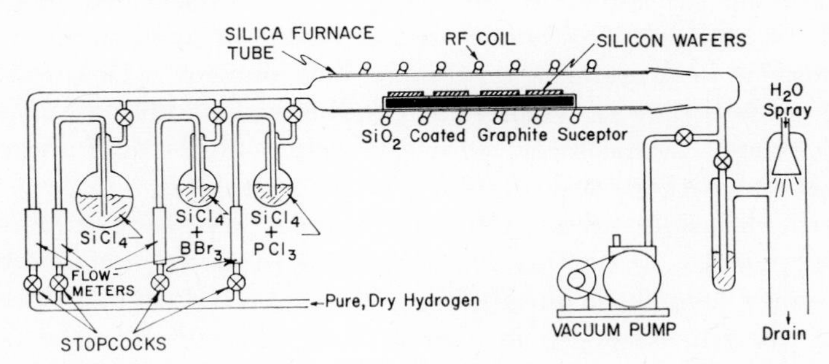

FIG. 5-16. A typical epitaxial system suitable for the preparation of doped-silicon layers.

saturate it with the vapor. Part of the flow of pure, dry hydrogen goes through the silicon tetrachloride reservoir, with the majority going directly to the reactor. By adjusting the relative flows, the partial pressure of the silane can be controlled to any desired value.

This particular apparatus also has provisions for other silane bottles, which are doped with boron tribromide and phosphorus trichloride. By putting a portion of the hydrogen flow through these doped sources of silane, it is possible to prepare films having *n*- or *p*-type impurities. At the exhaust, provision is made to assure a positive flow through the system by inclusion of an oil-bubbler trap. A water spray is shown for disposal of the silane. The hydrogen should be adequately vented, since relatively large quantities are employed. Both silicon tetrachloride and trichlorosilane hydrolyze violently. Care must be exercised in handling these materials, to be sure that they do not come into contact with any appreciable quantities of water.

In general, the properties of the film with which one must be concerned can be divided into three categories: thickness, doping profile, and perfection. The variables influencing each of these properties can be considered independently.

The thickness of the film is primarily determined by the growth rate, which depends upon gas composition and, to some extent, upon the temperature of the substrate and of the gas stream itself. In order to achieve a uniform growth rate over the entire area of silicon exposed, it is necessary that this area be subjected to a uniform gas composition. Because the reaction of hydrogen with the silicon tetrachloride proceeds rapidly and in several steps, this requirement of uniform gas composition is difficult to maintain. In fact starting with silicon tetrachloride through a single small reaction zone, it has been shown that a large fraction of the effluent gas, of the order of 20%, can be converted to trichlorosilane, with appreciable concentrations of other less heavily substituted silanes as well.[22] This being the case, it is important that, for uniformity, only a small fraction of the total available starting material be consumed during its residence time in the reactor.

In addition, it is necessary that the flow be as uniform as possible across the surface area. Achieving the desired flow geometry is pretty much a matter of trial and error. Not only is it necessary to concern oneself with the uniformity from wafer to wafer, but it is equally important to see that adequately uniform growth is obtained across the surface of a given wafer. Depending upon the flow pattern, it is possible to get a thicker deposit in the center of the wafer than at the edges, or even the opposite condition of a thicker deposit at the edges of the wafer than in the center. A very common type of nonuniformity is a wedging of the film from one side of the wafer to the other in cases where the gas flows across the wafer surface. When all the flow parameters are optimized, it is possible to obtain growth uniform to better than 10% over an entire group of 20 wafers or more.

The impurity concentration in the films can be controlled by doping the silane source, as is suggested in Fig. 5-16. For some of the structures of interest in microcircuitry, this control will prove to be quite satisfactory. In growing relatively lightly doped films on heavily doped substrates, however there is a considerable transfer of dopant from the substrate into the growing films.[27]

It is possible to grow n-type films with impurity concentrations in excess of 10^{16} atoms/cm^3 by using intrinsic silanes upon very heavily doped substrates, of the order of 0.005 ohm-cm. This transfer of dopant from the substrate to film can also involve the remainder of the system. In particular, any silicon or exposed graphite can act as a reservoir of impurity. The effect can often be minimized by inserting a backing wafer under the substrate wafer itself. In this manner, silicon from the backing wafer transfers to the substrate wafer during heating in the halogen-containing atmosphere, and serves to lock the impurity atoms in the substrate.

An alternative method employs a quartz plate which fits tightly to the silicon surface on the back of the substrate. This prevents appreciable gas diffusion from the back and, hence, minimizes transfer. One can also use a thermally grown oxide layer on the silicon itself to minimize the effect of transfer from the substrate. This gas-phase transfer can be appreciably more important than straight diffusion of the substrate impurity into the growing film. In all cases where heavily doped substrates are employed, one should test for this transfer of dopant.

Other ways of introducing desired impurities into the growing film have been tried. Small quantities of PH_3 or B_2H_6 can be added to cylinders of hydrogen for use as doping materials.[28] One of the principal advantages of this direct gas feeding is that it results in a low-inertia system wherein one can change dopants rapidly, which is very important in the preparation of multilayer structures.

Another method of feeding consists of the metering of the liquid silane into a flash evaporator, where it is picked up by the hydrogen gas stream. This is a positive method, and eliminates the change in doping level of a reservoir of silane, as it is fractionally distilled by hydrogen bubbling through it in the sort of system shown in Fig. 5-16.

The concentration of impurity in the film is not, in general, the same as in the original silane, since the impurity compounds have reaction rates different from the silane itself. In general, however, over a wide range of compositions, the concentration in the film is proportional to the concentration in the silane if the growth rate is held constant.

Many defects can appear in the growing film. These can usually be related to their cause from their general appearance. *Chevrons*, as shown in the left portion of Fig. 5-17, indicate too rapid a growth rate for the particular temperature being employed. *Pyramids*, as shown on the right side of this figure, are a defect which results from a number of causes, including impurities in the system and flaws on the substrate. The number of dislocations in the film is never less than that in the original substrate. On occasion, it can be several orders of magnitude higher. There are usually stacking faults which can be observed by lightly etching the surface of the film with a pitting etch. These have been proposed as a useful feature for the measurement of film thickness.[29] Polycrystalline-film regions indicate that the temperature is much too low for the particular growth rate employed or that gross contamination exists.

The preparation of the surface prior to epitaxial growth is an extremely important consideration. Both mechanically polished and chemically etched substrates can be used. In either case, it is usually preferable to give an additional in situ etching, either by soaking at high temperature in a pure hydrogen atmosphere or by gas etching with hydrogen chloride. The surface of the film reproduces quite well that of the original substrate. Scratches come through as scratches, although after the growth of a relatively thick film, the sharp features have been smeared out. Any orange-peel effect on the substrate resulting from chemical polishing is reproduced in detail.

The thickness of epitaxial films can be measured by weight gain, pro-

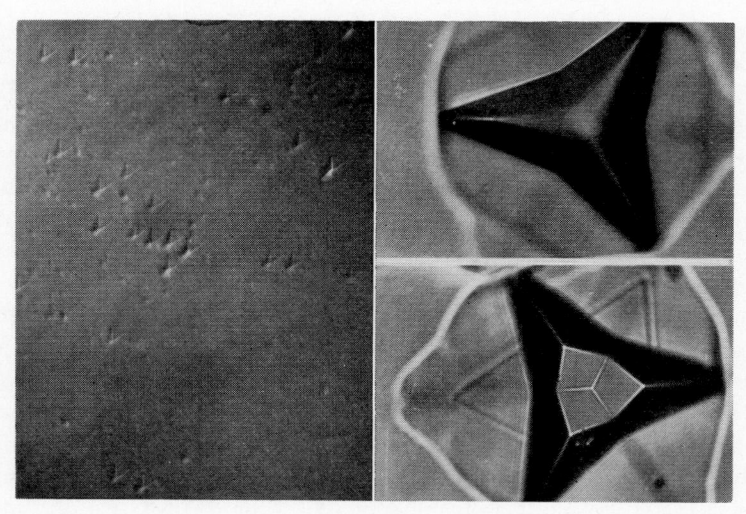

FIG. 5-17. Two typical defects often seen in silicon epitaxial films. Chevrons on the left result from too rapid growth. Pyramids on the right nucleate at the substrate-film interface. ($\times \sim 2000$)

viding the system is so arranged that growth occurs only on a known surface area; by beveling and staining, since the junction between the substrate and the film can usually be delineated by light acid-etching stains; and in the case of relatively highly doped films on heavily doped substrates of resistivity of about 0.01 ohm/cm or lower, the film thickness can be measured by observing Fabry-Perot fringes in the infrared.[30] These fringes result when the light reflected from the interface interferes with that reflected from the surface.

This last measurement technique will usually differ from that obtained by beveling and staining, since it really measures to the region in which the refractive index of the silicon is changing rapidly. Usually, because of diffusion, this is a micron or so into the film itself.

There is also the direct method of measurement wherein one measures the increase in thickness of the wafer with a precision mechanical measuring device. Of these several techniques, only the bevel and stain is destructive. Where there is a junction between the substrate and film, four-point probe measurements on the film give information concerning the average doping. However, where no such junction occurs, that is, where n-layers are grown on n-substrates or p upon p, this technique cannot be employed.

A good measure, and one that is directly relatable to the electrical properties desired, is obtained by measuring the breakdown voltage and capacitance of a diode-junction structure made near the top of the film. This technique is, however, time-consuming and destructive. It has been proposed to use measurements of the Seebeck voltage in order to obtain doping levels.[31] This, however, has not been reduced to a standard technique.

In any case, there is virtually always an impurity gradient in the films. This can be evaluated by measuring the capacitance of a diode as a function of voltage or by an interesting technique wherein one employs the diffusion of an impurity of known diffusion constant from a surface where a bevel has been cut through the film. By means of a second bevel and staining, one establishes a doping profile of the film in the region where the gradient is steep.

Silicon dioxide masks against epitaxial growth in certain temperature ranges. With proper control, one can get the deposit to go down selectively on the exposed areas of the silicon, while leaving the silicon dioxide film free of deposits and intact. As greater ability to make oxide-masked structures develops, and as diffusion and epitaxial growth are more closely interwoven, the range of structures which will be accessible for microcircuitry will expand tremendously. The importance of epitaxial growth in microcircuitry will certainly be much greater than in ordinary transistors, where the prime function is to reduce the effect of the large superfluous body of collector material necessary to give mechanical rigidity and ease of handling in the ordinary double-diffused transistors.

4. Surface-geometry Control

The existence of semiconductor integrated circuitry depends upon the ability to make fine-scale patterns in all three dimensions. Diffusion and epitaxial growth are used to control the depth of the structures. The surface geometry is controlled by masking operations. These are generally important in two areas: first, for making patterns in the oxide layer to serve for diffusion masks, and, second, for the deposition of the intraconnection metallization pattern.

It is also important that good geometry control be maintained for any thin-film elements deposited upon the substrate surface. The technology used to achieve this surface-geometry control was well developed for the manufacture of multiple-diffused transistor structures.

In the double-diffused mesa silicon transistor shown in Fig. 5-2, two precise indexed masking steps are required. The oxide-removal step prior to emitter diffusion in the figure must be registered with the metal-removal step. In addition, the masking for mesa etching must conform to the previous steps. However, this operation is usually less precisely controlled, and operates with larger tolerances.

In going to the planar-transistor structure, the number of indexed masking operations increases to four. As shown in Fig. 5-2, holes are first opened in the oxide before base diffusion. Indexed holes above the base regions are then opened before emitter diffusion. Additional holes are opened to expose the emitter and base contact region, and finally the metallization pattern necessary to achieve contact to the various layers is etched in index with previous maskings.

For the high-frequency transistors presently in production, the spacing between adjacent structural features is often as small as 0.0005 in. This means that alignment errors from one pattern to the next cannot exceed a few ten-thousandths of an inch.

In extending further to make semiconductor integrated circuitry, the number of indexed masking operations increases, as does the complexity of the patterns which must align. A simple integrated circuit might require five masking steps, while more complicated ones often employ six or seven. In the foreseeable future, as more pieces of the available technology are incorporated into single structures, the number of masking steps will undoubtedly rise still further. Thus, surface-geometry control is an extremely important part of the microcircuitry technology.

The most important techniques employed to achieve this geometry control are those of photolithography. These use a photosensitive material which can act as a mask against chemical etchants. The resists

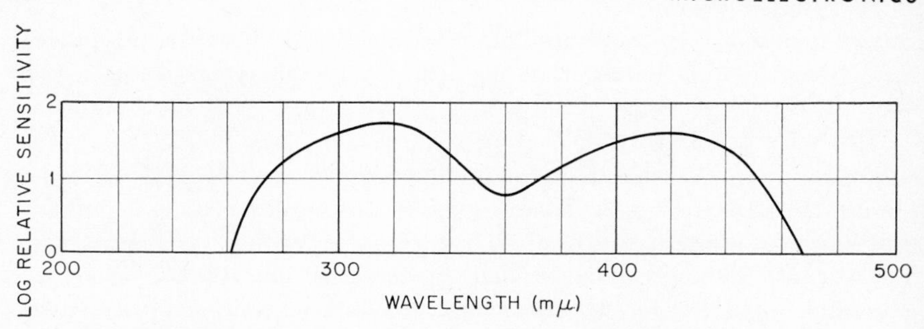

FIG. 5-18. Spectral sensitivity of KPR. (*Eastman Kodak Company.*[32])

are usually low-molecular-weight organic compounds which polymerize when subjected to ultraviolet radiation. The most common are commercially available compositions sold under such trade names as Kodak KPR, KMER, and KPL. Extensive application data on these materials are usually available from the manufacturer. For example, the spectral sensitivity of KPR,[32] one of the most common resist materials, is shown in Fig. 5-18.

The other method used in achieving geometry control is vacuum evaporation through a mask. Several variations of this exist. It is common in germanium mesa-transistor technology to employ double evaporation through a single aperture at different angles to achieve fine-scale structures. Other examples of evaporation through masks are given in Chap. 4 on thin-film microcircuitry. The remainder of this section will be concerned primarily with the technology of the photolithographic process.

Process. *Masks.* The first requirement for making fine-scale photolithographic patterns is the original mask that will be duplicated on the device surface. These are usually made from photographic emulsions either on glass plates or on flexible film bases. The artwork is done as a single pattern or small number of patterns at adequate magnification so that errors in drawing are minimized. This pattern is then reduced usually through at least two steps to the final dimensions desired. During this reduction, the pattern is transformed into an array of patterns either by a step-and-repeat procedure or by a multiple-lens camera. Figure 5-19 shows the original artwork and the final working masks for a typical pattern employed in microcircuitry.

The requirements placed upon the photographic equipment in order to achieve the desired precision of reduction are extreme. For example, in a pattern one-tenth of an inch across, a slight tilting in the plate holder corresponding to ½% variation in the focal length can result in a ½-mil size error, which is of the order of the maximum error band allowable from all considerations. Accordingly, rigid and precise equipment is needed for the production of quality masks. Figure 5-20 shows a typical copy camera employed in the preparation of photographic masks for microcircuitry. Several commercial suppliers for masks presently exist.

Care must be exercised to assure that masks have high contrast, and minimum pinholes in the dark areas or scratches and other blemishes in the clear regions. Ordinarily these masks are employed to make contact exposure of the photolacquers. They must be protected from damage and discarded when damage becomes appreciable.

Coating. After making the masks, the first step in performing the photolithographic process consists of coating the surface to be photoengraved with the unpolymerized photosensitive material. The general requirements of the coating are that it be uniform, adherent, and pinhole-free.

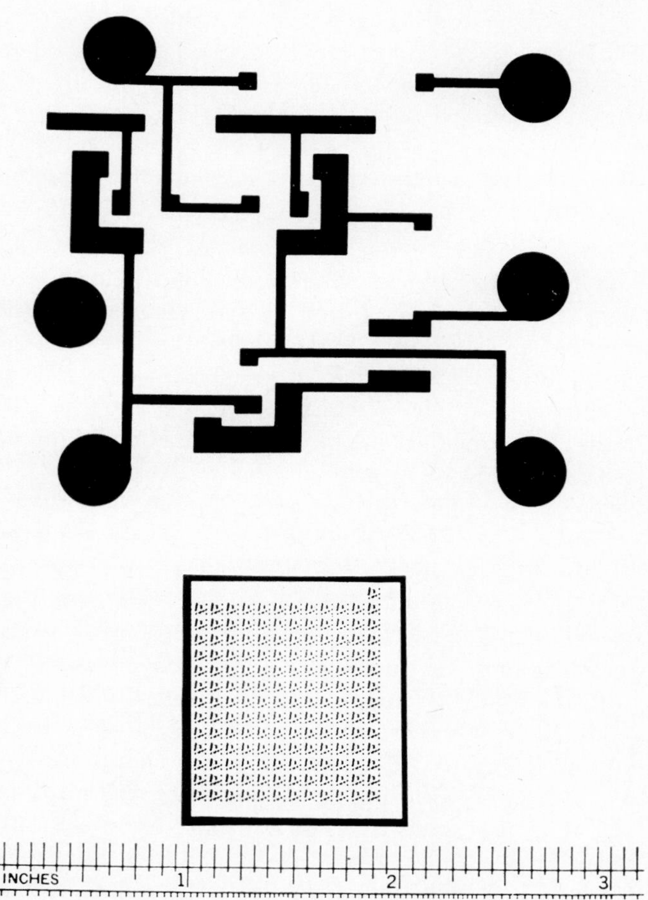

FIG. 5-19. An example of the original artwork and final working mask for a typical pattern employed in microcircuitry.

FIG. 5-20. A copy camera of the type useful in making high-precision photographic masks.

The most desirable thickness varies according to the particular engraving problem. For highest resolution, thin coatings are desirable. However, thinner coatings are more readily penetrated by etchants, and are more inclined toward pinhole-itis, a serious disease. The usual range over films employed in microcircuitry is about 0.5 to 2 μ. This is much thinner than that employed in conventional metal-etching applications, such as in the preparation of printed-circuit boards.

The film may be applied by spraying, dipping or by spreading with centrifugal force. For best results, the photosensitive materials should be filtered immediately before use. With the more viscous materials, such as KMER, a pressure filter is necessary. After coating, the film is dried thoroughly, and baked under infrared lamps or in a ventilated oven. These operations must be carried out in an environment with very little blue and ultraviolet light because of the photosensitivity of the material. Strong yellow light, however, can be employed.

Indexing and Exposure. The exposure of the photosensitive film can be done either by contact printing or by projecting the mask pattern. In either case, it is necessary to have a means for precise alignment of successive patterns. Two methods are employed. Either the masks are aligned precisely in a jig system, including index points against which the semiconductor wafer makes contact, or the system is arranged so that it is possible to observe the pattern on the wafer surface with light of wavelength outside the range of sensitivity of the film, and to adjust the position of the wafer for precise alignment before exposure to ultraviolet light.

Figure 5-21 shows a mask clamped in a frame suitable for contact printing. The mask is one of a set which has been aligned one with another. The two index points on the bottom of the jig frame make contact with a flat edge of the silicon wafer, while the circular arc on one side of the circle touches the flat side of the jig frame, thus assuring three-point contact. The wafer is pressed against the mask, either by a vacuum chucking arrangement or by other mechanical means. By this system, any damage to the index surfaces of the wafer, such as chipping during processing, results in loss of index. As in all contact-printing systems, the necessity of close contact between pattern and film results in damage, both to the film and to the mask from any dust particles or surface irregularities on either.

FIG. 5-21. A photographic mask in an indexed frame. The silicon wafer makes contact with the mask frame at the two semicircular bumps and at one flat edge.

Alignment by adjusting the wafer to the pattern can be accomplished by jigging under a microscope or vertically illuminated projector. Means must be established for separating the wafer and mask slightly during the time the alignment adjustment is being made, to minimize film damage.

This method has the advantage of allowing broken wafers and wafers with chipped edges to continue through the process. This is an especially important advantage in developmental work in microcircuitry where there are a large number of operations, increasing the probability of damage to the wafers. It is worth considering as a production technique. It is, however, appreciably slower than a well-jigged mechanical alignment scheme. In either the mechanical or the optical alignment schemes, the wafer and mask can be clamped together and removed to an exposure fixture. It is, however, convenient to build the exposure arrangement into the apparatus. This is readily done in the mechanical jigging by having an ultraviolet source provided with a shutter on the opposite side of the mask from the wafer.

A convenient arrangement has the optical axis at about a 45° angle to the horizontal. In this manner, gravity helps hold the wafer against index points during exposure. In the case of optical alignment under a microscope, it is possible to replace the usual vertical illuminator with the necessary high-intensity ultraviolet source for exposure.

This is usually best done by inserting an additional mirror into the system. It is, of course, necessary to remove the objective lens of the microscope in order to cover the entire wafer area. It is usually possible to modify the nosepiece of the microscope sufficiently to obtain uniform illumination over a circle at least one inch in diameter.

In projection printing, one images the mask through a high-quality lens onto the silicon wafer, which again is either located by preset mechanical index points, or an arrangement is made so that the image can be observed through an appropriate filter while the alignment is adjusted. Projection printing has the important advantage of preventing those problems caused by contact of the mask emulsion and photosensitive coating on the wafers. It does, however, require a very high-quality lens to obtain adequate resolution over the entire pattern.

The high-intensity ultraviolet light needed for the photolacquers can be obtained either from a carbon-arc lamp or from a high-pressure mercury arc. If the light from the source is collimated, or if it comes from essentially a point source, one obtains somewhat better definition than with diffused light, especially if contact between mask and wafer is not close. However, as in conventional photographic enlarging, collimated light, though it improves the resolution, increases the number of imperfections in the negative which reveal themselves in the print. The exposure time should be adjusted to achieve clean development in the unexposed regions and no lifting in those areas which have been exposed. These photolacquers are all extremely high-contrast materials. There is essentially no gray scale.

Development. In addition to proprietary developed solutions sold by vendors of the photosensitive materials, common laboratory solvents can be employed. These include any of the solvents for the monomers, such as methyl Cellosolve acetate, trichlorethylene, and xylene. These are used by dipping, spraying, or in a vapor-degreaser arrangement. Best definition with KPR has been obtained by employing a two-stage development process, wherein the film is first developed in an active solvent, such as trichlorethylene, followed by a less active one, such as xylene or isopropyl alcohol. Suppliers of the photosensitive lacquers are aware of the problems in the semiconductor industry, and can offer considerable application information.

Etching. For etching patterns in silicon dioxide, the usual etchant is buffered HF. A typical solution is 10 m NH_4HF_2 brought to a pH of 5 by the addition of HF. The exact composition can be varied appreciably, yielding faster etching rates for more acid compositions. Too low a pH can, however, result in lifting of the polymerized resist film. For metal films, other etchants in general are needed. With aluminum, a solution of 10% sodium hydroxide used with KPR is satisfactory. With other materials, KMER is usually more suitable since it is more resistant to acid etches and can even stand aqua regia for short times.

Stripping. Assuming the polymerized film has remained in place during etching, one is next faced with the task of its removal. Stripping procedures include softening the film by boiling in C_2HCL_3, followed by scrubbing in acetone, or by heating in mineral acids such as HNO_3 or H_2SO_4. Hot chromic acid, the classic "cleaning solution" for organic residues, will remove the film. However, it is well known that treatment with this solution leaves adsorbed chromium on glassware that is extremely difficult to clean off. There are proprietary pattern strippers of unknown composition which are somewhat effective in stripping the polymerized resist. These are not recommended because of the uncertainty in their composition and in the residues which might remain upon the etched surface. When stripping resist films from soft metal films, such as aluminum, care must be exercised so as not to damage the metal film. This is especially true when one employs the scrubbing technique.

Use of Resist Materials as Evaporation Masks. Occasionally for films it is more convenient to put the resist pattern on the wafer before deposition of the film. This would be true, for example, with a metal that was difficult to etch, such as Nichrome. In this case, evaporation covers the film and the exposed areas of the substrate. In the subsequent stripping operation, the resist film is removed with that metal which is on top of it, while those areas which were exposed during evaporation remain coated, hopefully, with the metal. The principal disadvantage to this technique is that the resist film limits the outgassing that the substrate can receive before evaporation. This can result in an adhesion problem.

Capabilities and Limitations. Definition by the photolithographic techniques of the order of 0.1 mil (2.5 μ) over an entire pattern is about the present state of the art. This means that a 0.001-in.-wide resistor is $\pm 10\%$ at best. Clean 0.1- mil lines have been made.[33] However, these employed a mask which was appreciably narrower than the resulting stripe.

5. Metallization for Intraconnections

The thin metallic film employed for intraconnections over the silicon dioxide layer must meet several general requirements. It must adhere to the silicon and to the silicon dioxide; it must not penetrate the silicon dioxide; it must make ohmic contacts to both *n*- and *p*-type silicon; it must be of adequately low sheet resistance and, in addition, it must supply a means for connecting leads for interconnection of the circuit blocks. Evaporated films of aluminum similar to those commonly used in transistor contact areas fulfill these requirements, providing that all the n-type areas to be contacted are heavily doped. Film thicknesses employed are usually in the range of 0.1 to 1 μ. The sheet resistivity of such films is of the order of 10^{-2} to 10^{-3} ohm/square.

The possible problem of reaction of aluminum with silicon dioxide because of the greater affinity of the aluminum for the oxygen does not prove to be a real problem at the operating temperatures experienced by the microcircuitry. This reaction at the interface does, however, yield excellent adherence. At temperatures above the melting point of aluminum, $\sim$660°C, the oxide is penetrated rapidly. In fact, at all temperatures above the aluminum-silicon eutectic temperature at 577°C, where a liquid phase can form, the rate of penetration is appreciable. Extended life tests, however, at 300°C with thin films of SiO_2 show that penetration at these temperatures is extremely slow.

The principal disadvantage to the aluminum metallizing is that the soft aluminum film is very susceptible to mechanical damage. It can be scratched easily during the normal processing operations—a problem which is much more severe in the microcircuitry than in ordinary transistors.

The aluminum is usually deposited in a conventional evaporator by heating on a refractory metal filament. Good adherence requires a well-cleaned, grease-free surface. Ionic bombardment with an inert gas, such as argon, prior to evaporation is helpful in obtaining reproducible results. After photoetching the pattern desired with sodium hydroxide etch, employing a photoresistive film, it is usually desirable to heat-treat the structure to the vicinity of the aluminum-silicon eutectic temperature briefly, in order to assure a firm bond between film and substrate.

The principal problem with aluminum aside from the ease with which it can be damaged mechanically is that it limits the temperature at which the microcircuit can be subjected. This limitation prevents the inclusion of certain potentially desirable packaging techniques. However, the well-established reliability of thermocompression bonds properly made to the aluminum film often outweighs this objection.

Other types of metallization have been considered. An interesting development in this direction was reported by J. Langdon et al.,[34] who used silver-chromium metallization, which was subsequently covered with a glass whose expansion coefficient matched that of the underlying silicon. Contacts were made by using holes through the glass to contact points, where appropriate metallic buttons were raised in place. Assembly techniques such as this offer the advantage of complete mechanical protection with essentially no external package. By interconnecting such structures directly on a mating metallized pattern—a sort of micro-printed-circuit board—one can hope to achieve very high packing densities at minimum expense.

6. Assembly and Packaging

The techniques employed to assemble and package the microcircuits have an important bearing on their usability and reliability. No single factor is more important in influencing the reliability of planar-semiconductor integrated circuits than the packaging of these circuits, since the residual failure modes are generally found to be mechanical. The close relation of packaging configuration and usability is obvious. While the input of the system manufacturer is necessary to achieve proper external configurations suitable for efficient system assembly, at least equally important in choosing a package design is careful consideration of the various materials and materials-assembly techniques, which go into its construction.

The basic requirements any package must fulfill are: first, that the circuit be secured mechanically; second, that the necessary connections required in the circuit's use be made available; and, third, that the circuit be protected as required from its environment. The various packaging schemes considered to date include the considerations of other boundary conditions, depending on the particular use to be emphasized. For example, some are oriented toward ease of systems assembly, while others emphasize small size and light weight. Two packaging schemes employed at present for semiconductor integrated circuits are worth describing in some detail. The technologies used for assembly in these cases are somewhat different, and so will be described in conjunction with a particular package configuration. A modified version of the standard transistor package has been employed where maximum reliability and usability are desired, since such a package makes maximum use of proved transistor technology and printed-circuit-board assembly.

An integrated circuit in such a package prior to sealing the "can" is shown in Fig. 5-22. The small, flat "bug" package is suitable where size and close packing are of paramount importance. Figure 5-23 shows such a structure. The multilead transistor package consists of a Kovar eyelet with a matched glass-to-metal seal, through which the leads are sealed, often with one lead welded to the bottom of the metal eyelet. In most cases these leads are of the same general configuration as the TO-5 transistor header with a 0.200-in. pin circle. Up to ten leads can be used, although a maximum of eight is more common. On occasions, the smaller TO-18 size header with eight leads maximum is employed. In either case, the silicon die is attached to the gold-plated Kovar header, using a gold solder or a gold alloy if electric contact is desired.

FIG. 5-22. Microcircuit mounted upon eight-lead transistor-type header. The eighth lead is spot-welded to the bottom of the metal eyelet. The top of the can has been cut away.

FIG. 5-23. A semiconductor integrated circuit in a flat package ⅛ in. wide by ¼ in. long. (*Texas Instruments, Inc.*)

To make contact to *n*-type silicon gold with about ½% antimony is appropriate. For *p*-type material, it is more difficult to make ohmic contact, although gold with several per cent gallium works with relatively heavy *p*-doping in the silicon itself. The solder-down operation is done in an inert atmosphere, either by a hand operation holding the silicon in tweezers, or through a jig operation employing a belt furnace. The operation proceeds easily, providing the silicon surface is free of oxide. The temperature during this operation is in the vicinity of 400°C. Gold-solder preform thicknesses of the order of 1 mil are typical.

Since the penetration into the silicon during the alloying is to a maximum of about 20% of the preform thickness when penetration is uniform, the circuit structure built within the top fraction of the die is undamaged. The solder joint is strong and of low resistance. Since the Kovar eyelet matches the thermal-coefficient expansion of silicon quite closely, there is a minimum amount of thermal strain introduced by this joint.

Such a soldering with gold-silicon alloy is well established in the transistor industry. The principal problems one is likely to encounter are the occurrence of voids and carrier injection into high-resistivity material. Since microcircuits, in general, do not handle high currents, these are not particularly important problems here.

Bonding of leads in this packaging is usually by thermocompression bonds. To make such a bond, the structure is heated to above 300°C, but below the gold-silicon eutectic temperature of approximately 370°C. About 350°C is the preferred operating temperature. Soft metal, usually gold but occasionally aluminum or silver, is compressed against the metallization. The tool employed for compression can be a wedge, but best bonds are obtained when a capillary tube is employed to make a nailhead bond.

In the latter case the gold wire is fed through a small glass capillary tube. The hole in the capillary should be just large enough to let the wire pass freely. A ball is made on the end of the wire by melting a portion of the wire with a small flame or heated platinum wire. Typically for 2-mil gold wire, the diameter of the gold ball is approximately 4 or 5 mils. The ball is positioned over the bonding pad over the device metallization, and pressed firmly against the pad so as to result in about a 50% deformation of the ball. Under these conditions a firm bond between the aluminum metallization and the gold wire is achieved. The other end of the wire can then be thermocompression-bonded to the gold-plated Kovar leads or it can be resistance-welded.

The device is hermetically sealed in an inert atmosphere by resistance welding of a flange on the Kovar eyelet to a metal cap. Well-made seals of this type show no leakage to the most sensitive leak-testing techniques available today. By pressurization with radioactive krypton and counting under a scintillation counter, devices of this type have been shown to be producible on a production basis with leak rates less than 10^{-12} standard cm^3/sec. Reproducible achievement of this seal has been the result of considerable development in the glass-to-metal sealing techniques, as well as the precise geometry and conditions for achieving reproducible welds.

Advantages of this type of package are that it employs only proved techniques carried over directly from transistor technology. The package and techniques have been evaluated through extensive life testing. In addition, conventional printed-circuit-board assembly techniques can be employed, although for best efficiency it is necessary to use two layers of conductors.

The disadvantages of this package are that it limits the number of leads to eight, or a maximum of ten, and it does not use space efficiently, since the integrated circuit die represents only a small fraction of the total volume of the package. This latter limitation can be alleviated appreciably by going to a multilead variation of the TO-47 package. In this case, the Kovar eyelet is replaced with a solid Kovar slug, and over-all package heights of 0.060 in. can be achieved.

In order to get around these shortcomings, the flat package with coplanar leads has been used. The one illustrated in Fig. 5-23 is a ¼- by ⅛-in. rectangle approximately 0.040 in. thick. As can be seen in the photomicrograph, this configuration can have a maximum of 14 leads—five coming out each side and two out each end. The leads are ordinarily on 0.050-in. centers. Such a package obviously makes much more efficient use of volume.

Assembly in this package varies. The die can be attached to a metallized ceramic base plate with metal solder, similar to the die-attach operation in the modified transistor package; or, in cases where electric contact to the bottom of the die is not desired, a low-melting glass can be employed to glue the integrated circuit to the package base. Lead attachment again is usually by thermocompression bonding, although, because of the limited vertical space available, use of the superior ball-bonding technique is severely limited. Here a wedge-shaped tool is more common for compressing the wire from the side upon the metallized areas. Because of the limited vertical dimension, care must be exercised to assure that no short-circuiting occurs between wires or between a wire and an undesired portion of the semiconductor die or package.

The sealing of these packages presents a rather formidable problem. Attempts to make such packages with flanges for resistance welding are thwarted by the lead positions. Solder sealing of the top has never proved to be a reproducible operation. The best technique available, at present, appears to be sealing with a low-melting glass. Unfortunately, the internal structure limits the maximum temperature to which the structure can be subjected, and so the choice of glasses for sealing is severely restricted. Both metal and ceramic covers have been used. The metal cover has the disadvantage that it can short-circuit the internal wires if extra insulation is not provided. The ceramic top adds more to the over-all package thickness, but at the present appears to be the best all-around solution

The advantages of this package are small size, light weight, and suitability for stacking into modular configurations. The disadvantages are that it is often expensive and that the new assembly techniques required introduce real and potential reliability problems, each of which must be investigated extensively before one can assure the same reliability as is attainable with the transistor-type package. System-assembly techniques to take full advantage of the size and configuration of this package are expensive, and appear useful only for small special-purpose systems. However, increased acceptance of this general configuration suggests that appreciable progress in its use is being made. An additional disadvantage of this particular package is that the ¼- by ⅛-in. size is too small for the largest microcircuits available today. As our ability to make still larger integrated circuits improves, it will soon outstrip the available package area.

Of course, this technique is not limited to the ¼- by ⅛-in. dimension. Other flat-package configurations are useful for special purposes. For example, a circular flat package with radial leads offers the advantage that the further one goes from the package center, the greater the separations between adjacent leads.

Other schemes for packaging integrated circuits exist. For example, the inclusion of integrated circuits in micromodule wafers (see Chap. 3) has been considered. Such a combination of integrated circuitry with the micromodule assembly technique would result in an extremely high effective component density.

The ultimate scheme for packaging and interconnection involves the complete elimination of separate packages for individual components or circuits and the packaging of the system as a whole. Such efforts as mentioned in the end of the section on metallization, wherein the structure is coated with an insulator through which holes are formed at the appropriate spots for interconnection, is moving in exactly this direction.

5-3. DEVICE STRUCTURES

The performance of microcircuitry is determined by the structural features employed for achieving the desired electrical functions. This section will be devoted to examining various of these structures, and discussing how they might be made from the existing technology, as well as the limitations and compromises which must be considered.

In general, semiconductor integrated microcircuitry is made from discrete regions which are associated with particular electrical functions. Although there is not necessarily a one to one correspondence between the regions of an integrated circuit and the components one might employ to make such a circuit by conventional means, such a correspondence can often be made. Some distributed effects can be employed. For example, distributed resistor-capacitor networks are sometimes used. In addition, one is forced to consider many parasitics which occur as basic parts of the structure itself. These can be used to advantage, although one is usually more concerned with minimizing their effect upon performance. With both active and passive elements, one must also achieve, within the integrated circuit, the functions assigned to the printed-circuit board in conventional electronics: that is, those functions of interconnection, insulation, and mechanical support. An appreciable part of this section will be devoted to these benign electronic functions.

1. Transistors

Since the whole concept of semiconductor integrated circuitry arose from and is significant because of the transistor, we will discuss transistor structures first. Figure 5-2 has shown schematic cross sections of conventional mesa and planar transistors. In such a schematic cross section, the geometry is considerably distorted in order to show the emitter and collector regions. Actually, the emitter and base layers penetrate only a few per cent of the thickness of the wafer. The oxide and metal films are also of the order of 1% of the over-all thickness. All the drawings of transistorlike structures will have this distortion in order to illustrate the important features while still showing the entire wafer thickness.

The steps necessary to make the two transistors shown in Fig. 5-2 are given in Figs. 5-24 and 5-25. In the mesa device, the original wafer is diffused over the entire top surface with the desired impurity element—for an *npn* transistor, either boron or gallium. The oxide growing during diffusion is photoetched to form a pattern of apertures, through which

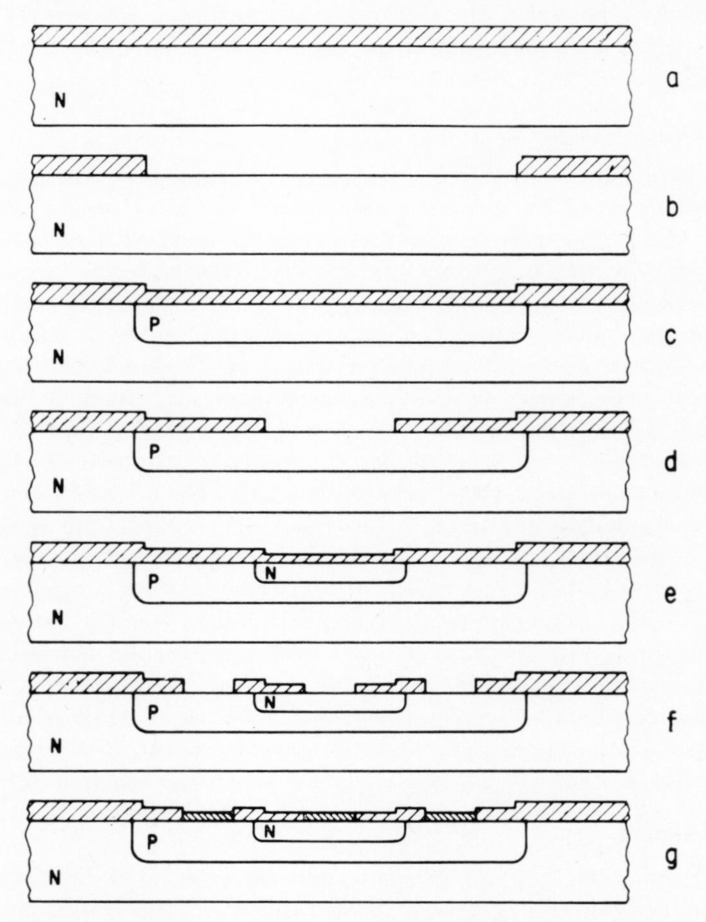

FIG. 5-24. Schematic cross sections of a single planar-transistor structure after various processing steps: (*a*) After oxidation; (*b*) after base oxide removal; (*c*) after based diffusion; (*d*) after emitter oxide removal; (*e*) after emitter diffusion; (*f*) after contact oxide removal; (*g*) after contact metallization.

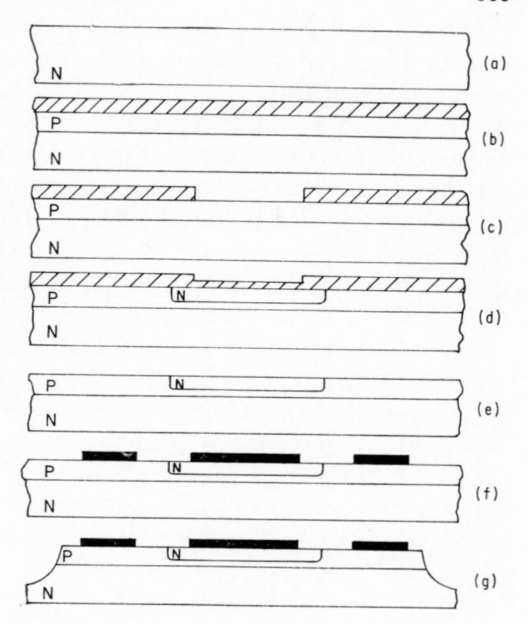

FIG. 5-25. Schematic cross section of a single double-diffused mesa transistor after various processing steps: (*a*) Original wafer; (*b*) based diffusion; (*c*) emitter oxide removal; (*d*) emitter diffusion; (*e*) oxide stripping; (*f*) contact metallization; (*g*) mesa formation.

the emitter is then diffused. The oxide is removed from the wafer, metal is evaporated, a second mask is applied and indexed with respect to the emitter dots, and metallized base and emitter contacts are made. The mesa is then cut by a third indexed operation, although usually with looser tolerances than with the first two.

For the planar transistor structure, the first step is an oxidation, followed by the opening of holes in the oxide for the base diffusion. An oxide-masked base diffusant such as boron is used, followed by an indexed oxide removal to open the aperture for the emitter diffusion. After emitter diffusion holes are opened above the base and emitter contact areas, and subsequent metallization and photoengraving leave the desired metallizing pattern in these areas. The advantages of the planar structure for microcircuitry are of sufficient magnitude that only the planar structure need be seriously considered for microcircuitry transistors. These advantages are associated with the facts that junctions are all covered by an insulating oxide layer and that the surface is relatively flat.

For integrated circuits, the same four masking operations and two diffusion steps can be made to yield a transistor structure identical with any isolated planar transistor. The differences in the performance one might get with such a transistor in an integrated circuit, as compared with the identical structure as an isolated component, are associated with the environment into which the transistor is placed.

Other applications of the technology to make transistor structures can also be considered, although they have not yet been employed. For example, planar transistor structures can be achieved by epitaxial growth of the base layer, followed by diffusion through the grown layer to limit the extent of the base region, followed, in turn, by a second diffusion to make the diffused emitter. In the final analysis, the structure itself is the important consideration, rather than the technology by which it is achieved.

2. Diodes

Diode structures follow easily from the transistor structures. Both planar and mesa diodes are made as isolated components, and again the planar structure is more applicable for integrated circuitry. Either the emitter-base or the collector-base diode of a transistor can be used for the two-terminal structure. In addition, it is possible to make zeners using, for example, the emitter-base diode of a transistor.

3. Isolation

Only a few very simple circuits can be made from transistors with all collectors common. In the usual array of transistors, such as that shown in Fig. 5-1, this common-collector configuration occurs naturally. This fact accounts for the popularity of Darlington amplifiers in integrated circuitry. To get flexibility in circuitry design, it is necessary to achieve electrical isolation of the individual component structures. In conventional electronics, this is done by cutting the device apart mechanically. In the chip approach, similar isolation by separation is also employed. In the integrated structure, achieving adequate isolation is one of the principal problems. Many schemes are possible. Nearly all depend upon the inclusion of extra junctions, so that regions which must be isolated are separated in operation by a reversed-biased diode.

The simplest scheme, in principle, is to produce a series of islands of one conductivity-type material in a wafer of other conductivity type, and to insist that the substrate wafer is always so biased with respect to the rest of the circuit that the junctions separating the islands from the wafer are never forward-biased. This can be achieved in several ways, each with its own advantages and disadvantages. An example of how this might be achieved is shown in Fig. 5-26.

An original wafer of p-type material is oxidized, and holes are photoengraved through the oxide layer. The diffusion of n-type islands is performed with a surface concentration held adequately low so that the subsequent diffusion of device structures into the islands can be accomplished. In the particular example in Fig. 5-26, the steps necessary to build a transistor in the island are shown.

Again the base and emitter diffusions of the transistor can correspond to those of a conventional planar device. One can appreciate the effects of this type of isolation by considering the limitations it introduces in the performance of the transistor structures. First, in order to achieve the diffusion of the isolation region, it is necessary that a concentration gradient be established, resulting in relatively light doping near the junction compared with that near the surface. Since all the collector current must flow laterally through this lightly doped region to the collector

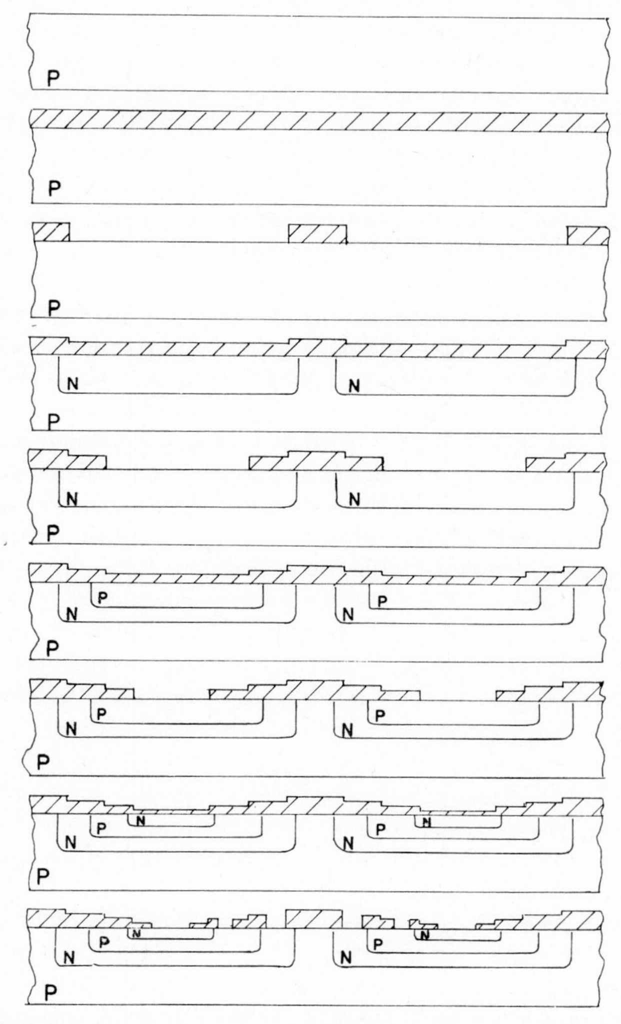

FIG. 5-26. Schematic cross section showing an example of how two transistor structures isolated by reverse-bias junctions can be achieved in a single block of silicon by the use of three diffusion steps. The crosshatched area is silicon dioxide.

contact made in the surface, a very large collector spreading resistance is added. This might well run of the order of thousands of ohms, and can be reduced to the order of hundreds of ohms only with considerable difficulty.

Second, the isolation junction acts as a capacitor between the collector and substrate, which couples this region to every other isolated region in the structure. Third, because of the necessary surface concentration for the isolation diffusion, the concentration gradient across the base-collector junction at the surface is appreciably higher than would be necessary for the isolated device, resulting in a lower base-collector breakdown voltage than would otherwise be required for the equivalent transistor. Hence, the electrical parameters associated with the separate transistor structure itself have been considerably degraded. The structure, however, would still be useful for relatively low-voltage low-current circuitry. An example of circuitry employing this type of isolation is the series 51 Solid Circuit family produced by Texas Instruments, Inc.[35]

An alternative method of achieving isolation is shown in Fig. 5-27. Two versions of this basic technique are illustrated. In this case, the isolated islands in which the components are subsequently constructed consist of the original wafer doping. Isolation is achieved by diffusion of the opposite polarity completely through the wafer, so that the diffusion fronts intersect.

At the left side of Fig. 5-27, a scheme is shown by which masking on both sides of the wafer is employed, yielding isolated islands completely penetrating the wafer, while on the right side, a structure is shown wherein diffusion takes place from the entire back surface, resulting in an array of isolated islands, which penetrate roughly halfway through the wafer.

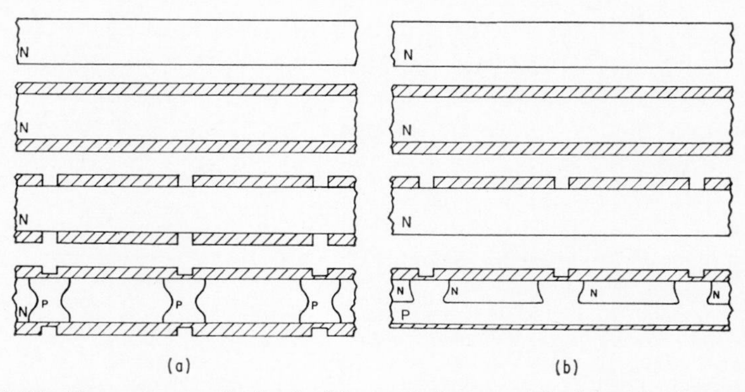

FIG. 5-27. Alternative methods of achieving isolation. The left side shows a scheme involving oxide masking on both sides of the wafer for preparation of a grid through the wafer. In the right portion, the oxide masking is employed only on the top surface, while diffusion takes place from the entire back surface. These isolated regions can then be used to produce the desired device structures.

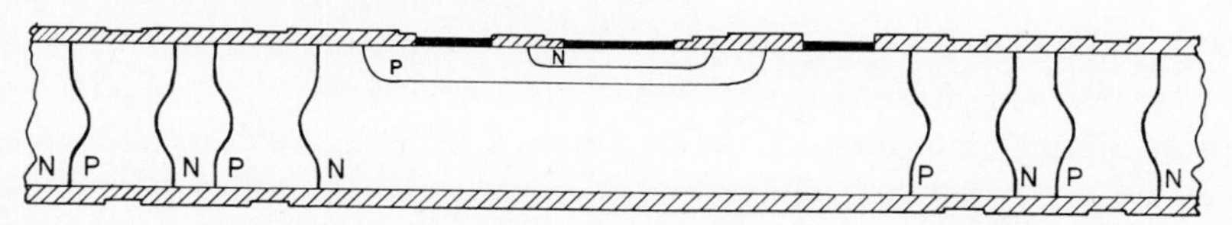

FIG. 5-28. Multiple-grid isolation suitable for the reduction of isolation leakage and capacitance.

The advantages of this technique over that above result because the transistor is made in a region of the original bulk doping as it would be for a separate component. This results in breakdown voltages comparable to that of the separate device; and, since the doping under the collector junction of a transistor can be appreciably higher, it is easy to obtain lower collector spreading resistances. This is especially true in the structures shown on the left, if one is willing to make collector contact on the opposite side of the wafer.

Disadvantages of this structure are that the diffusion through the wafer requires thin wafers and long diffusion times with high surface concentrations, resulting in high isolation capacitance. Also, the large junction area associated with these isolated structures makes it somewhat difficult to achieve good diode characteristics, resulting in relatively poor isolation. However, in the cases where capacitance and degree of isolation are important, it is possible to employ a multiple grid, such as that shown in Fig. 5-28, wherein more junctions are employed in series to decrease both the leakage and the capacitance, at the expense of an increase in area necessary for a given isolated structure.

By employing epitaxial growth, it is possible to make an isolated structure combining the advantages of both of the structures described above. Such a structure is shown in Fig. 5-29. By starting with a high resistivity *p*-substrate, one can minimize the capacitance per unit area of the isolation junction. The *n*-type film can be grown to a desired concentration or, to obtain maximum advantage, can be grown with a concentration somewhat heavier initially and lighter near the surface.

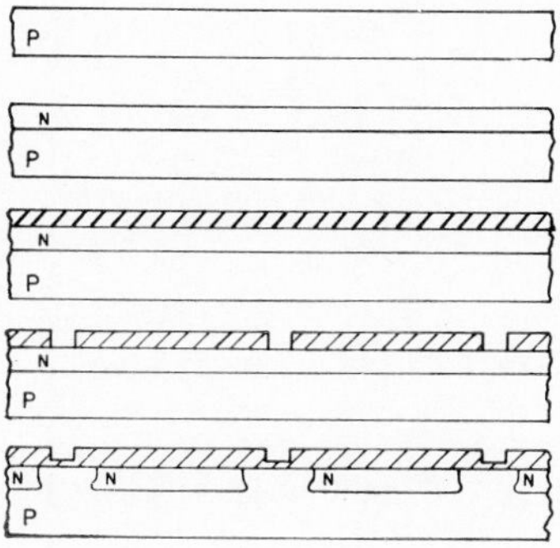

FIG. 5-29. Isolation structure employing an epitaxially grown layer. The five steps top to bottom show schematic cross sections after substrate preparation, epitaxial film growth, oxidation, oxide etching, and isolation diffusion, respectively.

The limitation on the concentration at the bottom of the film is that one must be able to diffuse through it to achieve isolation. In Fig. 5-30, typical impurity profiles for the three different isolation techniques described are compared, each with the same transistor structure included. In the diffused collector, one sees that there is relatively low doping in the collector region, corresponding to the high collector spreading resistance described above. With the diffused isolation, appreciably more doping is available in the collector region, although the isolation junction has high capacitance per unit area, because of the necessarily high diffusant concentrations necessary to completely penetrate the wafer. In the epitaxial collector with diffused isolation, however, conductivity in excess of that for the diffused isolation can be achieved, combined with the isolation capacitance per unit area of the diffused collector structure. The price one must pay for those advantages is the inclusion of epitaxial growth in the processing.

A variety of other schemes for achieving isolation have been tried. These include mesa etching in various configurations, and some attempts to remove the silicon between structures mechanically without the loss of index. For example, structures have been made wherein the regions of semiconductor are isolated by etching through from the back side with

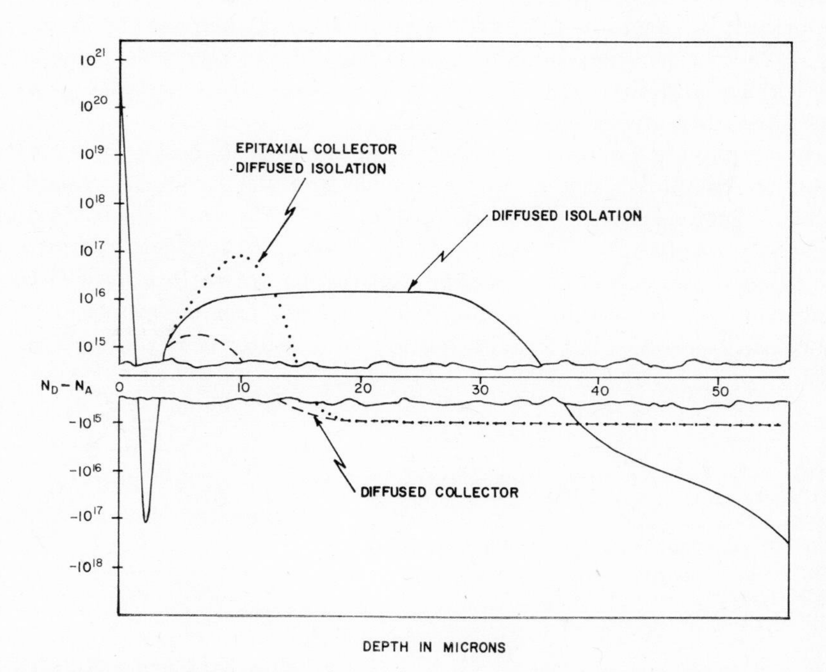

FIG. 5-30. Comparison of typical impurity profiles for three different isolation techniques.

selective etch that does not attack the silicon dioxide on top of the wafer. While the performance of such isolation is excellent, the process difficulties are still beyond the state of the art.

As the technology develops other isolation structures will arise. For example, selective growth of regions by epitaxy similar to oxide masking of the diffusion will open a variety of possibilities. Similarly, integrated circuits wherein the original substrate constitutes a common emitter rather than a common collector are likely to become producible structures. A large fraction of the improvements to be made in integrated circuitry in the next few years will be the direct results of improved schemes for achieving electrical isolation within a single block of semiconductor material.

4. Intraconnections

Equally important as the ability to isolate structures from one another is the ability to connect these structures in the desired circuit configurations. In the earliest integrated circuitry, such as the example shown in Fig. 5-31, this connection was accomplished by means of thermocom-

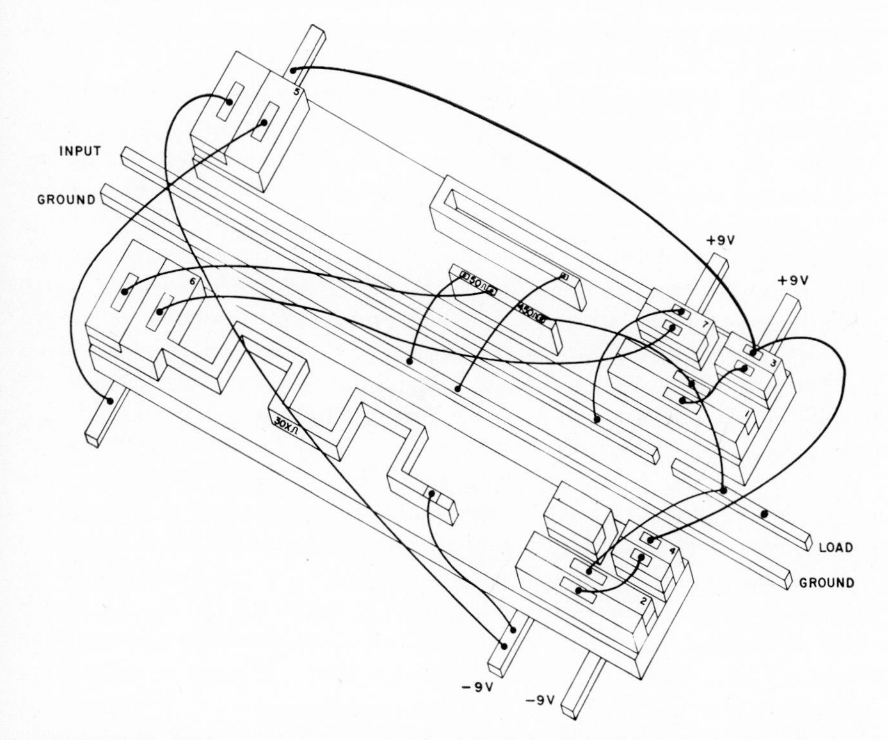

FIG. 5-31. Diagram of an early integrated circuit employing mesa structures and thermocompression-bonded wire for intraconnection. (*Texas Instruments, Inc.*)

pression bonding of fine wire, similar to that employed in standard transistor fabrication, wherever it was necessary to use something other than the contiguous silicon substrate itself.

This type of intraconnection discards a large portion of the potential cost advantage adherent to the integrated-circuit approach over the separate components and, in addition, introduces considerable opportunity for failure, since the most common failure mode of modern semiconductor devices is the opening of a lead. It was to a large extent, however, dictated by the mesa structure. The advent of the planar structure made it possible to intraconnect by means of thin films passing over the silicon dioxide insulating layer, making contact to the underlying substrate where appropriate. With oxide layers of the order of one micron in thickness, these intraconnections have a capacitance of the order of 0.016 pf/mil^2.

Aluminum is the most commonly employed intraconnection metal. It sticks well, does not penetrate the oxide at operating temperatures, and can be made to make ohmic contacts to the underlying silicon. The evaporated films are usually a few tenths of a micron thick and have a sheet resistance <0.01 ohm/square. Therefore, the resistance contributed by an intraconnection lead across even a large integrated circuit is only a fraction of an ohm, even for stripe widths as small as 1 mil.

In the intraconnection of complicated circuits, it is often necessary to cross leads. This can be done readily by utilizing the underlying silicon. Circuits of considerable complexity can be designed where all of the lead crossovers are done making use of structural features necessary from other considerations. For example, it is often convenient to run leads across a diffused resistor. Such features as split collector contacts can also accomplish the lead-crossing function. If it is necessary to add a completely separate lead crossing, one can make a transistorlike structure in a separate isolated region, and make use of the relatively low spreading resistance in the emitter diffusion. The 2 ohms/square sheet resistance usually employed for this diffusion allows one to make lead crossing with only a small resistance contribution, while maintaining the size reasonable.

Intraconnection in the chip approach is the major problem. The usual technique makes use of thermocompression bonding to each of the components. This is actually a microassembly job of circuit components without their normal packages. Considerable work has been done in an attempt to locate the separate chips precisely, and to fill the voids between these chips in a manner so that the resulting agglomerate is a suitable structure for evaporated metal intraconnection. When the technology is adequately developed to allow this or a similar scheme to be carried out economically, the single-chip approach will offer a new degree of flexibility with minimum interference from parasitics.

5. Resistors

In choosing a resistor structure, one must consider the range of nominal values of resistance accessible, the precision with which a nominal value can be achieved, the possibility of matching sets of resistors, the variation of the resistance values with time and temperature, the heat-dissipation properties of the structure, and the increase in construction cost necessitated by any additional steps in the processing. All the technologies appropriate at present make use of a resistive film of one type or another. The resistance value is, accordingly, the product of the length-to-width ratio of the structure and the sheet resistivity.

Because of the very high cost of the substrate in semiconductor integrated circuitry, compared with that for other types of resistor networks, an additional consideration is introduced. It is important that the resistive structures be as small as is compatible with the performance requirements. For this reason, it is necessary to use high sheet resistivities and very fine structures to achieve reasonably high resistance values. We will consider both structures made within the semiconductor material, and by films upon a semiconductor substrate.

Semiconductor Resistors. The bulk resistivity of the semiconductor can be employed to make resistors. This is easily done when the grid type of isolation shown in Fig. 5-27 is used. In this case, the nominal resistance is given by

$$R = \frac{l}{wd} \rho$$

where l, w, and d are respectively the length, width, and thickness of the layer, and ρ is the resistivity of the semiconductor material. A typical structure of the type shown in Fig. 5-27*b*, where a wafer might be of the order of 80 μ thick, resulting in a resistor with $d \sim 40$ μ, gives sheet resistances of the order of 250ρ ohms/square. Because the surface geometry is controlled by the difference between the original separation of grid lines and twice the lateral diffusion depth during the isolation diffusion, it is necessary to employ relatively wide lines in order to achieve much precision by this technique.

If high-resistivity material is used in order to get high-value resistors, one can easily run into incompatibility problems with transistors in the same structure. In fact, the sheet resistivity of the resistors employing this structure can be traded with the collector saturation resistance of adjacent transistors.

The temperature coefficient of such resistors is usually high and positive. By employing the epitaxial isolation structure as shown in Fig. 5-29, high sheet resistivities can be obtained by the use of relatively thin structures, but the same problem with respect to other components occurs. Resistors made from the bulk semiconductor are not, in general, very useful.

A very convenient technique for making resistors is to employ diffused stripes, since this can often be accomplished with no additional processing steps. For the diffused-collector isolation structures shown in Fig. 5-26, one can use either the collector diffusion or the base diffusion, while in the grid-diffused structure the base diffusion is usually convenient to employ. In either case, the sheet resistivity is of the order of 100 to 200 ohms/square.

Because of the oxide-masking capabilities, excellent geometry control can be maintained. Common line widths for such resistors are 1 or 2 mils with equal spacings. By employing 1-mil lines with 1-mil spaces with the aforementioned sheet resistivities, a resistor in the 20- to 40-kilohm range can be made in a 20-mil square. By decreasing the line width to ½ mil with ½-mil spacings, this value can be quadrupled.

Because of the diffusion profile, the doping in such a resistor is not constant. Most of the conductivity is contributed by the relatively highly doped region near the surface. When the base diffusion is employed, this doping is usually of the order of 10^{18} impurities/cm^3. At such a doping level, the resistivity has a high temperature coefficient. The resistance-versus-temperature curve for such a resistor is shown in Fig. 5-32. For temperatures above 25°C, the temperature coefficient is about 0.2% per deg and somewhat less for lower temperatures. The temperature coefficient is strongly dependent upon temperature, concentration, and the specific impurity.

By using gallium, for example, instead of boron a resistor like the one in Fig. 5-32 has a negative coefficient below about 0°C.[36] It is possible to change the temperature coefficient appreciably by the addition of deep-level impurities. However, these usually interfere sufficiently with the remainder of the microcircuitry so that they have not been employed for this purpose. The different behavior obtained with gallium-diffused structures results from their relatively high ionization energy.

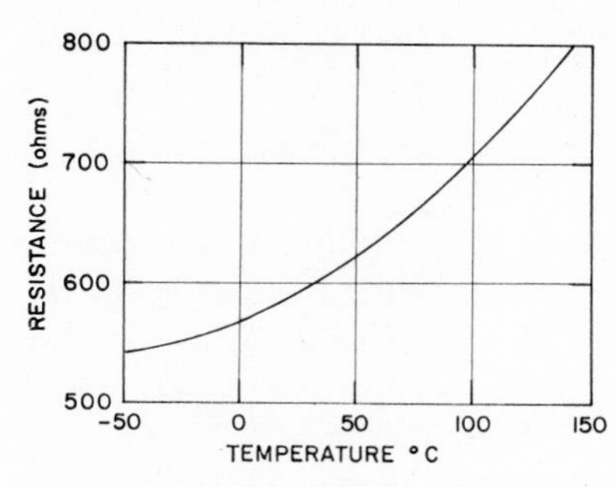

FIG. 5-32. Resistance versus temperature for a typical boron-diffused resistor.

The precision attainable is affected both by the variations in geometry and the uniformity of the diffusion. Geometry effects usually appear as a general increase or a general decrease in the line widths in a given area; that is, as a scale-factor variation rather than as a random deviation about the desired line width. Inhomogeneities in temperature or flow pattern during diffusion, or variations in the original semiconductor wafer, account for variations in the sheet conductivity. For both reasons, resistors in closely adjacent areas should be expected to show similar deviations from the desired nominal value.

It is difficult to achieve absolute resistor precision better than about 10% with high yields. A typical curve of resistor-value distribution designed for a nominal value of 580 ohms is shown in Fig. 5-33, representing a large number of diffusion runs over an extended production period.[37] The median value of 570 ohms misses the design value by 2%. The standard deviation in this case is 35 ohms, or about 5%. However, adjacent structures are matched more accurately than if they had been selected at random from such a distribution. Figure 5-34 shows a distribution curve of the ratio of pairs of resistors.[38] One sees that about 50% of such pairs are matched to within 1%. In spite of the large temperature coefficients associated with these diffused resistors, pairs of adjacent resistors track very well with temperature. Figure 5-35 shows the tracking of such resistor curves.

In cases where very high values of resistance are required, one can consider the use of the base spreading resistance by making a structure such as is shown in Fig. 5-36. This is similar to an ordinary diffused resistor using the base diffusion, except that the emitter diffusion is included over all of the resistor except near the contacts.

Typical sheet resistivities from this structure are 5 to 50 kilohms/square. Because this is achieved by taking the difference between two

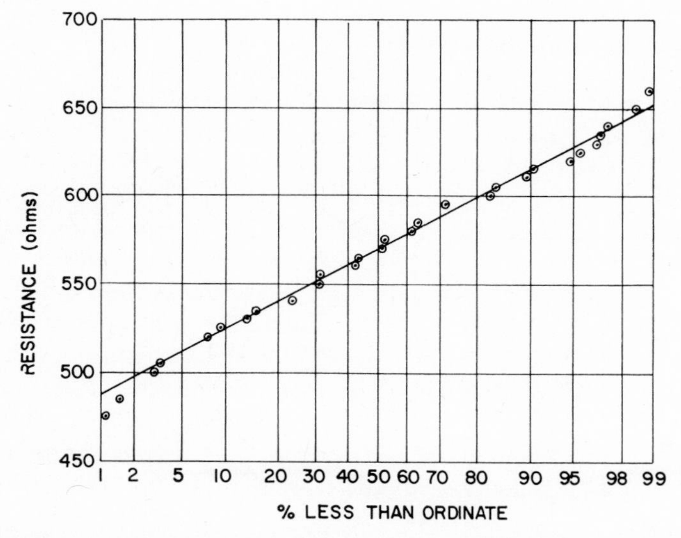

FIG. 5-33. Distribution curve of a diffused resistor representing an extended period of time in production. The nominal design value was 580 ohms.

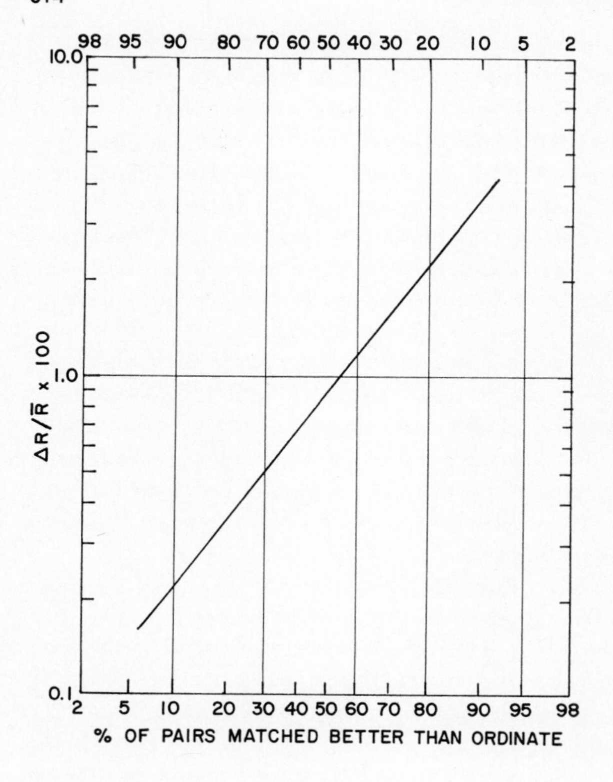

FIG. 5-34. Distribution of the ratio of pairs of adjacent diffused resistors. The ordinate is the ratio of the difference of values to the average value, expressed as per cent.

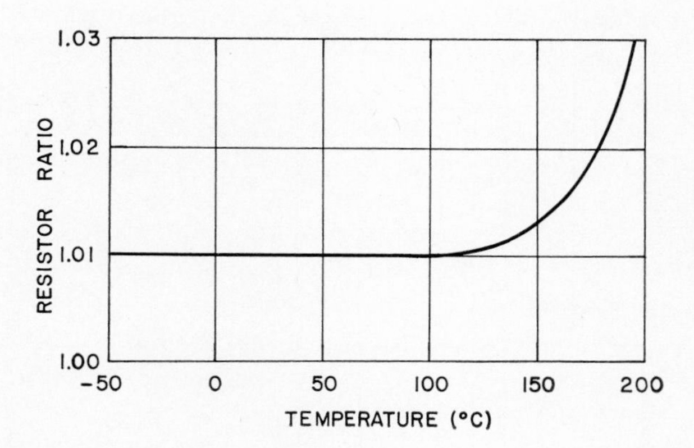

FIG. 5-35. Ratio of resistance versus temperature for a typical pair of diffused resistors.

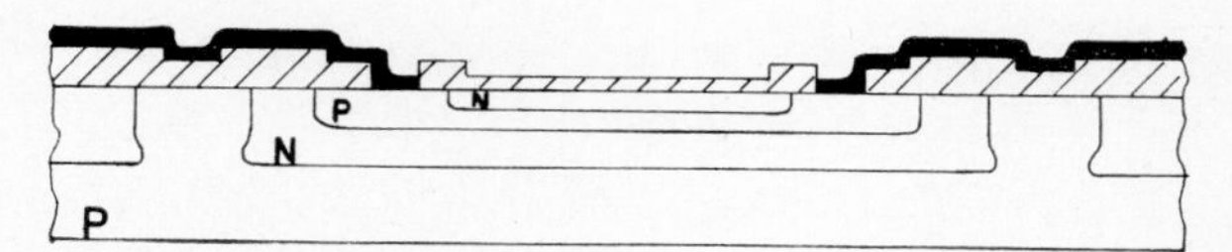

FIG. 5-36. Schematic cross section through a resistor structure making use of the equivalent of the transistor base-spreading resistance.

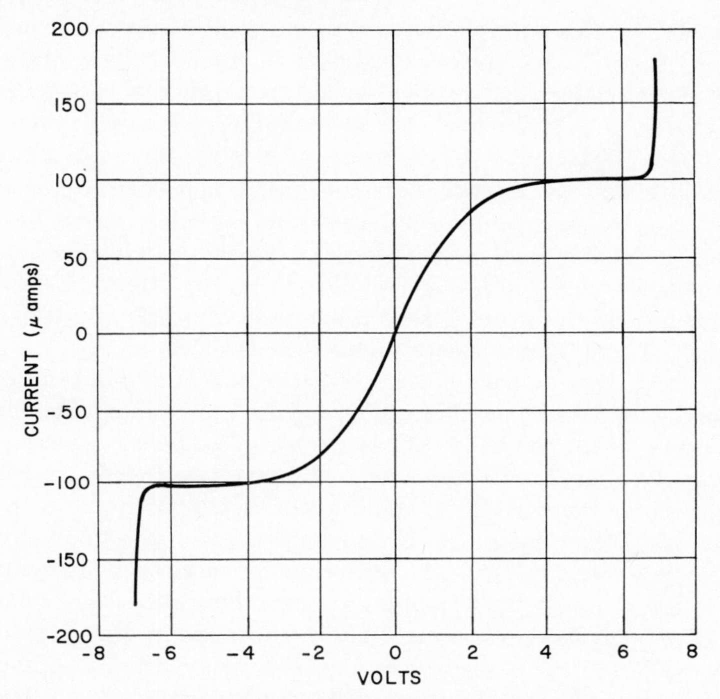

FIG. 5-37. *VI* characteristics of a resistor, such as shown in Fig. 5-36. The resistance at $V = 0$ is approximately 10 kilohms.

large numbers, it is not easy to control precisely. Fortunately, by using the emitter diffusion and the base diffusion, one gets a degree of control by monitoring the current gain of any transistor structures made at the same time.

For the usual spreads necessary to obtain useful yields, the sheet conductivity cannot be controlled much better than to within a factor of 2 from the desired value. In addition, resistors of this type are not truly ohmic since there is some self-pinching effect, and they are limited in applied voltage to the emitter-base breakdown voltage of the transistors. Figure 5-37 shows a voltage-current plot on such a resistor.[39]

Film Resistors. Many of the techniques described in Chap. 4 on thin films can be combined with the semiconductor circuitry. It is usually more convenient to make film resistors for semiconductor integrated circuitry by the use of resist-masking techniques, rather than to evaporate through a shadowing mask, because of compatibility with the remaining technology. Resist techniques take two forms: (1) coating the entire wafer with the desired film, and etching to leave the desired pattern; and (2) putting a resist pattern on the wafer before applying the resistive film, so that the excess metal will come off when the resist is stripped.

Any of the films suitable for use on glass substrates are worth considering. The added boundary conditions imposed by the semiconductor processing operations and by the limited area available require that the applicability of a given film to semiconductor integrated circuitry be separately confirmed. It has not yet been established which are compatible, although some preliminary conclusions can be drawn.

Nichrome, for example, is useful up to approximately 200 to 300 ohms/square. It is fairly stable, especially when sealed in an inert atmosphere, and it adheres nicely to the silicon dioxide layer. Figure 5-38 shows a photomicrograph of an integrated differential amplifier using Nichrome resistors. The refractory metals need to be proved.

The advantages of the thin-film resistors in this application over the semiconductor resistors are lower temperature coefficients, lower shunting capacitance, some possibility of saving area, and better isolation. The disadvantages are the inclusion of extra processing steps, at least one evaporation and one masking, and lower power capability.

Other films than metals have been considered. Structures have been made using tin oxide.[40] The inclusion of cermets by simultaneous evaporation of metal and insulators appears interesting.[41] The achievement of a high-sheet-resistance, stable, controllable film resistor compatible with the rest of the processing technology will be an important extension of the semiconductor integrated circuitry.

The tolerance of present thin-film resistors is of the same order as that for the semiconductor resistors, although one can reasonably anticipate that use in production will result in an appreciable improvement. Because of the ability to monitor the sheet conductance during evaporation, variations from this source can be minimized. In addition, with thin-film resistors, one has the possibility of adjusting the individual

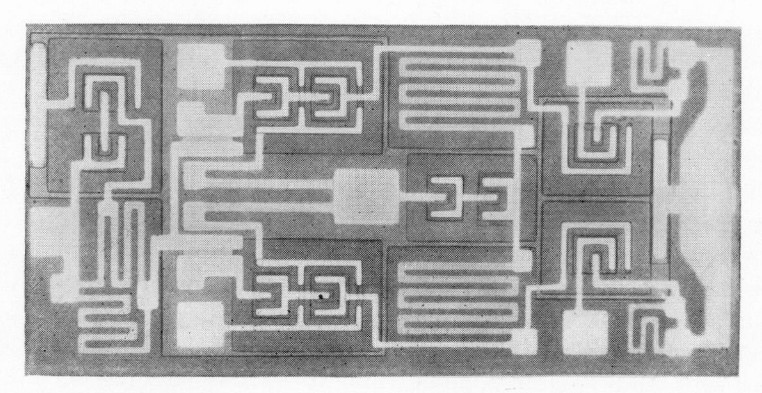

FIG. 5-38. Photomicrograph of an integrated differential-amplifier circuit employing thin-film resistors. The white metallization is aluminum intraconnections, while the gray stripes are Nichrome films.

structures subsequent to their preparation by any one of a variety of techniques which have been described.[42] This can be important for making circuits requiring precise matching of resistance values.

The capacitance distributed along the film resistors between film and substrate is of the order of 0.02 pf/mil^2 for the usual oxide thicknesses of approximately 1 micron. This is about an order of magnitude less than the capacitance per unit area usually associated with the diffused silicon resistors. The power dissipation of the film resistors is seldom a problem, since the over-all power dissipation is usually limited by the entire package.

Silicon is an excellent thermal conductor, and the thin silicon oxide layer increases the thermal resistance relatively slightly. Each thin-film structure, however, must be carefully evaluated with respect to its stability when dissipating power, since the deterioration of such thin-film structures has been shown repeatedly to be more severe under operating conditions than when subjected to what is designed to be an equivalent thermal environment.

6. Capacitors

A most obvious structure to use to achieve the function of a capacitor in semiconductor integrated circuitry is the *pn* junction itself. Although such a structure follows rather naturally out of the technology used to produce the other components, it does not, in general, have especially good electrical characteristics. The junction capacitance is a strong function of the applied voltage, and depends upon the impurity-concentration profile. For a typical emitter-base junction, one gets approximately 0.4 pf/mil^2 at 5-volt bias.

Because of the rectifying nature of these junctions, the capacitors can only be used with voltage applied so as to reverse-bias the junction. Somewhat less voltage dependence and operation in either polarity can be achieved by two junctions back to back, although this results in a factor of two decrease in the capacitance attainable in a given area.

In Fig. 5-39, several possible configurations of junction capacitors are shown. In Fig. 5-39*a* the simple junction structure is depicted. Figure 5-39*b* and *c* show two alternatives for the back-to-back junction configurations, while Fig. 5-39*d* shows how the emitter-base and collector-base junctions in a transistor structure can be combined so as to increase the capacitance achievable in a given area.

In all cases, one terminal has a fairly high resistance contributed by the low sheet conductivity of the semiconductor layer. If this series resistance is important in a given application, it can be minimized by going to many-fingered, interdigitated structures, such as those employed in high-frequency power transistors. As in other capacitor structures, the

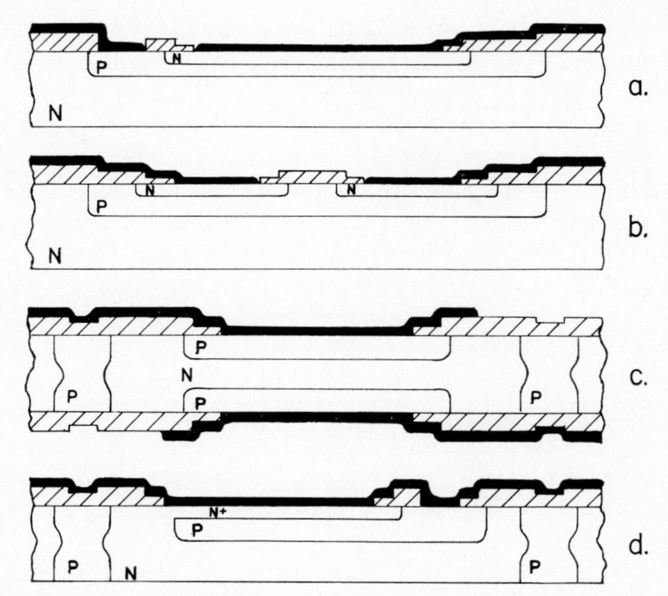

FIG. 5-39. Junction capacitor structures: (*a*) Double diffused; (*b*) double diffused back to back; (*c*) back to back bilateral; (*d*) combined emitter-base and collector-base junctions.

maximum useful voltage can be traded for capacitance per unit area. In contrast to some other techniques, however, these junction capacitors can be broken down and recover their original characteristics when the voltage is decreased, providing that their power-handling capacities are not exceeded.

Another type of capacitor structure is shown in Fig. 5-40. In this structure, the dielectric is made from the silicon dioxide layer grown upon the semiconductor. The underlying silicon forms one electrode, while the other electrode is the conducting film. These capacitors have good

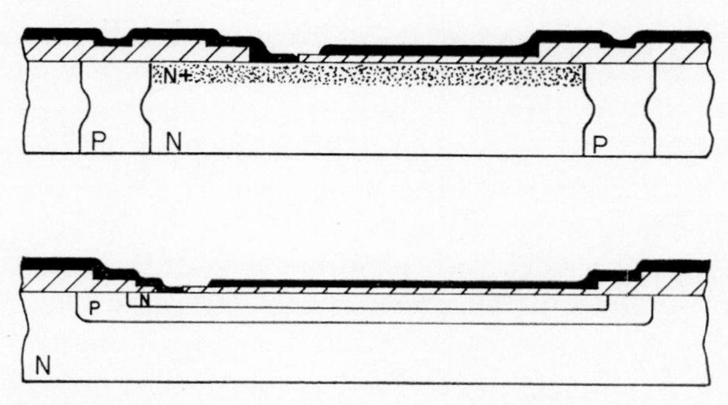

FIG. 5-40. Metal-oxide–semiconductor capacitor structures.

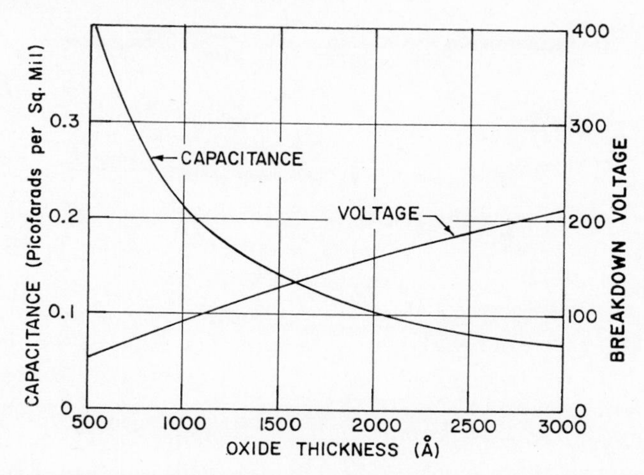

FIG. 5-41. Capacitance and breakdown voltage versus oxide thickness for metal-SiO_2-silicon capacitor structures.

linearity, and can be made with low series resistance. Again the capacitance achievable in a given area is limited to about 0.4 pf/mil^2. Figure 5-41 shows the capacitance per unit area and the dielectric breakdown voltage for metal-oxide–silicon capacitors versus the thickness of thermally grown silicon dioxide layers.

Working voltages should, of course, be restricted to something appreciably less than the dielectric breakdown. The incidence of pinholes in the silicon dioxide film limits the areas that can be employed economically in this type of capacitor. Values up to a few hundred picofarads are as large as can be combined with other semiconductor components at present with adequate yields.

Work to get higher dielectric-constant films has been reported. By combining titanium dioxide with the silicon dioxide, dielectric constants as high as 30 have been achieved,[43] but such structures have not yet been shown to be adequately stable. Extension to high-capacitance values in semiconductor integrated circuitry seems to require the use of thin-film structures, such as those described in Chap. 4. Until these have been successfully integrated, high capacitances should be avoided wherever possible or included upon a separate substrate where absolutely required.

7. Other Structures

Many other structures are possible, some of which will be extremely important. In this section we will describe a few of these, suggesting some of the advantages they might be employed to achieve.

As a first example, consider the unipolar field-effect transistor,[44] which is desirable in several applications requiring high input impedance or as a

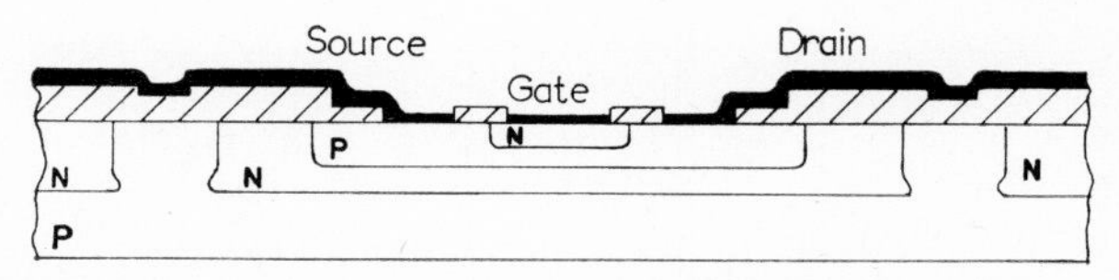

FIG. 5-42. Isolated unipolar field-effect transistor structure.

switch with very low offset voltage. Field-effect structures can be made in a variety of ways. In Fig. 5-42, a structure similar to the resistor employing the base spreading resistance is illustrated; this will act as a field-effect device.

Referring to the characteristics of the pinch-off base spreading resistors shown in Fig. 5-37, one sees that a pinch-off voltage of the order of two volts can be expected for this structure. By employing a reasonably long narrow channel consistent with usual processing technology, say, for example, a channel that has a length-to-width ratio of 20, one can reasonably expect to get source-to-drain saturation currents of the order of 1,000 μa. Combined with the 2 volts pinch-off voltage, this suggests transconductances in the vicinity of 500 μmhos. By further adjusting the geometry and diffusion schedule, one can get transconductances above 1,000 μmhos and on-impedances less than 1,000 ohms. In addition, the technology by which this structure is prepared is completely consistent with the transistor technology, making circuitry requiring a combination of unipolar and bipolar transistors relatively straightforward.

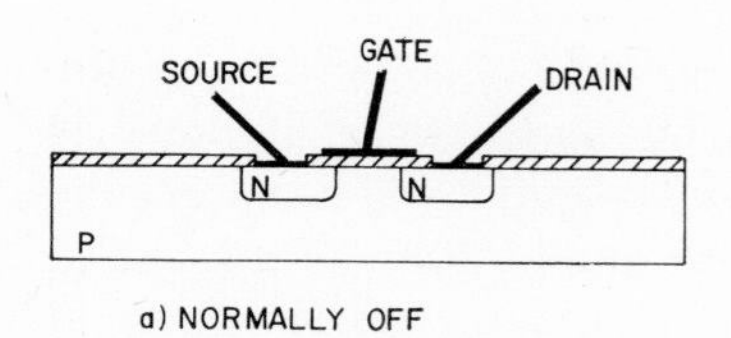

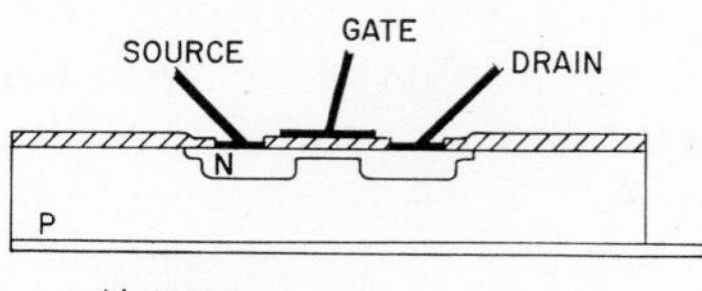

FIG. 5-43. Schematic cross sections of surface-controlled field-effect transistors. In the normally off structure a channel is induced between the n-type regions by applied gate voltage. The normally on configuration is usually used with gate voltages to decrease the channel conductance.

Field-effect structures in which the gate is separated from the channel by an insulating layer can also be made. Typical cross sections for these surface-controlled field-effect transistors are given in Fig. 5-43. In the top structure, n-type contact regions are diffused into a p-type substrate to form the source and drain contacts. A gate contact overlying the oxide completely covering the region between the two n-type contact areas is employed. When the gate potential in this polarity structure is adequately positive, electrons are attracted to the surface of the silicon, resulting in the formation of an inversion layer, which connects the two n-type contact regions.

Such a field-effect device is normally nonconducting, although the effects of surface states at the silicon oxide–silicon interface can result in the formation of an inversion layer at zero gate voltage. The d-c input impedance of such an insulated gate is extremely high, of the order of 10^{15} ohms for a typical structure.

The other variation of the surface-controlled field effect shown in Fig. 5-43 is constructed to have a channel connecting the contacts at zero gate bias. In this case, where a junction exists to the channel, the semiconductor body can also be used as a gate electrode as with the ordinary field-effect transistor. Here the top gate overlying the oxide need not extend completely from source-contact region to drain contact in order to control the channel conductance.

The future of such surface-controlled structures looks extremely interesting. However, there are problems associated with the stability of the surface states that can result in appreciable drifts of the characteristics. Additional understanding of this interface is required before wide use of this type of structure can be considered.

Another device used for integration is the surface-controlled transistor tetrode, described by Sah.[45] A cross section of this structure is shown in Fig. 5-44. Essentially it consists of a planar transistor, either *npn* or *pnp*, modified by the addition of a grid electrode overlying the region where the emitter-base junction intersects the semiconductor surface. By varying the potential on the surface at this point, the current gain of the transistor can be modulated. The effect is the result of modulating the surface recombination velocity and, in some instances, the formation of a channel over either the base or the emitter under the grid electrode.

This structure also offers extremely high input impedance, of the same order as that for the surface-controlled field-effect transistor. In addition, it is relatively easy to achieve transconductances of several thousand micromhos. Again, the technology by which it is produced is completely compatible with that employed in making ordinary integrated bipolar transistors. Certainly, a device with the properties possessed by this structure will have applications, at least in special situations.

One of the earliest examples of a true integrated circuit was given by D'Asaro[46] when he described a stepping switch employing *pnpn* structures. By making use of the voltage drops which occur in an asymmetric four-layer device structure, he was able to control whether the next stage would trigger or not, upon the application of properly timed clock pulses by the condition of the first stage.

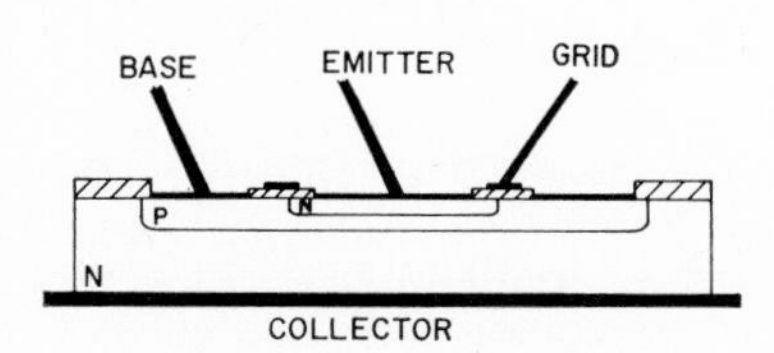

FIG. 5-44. Surface-controlled field-effect transistor. The grid electrode acts as a high-impedance input which modulates the current gain of the transistor structures.

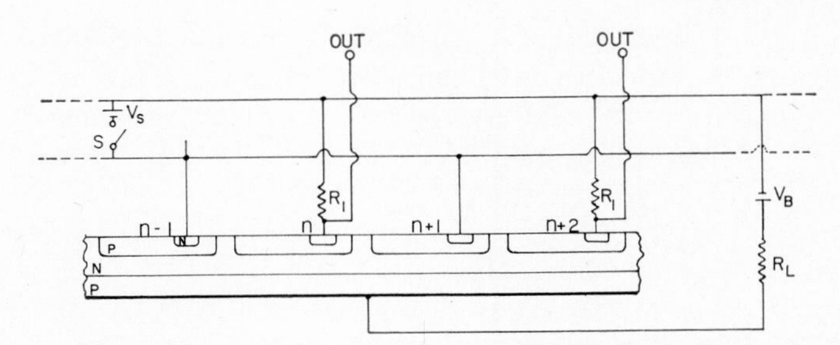

FIG. 5-45. Schematic representation of the connection of an integrated *pnpn* structure to construct a pulse counter. (*D'Asaro.*[46])

A schematic representation of the circuit he used as a pulse counter is shown in Fig. 5-45. The operation of this structure depends on the fact that, in the common *n*-layer (as drawn), the voltage is higher under the emitter of the structure that is conducting than it is under the opposite end of the individual *p*-regions. Thus, the common *pn* junction is more heavily forward-biased in the next structure to the right than it is in the structure to the left.

For a correct combination of currents the $(n + 1)$ device will conduct when switch S is closed, because the bottom junction will be sufficiently forward-biased to remove the high-impedance state of the four-layer structure. Upon opening switch S, the on-state moves to position $(n + 2)$. Thus, in this structure, each time switch S is opened and closed, i.e., each time the line is pulsed, the conducting state moves two positions to the right. The structures described by D'Asaro would trigger the device to the right and not trigger the one to the left for a range of currents through a structure from less than 3 to greater than 30 ma.

Such a structure represents an extremely simple technique for achieving a shift-register function, something which, by means of conventional circuitry using only bipolar transistors, requires a considerable number of components. Though this is one of the earliest examples of a true integrated circuit, it still remains one of the best examples of capitalizing on a unique interaction in order to achieve a desirable electronic function.

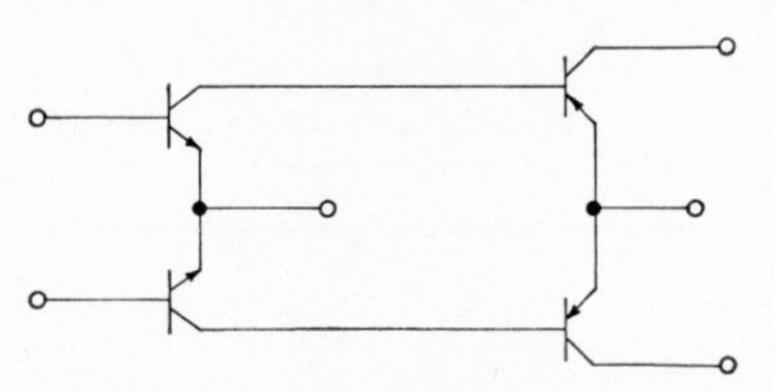

FIG. 5-46. Complementary circuit configuration useful in analog circuitry.

Another structure of considerable interest, although not really employing devices other than those discussed previously, is that where both *npn* and *pnp* transistors are included in the same circuit. In many instances in analog circuitry, configurations such as that shown in Fig. 5-46 are

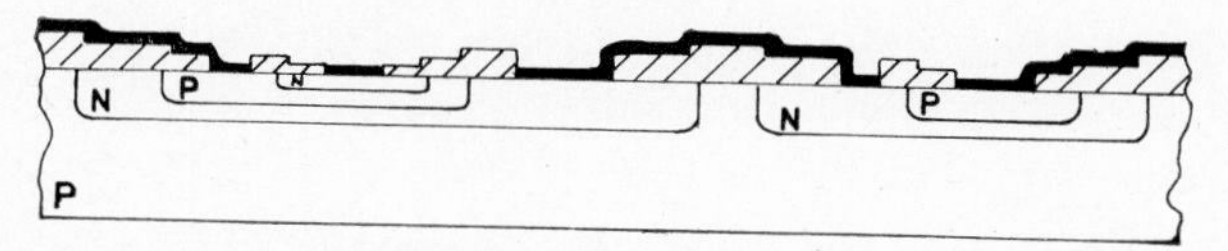

FIG. 5-47. Schematic cross section of a structure with *npn* and *pnp* transistors in the same substrate.

useful, in order to use differential input and to obtain output not displaced with respect to ground potential.[47]

Many other examples where complementary circuitry can be used to advantage exist. A simple method of achieving *npn* and *pnp* transistors in the same structure is shown in Fig. 5-47. Here in the case where diffused isolation is employed for the *npn* transistor, the *pnp* transistor is prepared by using the base diffusion of the *npn* as its emitter and the collector diffusion of the *npn* as the base. The use of these diffusions for the *pnp* transistor cannot be expected to result in an optimum structure since they must be adjusted in order to achieve reasonable operating characteristics for the *npn*.

In situations where low beta can be tolerated in the *pnp* device, this represents a straightforward technique for achieving complementary circuitry. In cases where it is important that the characteristics of the two devices be separately optimized, it is necessary to resort to more complicated processing schemes.

A wafer having both *p*-regions and *n*-regions suitable for the collectors of the *pnp* and *npn* transistors, respectively, can be obtained in a variety of ways. For example, in Fig. 5-48 we have shown two methods by which such a structure might be achieved by epitaxial growth. In one of these structures, the growth covers only a portion of a *p*-type substrate. This can be done either by masked growing techniques or by etching through the grown layer. In the second configuration, the epitaxial growth is used to grow a layer over the entire surface of the substrate,

FIG. 5-48. Schematic cross section of two structures achievable by epitaxial growth suitable for the construction of optimum *npn* and *pnp* transistors in the same semiconductor substrate.

layer and substrate being of opposite conductivity type. In the second case, the *pnp* transistor will be produced on one side of the wafer, while the *npn* transistor is made on the opposite side. In the first case, of course, these transistors can be prepared side by side.

Other structures employing diffusion at low surface concentration can be used to form a similar grid. In order to separately optimize the *npn* and *pnp* transistors, it is necessary to first perform those diffusions which can be done with extremely slowly diffusing impurities.

For example, the first diffusion might use antimony or arsenic as the *n*-type impurity, in order to produce the base region for *pnp* transistors. A second diffusion might diffuse boron as the base of the *npn* transistor, and, simultaneously, it could be diffused into the previously prepared antimony or arsenic region in order to form an emitter for the *pnp* device.

Since this boron will probably be too lightly doped to be an efficient emitter, it might be followed with a relatively heavy boron predeposition over the *pnp* emitter and a phosphorus predeposition on the surface of the *npn*. The base width and hence, the final beta of the *pnp*, will be determined primarily by the difference in the diffusion depths of the first antimony or arsenic layer and the first boron diffusion. Since the impurity gradient at the emitter-base junction in this structure is relatively small, it will not diffuse much during the time that a phosphorus emitter is diffused into the *npn* structure. On the other hand, the heavy boron concentration placed near the surface of the emitter for the *pnp* transistor will assure adequate emitter efficiency.

One sees that in order to prepare such a structure several additional diffusion and masking operations are required. The circuit design should be considered carefully before one decides that the best solution to his problem resides in making high-quality *npn* and *pnp* devices in the same semiconductor substrate.

As a final example of an interesting structure useful in microcircuitry, we would like to consider the use of thermal feedback.[48] In this example, the temperature variation of the emitter-base voltage of a transistor is used as a thermal sensor, to control the amount of current being passed through a transistor structure and, hence, the amount of power dissipated in the collector spreading resistance. It has been shown that in this manner one can actually thermostat the piece of silicon to within a few degrees centigrade, in spite of ambient variations from -55 to $+100°C$. In effect, one has succeeded in making a miniature thermostated oven.

A diagram of a circuit capable of performing this function is shown in Fig. 5-49. As the temperature of the circuit falls, the emitter-base voltage of the first transistor necessary for conduction increases, resulting in lower conductance of the first transistor and, accordingly, making the base of the second transistor more positive. As the second transistor turns on, relatively large current flows, dissipating power in the internal resistances of the structure. This heats the silicon block until the first transistor conducts and shuts off the second transistor.

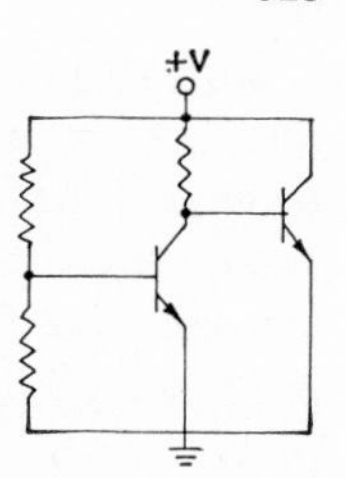

FIG. 5-49. Circuit diagram of a structure suitable for maintaining constant temperature in an integrated circuit.

The usefulness of this sort of structure in stabilizing temperature-sensitive circuits such as precision amplifiers or precise voltage-reference sources is apparent. A voltage reference stabilized in this manner has shown stability to one part in 10^5 over an ambient temperature range of 150°C.[49] Here is an example of improved performance achievable by a structure uniquely possible in semiconductor integrated circuitry.

The structures described in this chapter are by no means all inclusive. Many others will be of importance. In particular, the use of unique interaction effects, available only when the various circuit elements are formed within parts of the same monolithic block, will increase.

5-4. CIRCUIT CONSIDERATIONS

The same general considerations that go into the design of conventional circuits are required in the design of integrated circuits. However, some additional boundary conditions must also be considered. As with conventional circuitry, the prime considerations are that the resultant design will perform the required function with adequate reliability, and that it will do this at the lowest possible cost. The principal changes in boundary conditions which occur when one considers the present technology for circuit integration are the following:

First, the relative costs of components change. Costs of components are now primarily determined by the area they consume on the semiconductor wafer. Thus, a transistor is approximately as expensive as a diode, which, in turn, corresponds to a resistor of about 4 kilohms $\pm$ about 30% or one-quarter that value $\pm 20\%$.

Second, the values of available components are restrictive. Values of resistors exceeding about 50 kilohms are prohibitive, and capacitors above a few hundred picofarads must be assembled separately.

Third, the absolute values of resistors are poor, although the ratios between resistors in the same structure can be held fairly closely and track well with temperature, as was discussed in more detail in the section covering resistors.

Fourth, all the components of the integrated circuit are coupled by various parasitic capacitances and conductances. This is a function of the close spacing and the techniques employed for isolation.

In general, in the design of integrated circuitry, prime consideration must be given to the size of the resulting die and to the yield which can be expected, for die size and yield are of prime importance in determining the economics of integration. In order to assure highest possible yields, the voltages to which the circuits are subjected should be kept as low as possible. The effect of voltage upon yield will be discussed more completely in Sec. 5-6.

A useful technique to aid the circuit designer in taking into account the various peculiarities imposed by integration utilizes special separate components for circuit breadboarding. The various component structures available in the integrated circuits can be prepared separately by the technologies used for the final integrated version.

Thus, for example, transistors are made within isolated regions as they would appear in the final structure. In addition to the emitter, base, and collector leads, the isolation region is also made available. These transistors can be packaged in regular four-lead packages.

Similarly, diffused resistors or metal-film resistors can be made as separate components surrounded by the environment which simulates what will surround them in the integrated circuit. In this manner, a circuit can be assembled wherein each component appears in the environment it will occupy in the final circuit, and all isolation regions can be connected in an appropriate manner. By making combinations of transistors in the same isolation region and by making more than one resistor structure in a given block of silicon, the effects appropriate to common-collector transistors and to pairs of resistors made simultaneously on the same substrate can be included.

Circuits breadboarded with such components can be evaluated before laying out the masks and constructed in fully integrated forms. Such prior evaluation has proved to be quite representative of the final integrated structure, although usually slightly on the conservative side, since additional parasitics are introduced by having separate packages, and the advantageous effects of the uniform temperature which exists in the common integrated circuits are not shown to full advantage with the separate integrated circuit components.

This technique results in a procedure by which integrated circuits can be designed and integrated with confidence that they will perform the desired function. Such a procedure is of considerable aid in giving the circuit designer the ability to include the effects of those peculiarities of semiconductor integrated circuitry.

1. Digital Circuitry

Digital circuits present the ideal instance for integration. The binary operation of such circuits places minimum strain upon the tolerances of individual components. The transistor itself is used in its optimum application when it is employed as a switch. Logic systems often use large numbers of identical circuit functions important in the most economic use of integration, and the low voltage levels usually employed are consistent with the requirements for high yields. It is not surprising that the first useful applications of integrated circuitry have been in the digital area.

Most of the common types of logic circuitry are suitable for integration. However, the effects of the new boundary conditions may change previous ideas as to which circuit schemes are most appropriate. Both the type of logic and the circuitry desired must be considered in deciding on the most appropriate approach to a given problem.

Some circuitry is most useful for high speeds, other for lowest power, and still other configurations might be most appropriate in situations where the highest possible logical gain is important. As examples of the trade-offs which might be considered in choosing the circuit configuration, the basic three input gates for several types of logic circuitry presently used in digital computers are shown in Fig. 5-50. By comparing these, some idea of the relative usefulness can be made.

For example, in Fig. 5-51 a plot of the relative speeds that can be expected for several of these configurations as a function of the power per node is plotted. Such a plot assumes that all of the gates are made with identical technology, and that a transistor of relatively low storage time is available. As the technology is altered, the positions of the curves is altered.

For example, if a transistor with appreciably lower storage time were included, the advantage of current-mode logic at high power levels over direct-coupled logic would largely disappear. The component values included in Fig. 5-50 are representative of those which might be employed when the particular type of circuitry is used somewhere near its most advantageous condition. Of course, these values change appreciably as the power level and speed of these circuits is altered.

A comparison of the TRL gate in Fig. 5-50*a* and the DTL gate in Fig. 5-50*b* is a good illustration of the effect of relative component costs upon circuit design. The TRL gate uses resistors in series with the inputs while the DTL gate employs diodes. With conventional circuitry, the resistors are often cheaper than the diodes, making TRL somewhat less expensive than DTL. This was especially true in the past.

In the integrated case, however, the diodes required in the input of the DTL actually consume less area than the resistors required for TRL. Hence, DTL is a cheaper circuit to make than TRL. Since diode-transistor logic also offers higher speed and greater digital gain than TRL, there is no reason to even consider TRL seriously for integrated logic

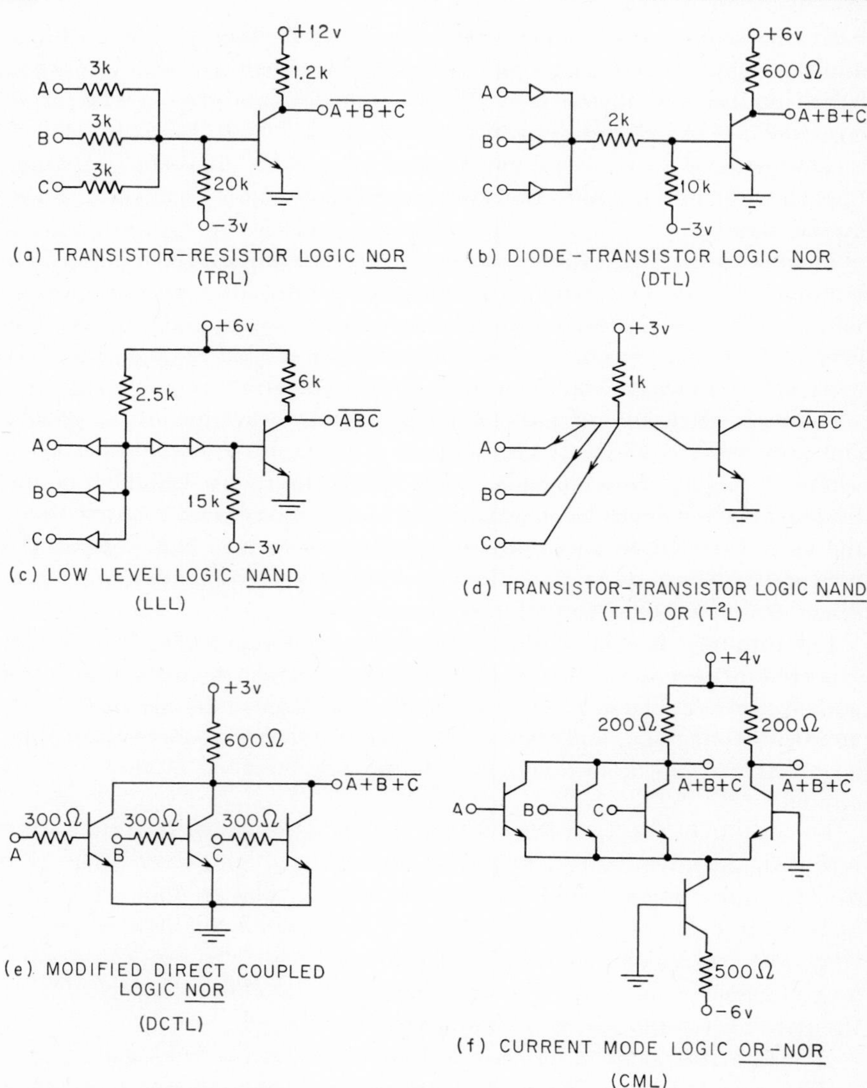

FIG. 5-50. Various forms of logic gates suitable for integration. Values given are typical of those commonly employed.

circuitry. DTL offers large voltage swings with corresponding high noise immunity.

It is, however, quite slow, as can be seen from the relative speed plot in Fig. 5-51. The pulldown resistor is relatively large, and the need for both a positive and a negative power supply can be a considerable disadvantage, especially in a densely packed system, where one does not wish to have any unnecessary non-signal-carrying conductors. In addition, DTL ordinarily employs relatively high-voltage power supplies which, as was mentioned above, is not especially desirable in integrated circuitry.

The low-level logic NAND gate shown in Fig. 5-50*c* offers the advantage of extremely high fan-in. With the two diodes in series, as shown, it has good noise immunity. This can be further improved by the substitution of a zener diode. However, this is at a sacrifice in power and speed. It is difficult at present to make an adequately low-voltage zener diode in integrated circuitry, and so the structure shown, employing forward diode drops, is more appropriate. The large resistor in the pulldown circuit is a problem, and again the requirement of two power supplies with the accompanying disadvantages of extra power buses throughout the system is a drawback.

The T^2L gate shown in Fig. 5-50*d* offers the same NAND function afforded by low-level logic, with some of the drawbacks relative to its integration removed. The input transistor is drawn to represent the manner in which it would ordinarily be made in integrated circuitry. Three emitters would be placed in a common-base region, since the load transistors are connected in a common-base common-collector manner.

This circuitry is extremely simple, fast, and easy to integrate. The gate shown requires only one relatively small resistor in addition to the two transistors, including the one with multiple emitters. Unfortunately,

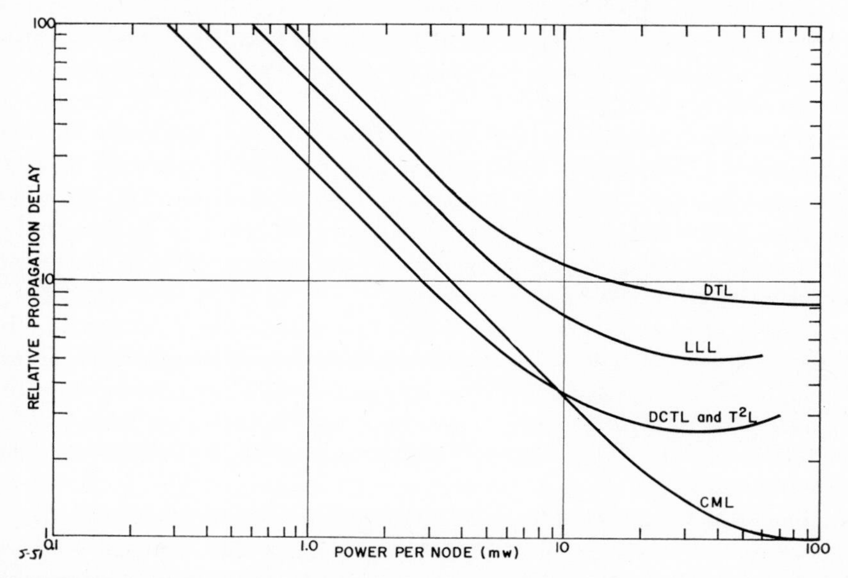

FIG. 5-51. Relative propagation delay versus power for fan-in and fan-out of unity, comparing various integrated circuits made with a particular technology.

this circuitry has some disadvantages. Because of the additional drop across the gating transistor, the noise immunity is less than one gets even for straight DCTL, since the down voltage on the base of the inverter is always held higher than the saturation voltage of the previous inverter by the voltage drop across the gating transistor. It also suffers from a current-hogging problem, considerably limiting fan-in and fan-out capabilities. If the additional components necessary to cure these problems are added, the advantages of this type of circuitry disappear rapidly.

The modified DCTL shown in Fig. 5-50*e*, wherein a resistor is inserted in the base of each transistor to prevent current hogging is simple and fast. All resistors are small, voltages are low, and only one power supply is required. The logical gain of such circuitry can be traded for noise immunity. Proper choice of base resistances with low-saturation resistance transistors gives good noise margin over a wide temperature range with fan-in of 3 and fan-out of 5. While such circuitry with conventional components is quite expensive because of the large number of transistors employed per logical operation, it is quite suitable for integrated circuits.

The current-mode logic gate shown in Fig. 5-50*f* is most useful where speed is of paramount importance. With the technology one can expect to have available in the foreseeable future, such nonsaturating current-steering circuitry presumably will always produce the minimum average propagation times at high power. As the gate is drawn in the figure, there is a level-setting problem which can be solved by using alternate stages of complementary polarity or by the inclusion of a zener diode-resistor network. Neither of these techniques is especially desirable for integrated circuitry.

A possible solution to the level-setting problem employs emitter followers.[50] This results in an appreciable increase in the area needed, and also requires a separate reference bus to each circuit. Another advantage of this CML circuitry is that both the OR and NOR outputs are available. The use of both outputs in perhaps 20% of the logic nodes in a system can be advantageous.

The circuitry can be so designed that the relative values of resistors in a particular logic node are important rather than their absolute values, which again is compatible with the integrated-circuitry boundary conditions. This type of circuitry is presently restricted to high power, since at low power the values of the resistors exceed those readily available. The level-setting is a severe limitation. All solutions to this level-setting problem have resulted in a considerable increase in the area required.

Of course, consideration of just the three-input gate does not give the whole story. If only gate structures are available as integrated circuits, the advantages inherent in the technique are hardly apparent. Only when larger logical assemblies are considered, do the real advantages become apparent. As one considers larger functions, the effects of area and yields become of increasing importance. This should be considered carefully in choosing the logic form desired. For example, using modified DCTL of the type depicted in Fig. 5-50*e*, a complete shift-register stage as shown in the photomicrograph in Fig. 5-52 can be made in an area 0.075 by 0.075 in. The equivalent function using CML, even at a much higher power level, takes approximately three times the area when the required level setting is included.

With evolution of the technology, perhaps the relative advantages of one structure over another in this respect will change. For example, with the inclusion of a good technology for making high values of resistances in small areas, the disadvantages to those circuits employing high resistors will disappear and such structures as low-level logic must again be considered.

In any case, all the logic forms in Fig. 5-50 are suitable for integration. When new systems are under consideration, the logic forms should be considered from the point of view of taking maximum advantage of semiconductor integrated circuitry. For inclusion into an existing system,

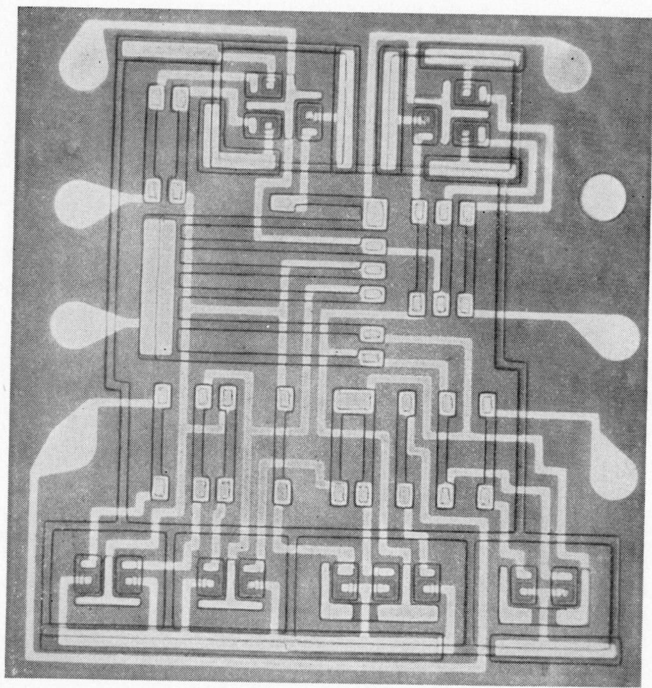

FIG. 5-52. Photomicrograph of a complete shift-register stage using diffused resistors. This circuit has 15 transistors and 21 resistors. It employs grid-type diffused isolation.

however, it may be advantageous to make integrated circuitry, using logic forms which existed in the system when originally constructed from conventional components, even though these may differ from those that take maximum advantage of integration. In these instances, one can still achieve the advantage of size, weight, and elimination of interconnections.

The use of the integrated-circuit components for breadboarding of any of these logic configurations is a useful aid in designing and evaluating a circuit before the investment necessary to make it in a fully integrated form is committed. One can expect larger and larger logic blocks to be available in the future, as well as other digital circuitry, such as coding and decoding matrices.

2. Linear Circuits

The limitations imposed by the restricted values and precision of passive elements become extremely important in the area of linear circuitry. The lack of quality reactive components, especially the almost complete lack of inductance available in integrated circuits, is a severe limitation on any structure requiring sharp tuning. The lack of precision available in the resistors is a limitation on many analog-computing functions. Although a considerable amount of work has been supported in the area of linear integrated circuits, very little of it has resulted in the construction of practical, useful configurations.

The single real advantage which has been exploited to date in the area of integrated analog circuitry has made use of the fact that components built in the same piece of semiconductor material are very close to the same temperature at all times and track well with temperature. Thus, it has been possible to make differential amplifiers with excellent performance. The thermal feedback amplifier described in the section on other device structures also makes use of this property.

Another area where existing technology can be used to make linear circuits which give advantages not available with conventional circuitry is in those applications where size is important. For example, it is possible to build an entire integrated sense amplifier into the head of a magnetic pickup, thus considerably reducing the noise problem inherent in running leads from the head to a remote amplifier. Also, the inclusion of some amplification in such structures as transducers close to the sensing elements themselves is often advantageous. Special applications, such as hearing aids, where size can be important and maximum fidelity is of relatively little concern, are natural places to consider linear integrated circuitry.

Such applications as high-quality i-f stages are severely limited by the lack of an appropriate compatible tuning element. The *RC* tuning that can be obtained with present resistor-capacitor networks is considerably inferior to that easily achieved by employing *LC* networks in conventional electronics. The inclusion of small, sharply resonant structures, such as piezoelectric crystals, may eventually result in good tuning in a small volume. However, anything done in this area to date has been more nearly the isolated chip approach than an integrated structure.

The economic application of truly integrated circuits to applications in linear electronics will be restricted to relatively special applications for the present. Considerable development, both with respect to the manner in which circuit functions are performed and the technology available, will be required before the majority of linear circuitry is accessible to economic integrated circuits.

5-5. AN EXAMPLE

As an example of the application of the considerations in the previous sections concerning the technology, structures, and circuits, it is worthwhile to examine in detail the design, manufacture, and evaluation of a semiconductor integrated circuit. For such an example, consider the logic circuit shown in Fig. 5-53. This is the circuit diagram for a gated flip-flop. It is one of a family of circuits called *micrologic elements*, manufactured by Fairchild Camera and Instrument Corporation.[51] This family consists of a buffer element, a counter adapter, a flip-flop, a three-input gate and a half-adder in addition to the gated flip-flop called a half shift register. The logic diagrams for the various circuits are shown in Fig. 5-54.

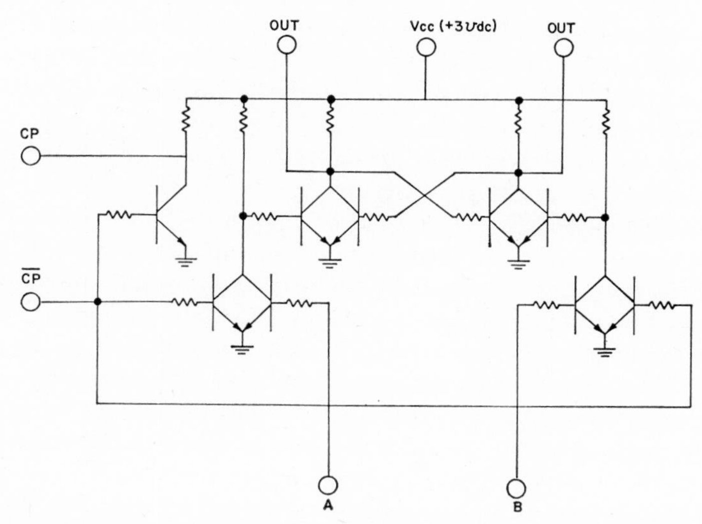

FIG. 5-53. Circuit diagram corresponding to a micrologic *S* element.

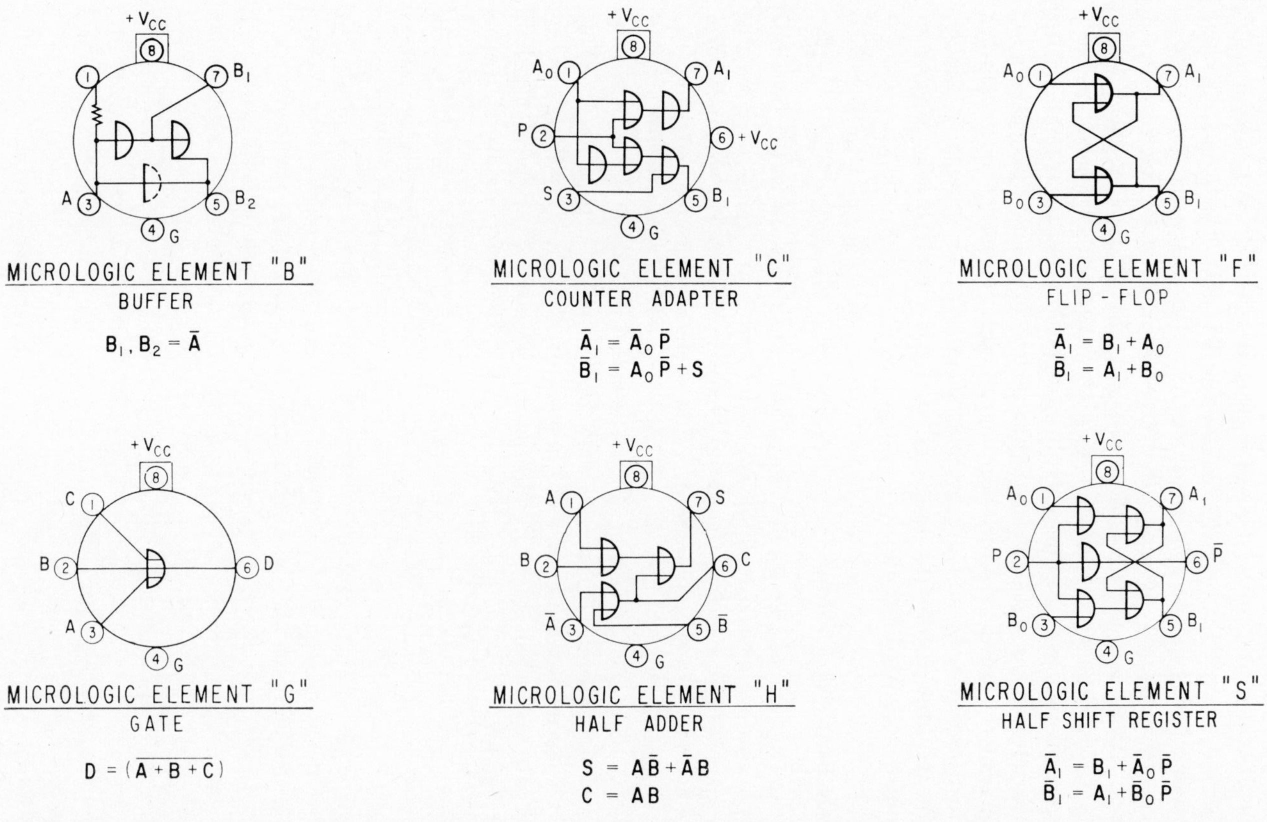

FIG. 5-54. Logic diagram and Boolean functions of micrologic elements. Each D-shaped symbol represents a NOR gate. The dotted symbol in the buffer element represents an emitter follower.

The circuit form used in this family is modified DCTL, as was shown in Fig. 5-50*e*. All the integrated circuits are made with the same technology and structural features; only the surface geometry is varied from circuit to circuit.

One sees that the gated flip-flop we have chosen for our example consists of 9 transistors and 14 resistors. The transistors exist as four common-collector pairs and one separate device. Hence, the transistors must fall into five separate isolated regions. In order to make transistors of the relatively low saturation resistance needed in the chosen circuitry, and in order to achieve high-speed performance, the diffused-grid isolation structure shown in Fig. 5-27 will be selected. The epitaxial isolation would be equally suitable, but the design of these integrated circuits preceded the epitaxial technology. The five collector load resistors required can be made by having several diffused stripes of one conductivity type of material in a region of the opposite conductivity type. Since only positive voltages are applied to the circuit, this structure which always separates resistors by two junctions will assure that these junctions do not get forward-biased.

Hence, all five collector resistors can be included in a single isolation region. In addition the nominal value of these resistors, of around 600 ohms each, is nicely compatible with the sheet resistance of approximately 150 ohms/square desirable for the base diffusion used in making the transistors. Accordingly, all five collector load resistors will be made in a single region simultaneously with the bases.

The nine base resistors could similarly be placed in an isolation region. However, if they were concentrated in a single region, it would become cumbersome to interconnect them to the various bases spread throughout the remainder of the circuit. Since the base-diffusion sheet resistance is convenient to employ for these resistors also, an obvious possibility is to distort the geometry of the base region so as to obtain the desired number of squares between the base contact and the edge of the emitter.

Care, however, must be exercised in proceeding in this direction. If the straightforward structure of a large rectangular base is made, additional effects come into play, which have an important bearing on the operation of the device. In this case, because of the relatively large base resistance between contact and emitter, when the transistor is driven into saturation, the region of the base under the base contact becomes heavily forward-biased.

In this configuration the transistor acts as if it had a diode shunting from the base contact to the collector. This could result in an additional current path, negating the effect of the base resistor to prevent current hogging by lowering the base voltage at which appreciable current flows. In order to make this "overlap diode" effect insignificant,

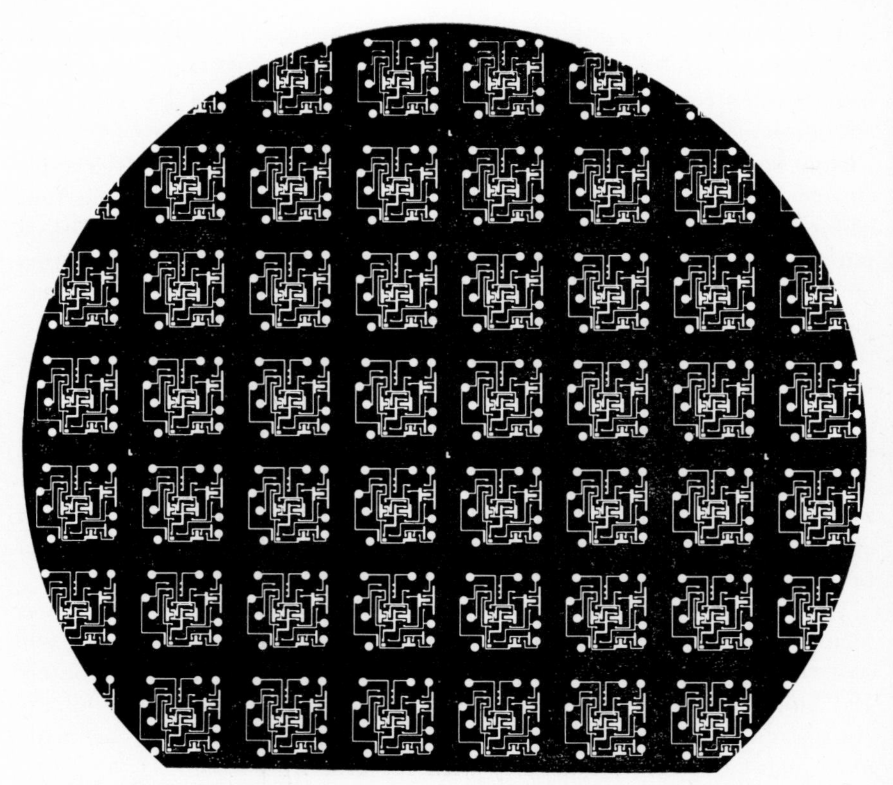

FIG. 5-55. A wafer of a completed microcircuit structure. Only the aluminum intraconnections for lead-bonding pads are visible with this illumination. Each of the structures is the 9-transistor, 14-resistor micrologic *S* element. The wafer is approximately ¾ in. in diameter.

it is necessary to insert in series with this shunting diode a resistance large compared to that of the base resistor itself.

This can be done by making the overlapped diode extremely small, increasing "collector" spreading resistance directly under the contact area. Accordingly, a contact dot as small as consistent with the available state of the technology is chosen for the end of the base resistor. This accounts for the peculiar transistor geometry to be seen in the photomicrographs of Fig. 5-56.

Next, we must choose the transistor surface geometry with an emitter size consistent with the current level at which the circuit will operate. The resistor length-to-width ratios are determined from the sheet conductivity of the base diffusion. The remaining problem is to lay out a geometric configuration which makes efficient use of the area, while resulting in as simple an intraconnection pattern as possible. A considerable simplification in layout can be achieved by running all grounded-emitter connections to the isolation region itself. In this manner, the back isolation diffusion acts as a common ground plane. This greatly simplifies the lead-crossing problem.

Figure 5-55 shows a completed wafer of these integrated circuits, illustrating the relationship of an individual circuit to a wafer of silicon.

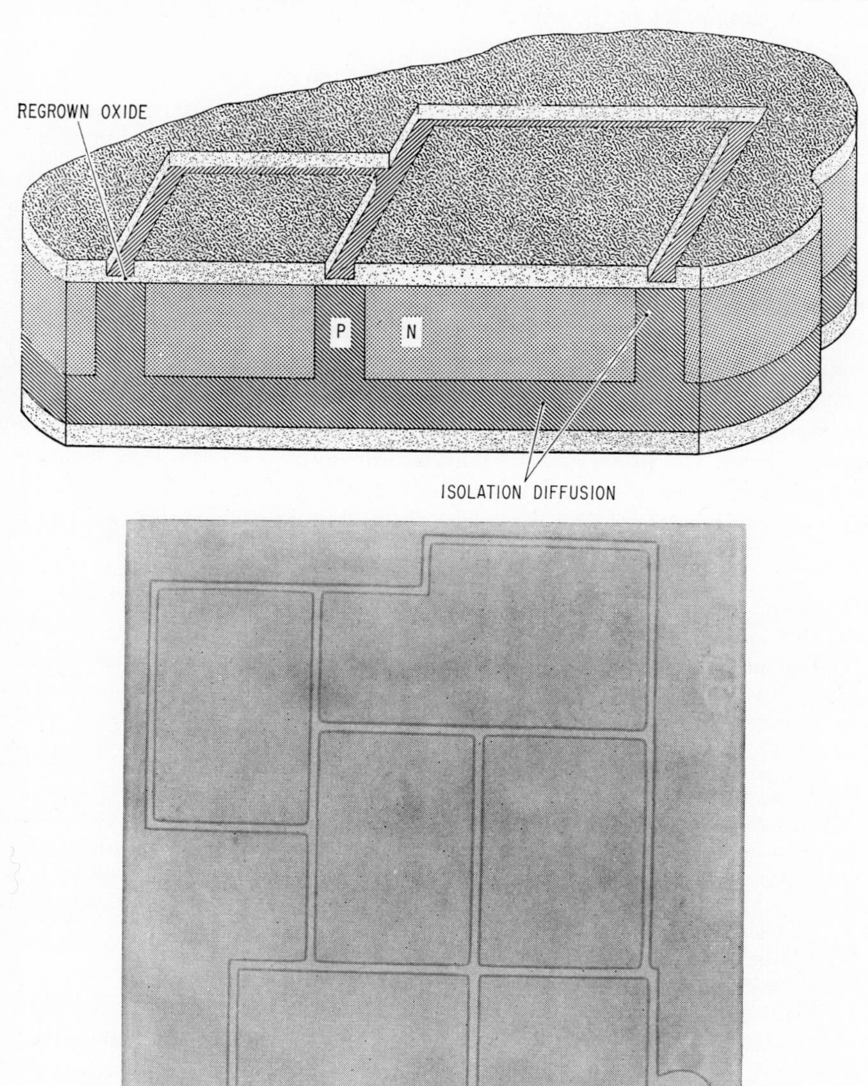

FIG. 5-56a. Photomicrograph and schematic cross section of a micrologic element after isolation diffusion.

Only the metal intraconnection pattern is visible. All the photomicrographs in Fig. 5-56 refer to a single structure on the wafer.

The photomicrograph in Fig. 5-56*a* shows a configuration of the isolation region suitable for this circuit. This consists of an array of 0.001-in. lines. The five collector resistors will be prepared in the upper right-hand rectangle, while the other five rectangles will contain the transistors and their associated base resistors. The circle at the lower right-hand corner of the isolation pattern is included so that a common ground contact might be made to the isolation region from the top surface of the device for packaging techniques, where it is inconvenient to employ a connection to the bottom of the semiconductor die.

1. Manufacturing Steps

Table 5-1 shows the process steps required to complete the micrologic element. After the crystal is grown and sliced, the wafers are lapped to a uniform thickness of 120 μ, to remove the surface damage introduced by diamond sawing. A controlled etching step reduces the thickness of the wafers to 80 ± 5 μ. This thickness is chosen as the best compromise between difficulty in handling the wafers without breakage through the remainder of the process and complications caused by the long isolation diffusion required to completely penetrate the wafer. This 80-μ dimension determines the 120-μ dimension after lapping, since approximately

TABLE 5-1. STEPS IN MICROLOGIC PROCESS

1. Grow crystal.
2. Slice crystal.
3. Lap wafers.
4. Etch wafers.
5. Oxidize wafers.
6. Photoengrave oxide (to expose isolation pattern).
7. Deposit boron (in isolation pattern).
8. Diffuse boron (through wafer).
9. Photoengrave oxide (to expose base and resistor patterns).
10. Deposit boron (in base and resistor pattern).
11. Diffuse boron (to desired base-collector junction depth).
12. Photoengrave oxide (to expose emitter and collector contact patterns).
13. Deposit phosphorus (in emitter and collector contact pattern).
14. Diffuse phosphorus (to desired emitter junction depth).
15. Strip oxide from back of wafer.
16. Evaporate gold on back of wafer.
17. Diffuse gold throughout wafer.
18. Photoengrave oxide (to expose regions for interconnection contacts).
19. Evaporate aluminum (over entire surface).
20. Photoengrave aluminum (to the desired interconnection pattern).
21. Heat-treat wafer (to assure good bonding and ohmic contacts).
22. Scribe and break wafer to dice.
23. Sort dice (optically and electrically).
24. Attach die to header.
25. Bond leads.
26. Weld leads.
27. Clean subassembly.
28. Vacuum-bake subassembly.
29. Seal micrologic element.
30. Tumble element.
31. Temperature-cycle element.
32. Leak-test element.
33. Centrifuge element.
34. Test element.

20 μ of material must be removed from each surface of the wafer to assure that all the damage introduced by the lapping operation has been removed.

Step 5, oxidation of the wafers, takes place in a wet-oxygen atmosphere at 1200°C. It is necessary to build up a relatively thick oxide at this point since the high boron concentration and long diffusion time which will be employed to achieve isolation put a severe strain on its masking ability. Approximately 0.8 μ of oxide is prepared.

Step 6, the photoengraving of the oxide to expose the isolation pattern, is a straightforward masking operation employing contact printing of the mask pattern onto the wafer surface and etching with fluoride-containing oxide-removal etch.

Step 7, boron deposition, is performed at 1200°C, using a boron trichloride or boron tribromide source. It is desirable to achieve very high surface concentration at this time, and so a maximum concentration of boron is deposited consistent with no formation of the impenetrable black skin. The oxide has been completely removed from the back of the wafer during process step 6. The boron deposition is done with the wafers in a standing position in the boat so that both sides receive the deposition.

Diffusion of boron in step 8 then proceeds for 24 hr at 1300°C in a slightly oxidizing atmosphere of dry oxygen. This time-temperature combination is adequate to assure that the diffusion fronts from the isolation grid pattern on the wafer top and from the entire back side meet and overlap significantly. It is important that the overlap be relatively large, since we depend upon the conductivity from the grid on the top side to the layer on the back surface for our emitter-grounding connection. If these regions do not overlap sufficiently, it is possible to introduce large emitter resistance values, which interfere with the operation of the circuit.

The next operation, step 9, is another masking step to expose the bottle-shaped transistor bases and the resistor patterns. The same oxide etch is used as before. In this case, it is necessary that the mask pattern be positioned precisely with respect to the initial pattern.

Boron is then deposited in step 10 and diffused in step 11 to the desired base-collector junction depths. The photomicrograph in Fig. 5-56*b* shows a structure on the wafer at this stage of the operation. The accompanying diagram shows a cross section through one of the resistors and a transistor.

Since the boron diffusion, step 11, was performed in an oxidizing atmosphere, a layer of SiO_2 has re-formed over the originally exposed area. This layer is of the order of 0.6 μ in thickness.

Step 12 is another oxide-photoengraving operation. In this case, holes through the oxide are opened for the emitter and in those regions where collector contacts will be made to the transistors.

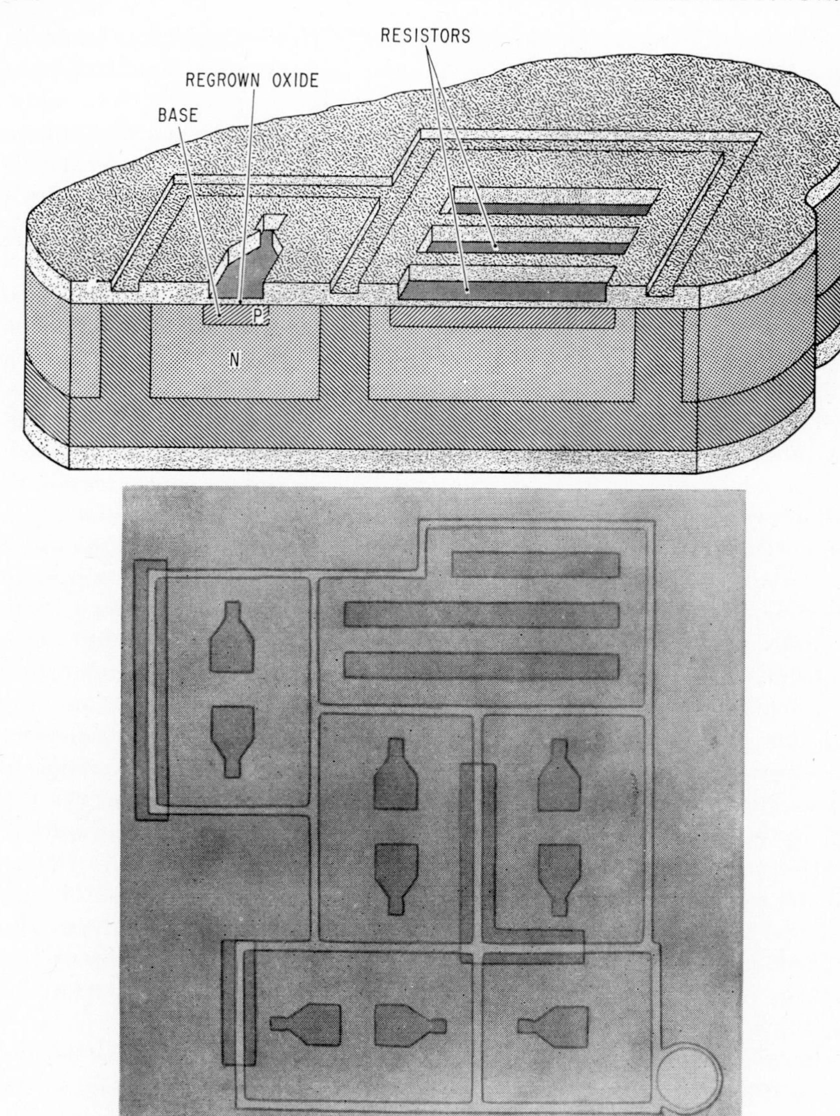

FIG. 5-56b. Photomicrograph and schematic cross section of a micrologic element after base and resistor diffusion.

Step 13 is the deposition of phosphorus from a P_2O_5 source, followed by the diffusion of the emitters in step 14 to the desired depth. For the transistor structures employed here, we desire an emitter junction depth of about 1.5 μ, with a base width of the same magnitude.

The inclusion of the heavy emitter diffusion, corresponding to a sheet resistivity of about 2 ohms/square over the collector contact areas, serves a twofold purpose. First, it facilitates the later preparation of ohmic contacts to these regions, and, second, it will decrease the effective collector saturation resistance of the devices. This is especially true since some of these regions will be used for lead crossings; for example, the one in the lower right isolation region is so used. The photomicrograph in Fig. 5-56*c* shows the device structure subsequent to this emitter and collector contact diffusion.

The next three steps, 15, 16, and 17, are those required to control the storage time of the transistors to a value consistent with the speed requirements of the circuit. In this case, gold is diffused from the back side of the wafer throughout the entire structure and, in particular, into the collector body, where it forms recombination centers for any holes injected during the operation of the transistors in saturation. This is a relatively common technique of minority-carrier storage control in silicon switching devices. The gold is evaporated on the bare back of the wafer, and diffusion proceeds at temperatures of the order of 1000 to 1050°C for times of the order of 10 min. Gold is extremely rapid as a diffuser, and this time-temperature cycle is adequate to assure appreciable concentrations in the desired regions of the structure, while not interfering with the diffusion profiles of the ordinary donor and acceptor impurities established previously.

The next step, number 18, is another oxide photoengraving operation that is employed to expose those points where contact must be made to the silicon for the thin-film intraconnections. The photomicrograph in Fig. 5-56*d* shows the structure after this operation. As well as emitter, base, and collector contacts to the transistors, holes have been opened over the resistor stripes.

Next, in step 19, the entire surface of the wafer is coated with an evaporated layer of aluminum which is photoengraved in step 20, employing a sodium hydroxide etch to leave the desired intraconnection pattern. The last photomicrograph, Fig. 5-56*e*, shows the wafer subsequent to this aluminum photoengraving operation.

Though aluminum adheres readily to both silicon dioxide and silicon, it is desirable to give the film a heat treatment to improve this adherence and to assure that good ohmic contacts are obtained. This heat treatment is in the vicinity of the aluminum-silicon eutectic temperature. If the eutectic temperature is exceeded, the time cycle must be held short, so that the aluminum does not significantly penetrate the silicon dioxide layer.

At this stage, the circuits are complete and need only be separated from one another, packaged, and tested. The rest of the operations in the flow table suggest the sequence of events employed for micrologic elements packaged in a standard Kovar eyelet-type header similar to the

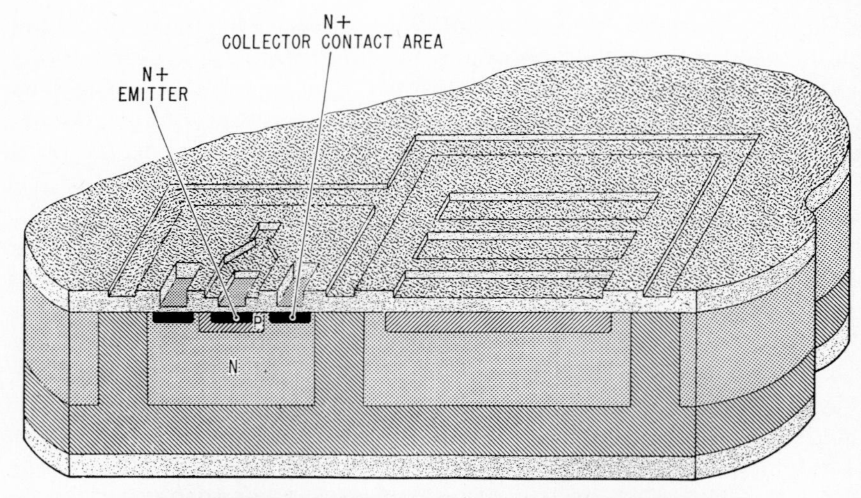

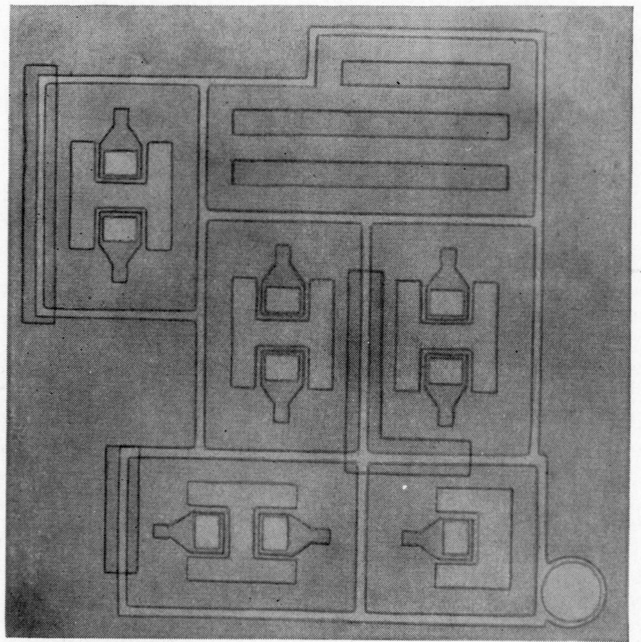

FIG. 5-56c. Photomicrograph and schematic cross section of a micrologic element after emitter diffusion.

TO-5 transistor package. For other packages, for example a flat ceramic package, the sequence of events might be changed slightly.

In summary, one sees that the operations required to make this particular semiconductor integrated circuit include some nine high-temperature furnace operations, two metal-film evaporations and five indexed masking operations. If each of these major steps is broken down into the substeps which must be considered in production, for example, if each masking operation is divided into cleaning, coating, exposing, developing, and stripping steps, as is commonly done in specifying these processes for production workers, then the total number of process steps prior to dicing of the wafer is well over 100. A sequential materials-processing operation of this sort can function only if most of the individual steps

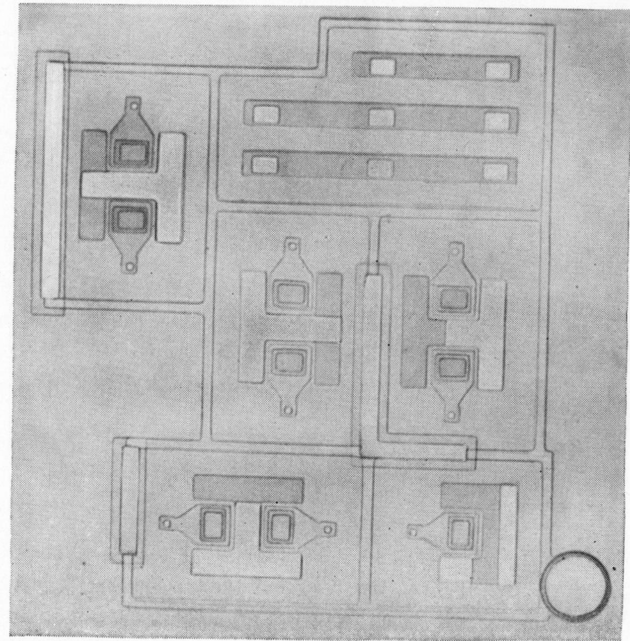

FIG. 5-56d. Photomicrograph and schematic cross section of a micrologic element after oxide removal for contacting.

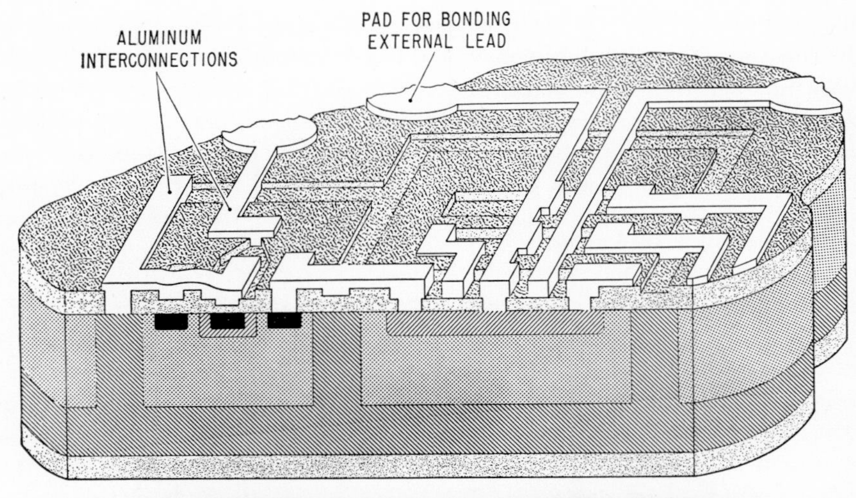

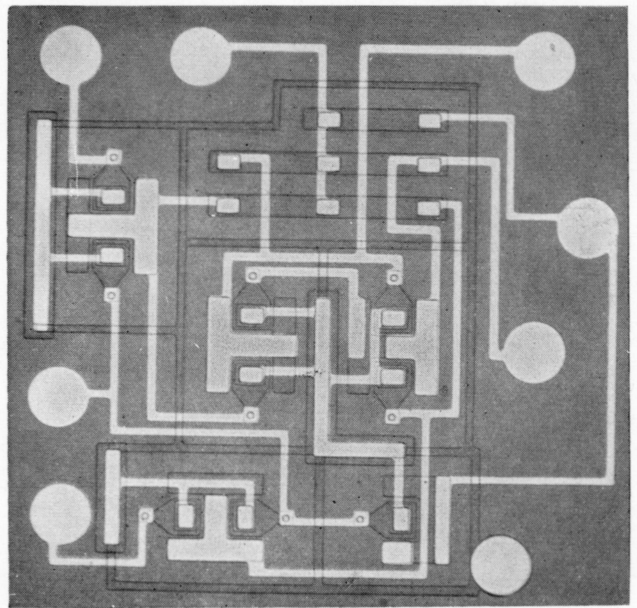

FIG. 5-56e. Photomicrograph and schematic cross section of a micrologic element after intraconnection metallization.

have yields very close to 100%. These operations can be this well controlled. In other circuit structures which it will be desirable to make in the near future, for example, those including epitaxial growth, thin-film resistors and capacitors, and *npn* and *pnp* transistors in the same substrate, the total number of processing operations can increase by an appreciable amount. Even so, it appears certain that the increased processing complexity will more than justify the additional production steps by offering greater electronic function for a given cost.

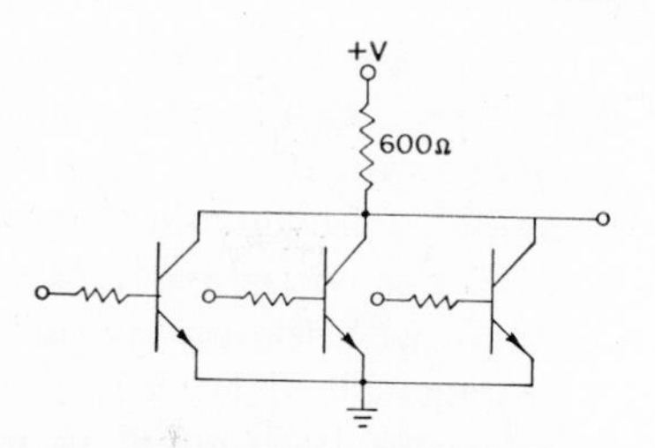

FIG. 5-57. Circuit diagram of a micrologic *G* element.

2. Properties and Characteristics

Let us now examine some of the properties of the microcircuit which we have made. Perhaps to simplify discussion, it will be more appropriate to discuss the gate or *G* element whose circuit diagram is shown in Fig. 5-57. First consider the isolation junction containing the three transistor structures. Figure 5-58 is a distribution curve of the breakdown voltages of isolation regions on micrologic gates prepared in this manner, measured at 10 μa current. One sees that the breakdown voltage is typically above 100 volts. Since the circuitry was designed to operate at 3 volts, it is obvious that adequately high isolation has been achieved. It can be confirmed that the collector resistances are near the nominal value, and experience over extended production runs suggests that $\pm 20\%$ is a reasonable estimate of the long-term accuracy which can be achieved by this production process.

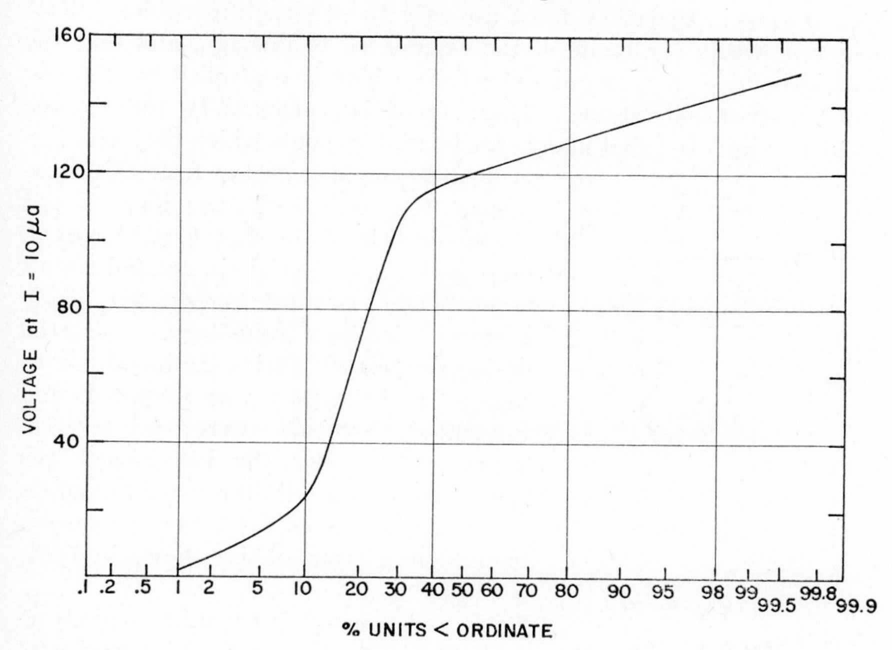

FIG. 5-58. Distribution of voltage across isolation region corresponding to 10 μa leakage current for a micrologic gate element.

The production distribution obtained for this resistor was shown in Fig. 5-33. Fan-in and fan-out measurements over the required temperature range show that the base resistor is performing the desired function to prevent the current-hogging problem, and that the small base contact has so reduced the effect of the overlap diode that it does not interfere with the proper operation of the base resistor.

Propagation delay times measured at room temperature with fan-in and fan-out both equal to unity show that average propagation delay per gate runs about 20 nanosec. The isolation capacitance for the three-input gate turns out to be about 60 pf at a bias of a few tenths of a volt, the level it experiences in operation. Thus, one estimates that of the 20-nanosec propagation delay, approximately half is contributed by the charging of the parasitic isolation capacitance. Since the 20-nanosec propagation delay is completely consistent with the speed objective of one megacycle clock rate for a system over the entire temperature range -55 to $+125$°C, the effect of the isolation is compatible with the desired operation.

It should be pointed out that when such circuits are being produced in quantity at reasonable yield, the production capability greatly exceeds that required to make a gate or a family of six microcircuits. What exists is a technology capable of making a large variety of circuits, limited only in that they must use only those structures which can be achieved within the limits of the established technology, and that the complexity does not exceed some limit which is controlled by yield or other cost considerations. Thus, there is a capability for making custom circuits to fulfill many specific requirements which may arise.

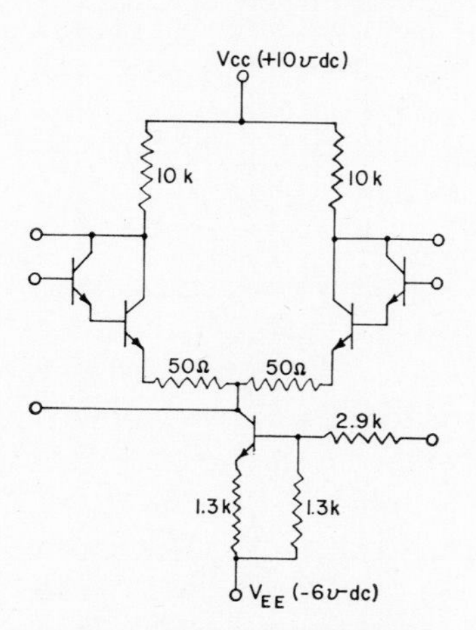

FIG. 5-59. Circuit diagram of Darlington differential-amplifier structure.

As an example to demonstrate this capability, the differential-amplifier circuit shown in Fig. 5-59 was bread-boarded, employing the special integrated-circuit components described in Sec. 5-4, made with the micrologic technology. Using the design established during the bread-boarding, a set of masks was prepared, and integrated versions were constructed. Three weeks after the breadboard was firmed up, completed differential amplifiers in wholly integrated form were produced, having the characteristics shown in Table 5-2.

The performance of the integrated structures was actually somewhat superior to that of the breadboard itself. The process

TABLE 5-2. TYPICAL PERFORMANCE DATA ON DARLINGTON DIFFERENTIAL AMPLIFIER SHOWN IN FIG. 5-59

Offset voltage referred to input (both inputs grounded)	2 mv, −55 to 125°C
Common-mode rejection	>80 db, −55 to 125°C
Gain (single-ended, 10 mv rms input)	25–28 db, −55 to 125°C
Bandwidth	500 kc

was held constant in this example. Even the gold diffusion for the limiting of lifetime in the collector bodies of the transistors was included. Here we have an example of a technology established primarily for its usefulness in the preparation of a family of digital circuits being employed to make a serviceable linear amplifier. As more bits and pieces of technology are added to the available production capability, the range and performance of circuits accessible in fully integrated form will expand rapidly. The idea of using special components consistent with the technology in the design of future integrated circuits will certainly decrease lead times, and result in rapid turnaround to achieve the lowest-cost optimum, special integrated structures.

5-6. COST AND RELIABILITY

Since it is generally agreed that the major promises of the semiconductor integrated-circuit approach to microcircuitry are reduced systems costs and improved reliability rather than merely decreased size and weight, it is important to examine in more detail why improved reliability and lower systems cost can be expected. First, let us consider cost. This can be divided into the cost of the individual integrated circuits, i.e., the "component" cost, and the costs of assembling these to make a system. The cost of the microcircuits themselves can be further subdivided into three areas: the production cost, the tooling cost, and the engineering cost necessary to design the circuits initially.

Consider the production cost. For this it is useful to divide the manufacturing process into two parts which we will call *fabrication* and *assembly*.

Fabrication will take in all the batch-processing steps necessary to make the circuit die and will include a first electrical test corresponding to step 23 in Table 5-1 for the micrologic process. At this test, a check for gross electrical problems such as short circuits and open circuits is made on the individual circuit structures by probing on the separate dice.

The assembly part of the process is that part from this die-sort operation through assembly and including final test. The total microcircuit production cost C_t can be considered as made up from the cost of a good die plus the cost of assembly as modified by the yield at final test. That is,

$$C_t = (C_d + C_a)Y_f^{-1} \tag{5-3}$$

where C_d is the cost of a good die, C_a is the assembly cost, which equals the total cost of all the operations in this portion of the process divided by the number of units finally tested, and Y_f is the yield at final test, defined as the ratio of the number of good devices to the total tested. But C_d can be expressed in terms of the cost of any die C'_d and the die-sort yield Y_d, as follows:

$$C_d = C'_d Y_d^{-1} \tag{5-4}$$

Define further a as the area of the die, A as the area of the wafer, and C_p as the cost of processing a wafer up through the die-sort operation. Then

$$C_d \cong C_p a (A Y_d)^{-1} \tag{5-5}$$

since for $a \ll A$, the usual case, the number of dice obtained from a wafer is A/a. Accordingly, the cost of the completed unit is given by

$$C_t \cong [C_p a (A Y_d)^{-1} + C_a] Y_f^{-1} \cong C_p a (A Y_d Y_f)^{-1} + C_a Y_f^{-1} \tag{5-6}$$

For small transistors, the second term in Eq. (5-6) usually dominates. That is, the assembly cost exceeds the fabrication cost. The crossover point where the contribution of these two terms is equal occurs somewhere in the range of a = 1,000 to 5,000 mils^2: that is, for a die size of about 30 to 70 mils on an edge. The exact point of crossover varies with the particular product and with the detailed specification which the device must meet.

For microcircuits, both C_p and C_a are increased over the transistor case. The increase in C_p results from the increased number of processing steps required during fabrication, whereas C_a reflects the increased number of leads which, in turn, affects the price of the package, the lead-bonding operation, and final test. C_p increases somewhat faster than directly proportional to the number of process steps, since more process steps introduce additional opportunity for loss of material during fabrication. Even if only the increased probability of dropping batches of wafers on the floor is considered, the rate of increase exceeds direct proportionality.

As an indication of how the number of steps for microcircuits compares with that for conventional transistors, the process flow diagram in Table 5-1 has approximately twice the number of operations that would be included in a similar flow chart for a standard planar transistor. Hence, the processing cost for a microcircuit wafer is something in excess of twice that for the transistor. C_a, on the other hand, generally increases at a rate less than proportional to the number of leads, since only a few of the operations during the assembly portion of the process are directly affected by the number of leads, and these usually only in approximately direct proportion.

Accordingly, the crossover point for microcircuits where the fabrication cost equals the assembly cost moves in the direction of smaller die size. On the other hand, the dice useful in microcircuitry are usually appreciably larger than those associated with ordinary low-power transistors. This results in the dice cost being the dominant cost for semiconductor microcircuits.

From Eq. (5-5) we see that die cost is inversely proportional to the die-sort yield Y_d. To appreciate what influences Y_d, it is helpful to examine the cause for shrinkage at die-sort testing. This test point is, in general, a place where no precise parameter measuring is done. It is not ordinarily employed as a point where the tails are cut off normal distribution curves. Rather it is used as a monitor for catastrophic problems.

Sometimes, entire batches can be found which are rejects, but even in good batches one finds that there is a certain occurrence of inoperable structures. In general, this occurrence appears very much as though it is linked to the probability of inclusion of a "bad spot" in a sensitive portion of the circuit structure.

These bad spots can occur from a variety of mechanisms; for example, a pinhole in the silicon dioxide layer under a metallized intraconnection can result in a short circuit to the substrate. Similarly, one of the diffusion defects commonly called "pipes," which occurs in the region of a transistor structure, can result in an emitter-collector short circuit. Problems of this kind are major causes for shrinkage at die sort.

In general, the problem can be represented by assuming that there is a certain density of bad spots, spread more or less at random over the surface of the semiconductor wafers. Inclusion of one of these in an area of the structure sensitive to this particular type of bad spot results in an inoperable device.

Hence, assuming random density of the spots, the chance of inclusion of one increases with the increasing area of the device. Accordingly, Y_d is a strong inverse function of device area. For die-sort yields less than 50%, which are typical of microcircuits, and of large transistor structures for that matter, the loss of yield due to bad-spots inclusion varies more rapidly than in inverse proportion to area.

There is the possibility where most of yield loss occurs because of parameter distribution problems that this picture can be altered appreciably. For example, a microcircuit depending on an extremely precise ratio between a pair of large resistors might actually show increased yield, with an increase in the width of the resistor stripes, so that minor variations in the line widths due to the masking operations would have less influence on the resistor values. In a case like this, however, it is still necessary to consider the total effect on die costs rather than die-sort yield alone. Unless increasing the resistor width increases the yield faster than it decreases the total number of dice per wafer, the total die cost will increase.

Using the above estimate that Y_d varies more rapidly than inversely proportional to the device area, we see from Eq. (5-5) that the cost of a good die varies faster than the square of the die area. This strong dependence upon die size should be borne in mind in any microcircuitry design considerations. Since area becomes the important contributor to cost, components consuming large areas are very expensive. For example, large resistance values made with low sheet resistances should be carefully considered. If one uses the base diffusion to make a 40,000-ohm $\pm$ 20% resistor, it consumes an area the equivalent of four completely isolated transistor structures. If greater precision is required in the same resistor so that the line width must be increased from 1 to 2 mils, another factor of four is introduced into the required area.

On the other hand, extra processing steps required to extend the available technology in the construction of microcircuitry are relatively inexpensive. For example, in order to add thin-film resistors to the micrologic technology, one need add only three additional processing steps—an evaporation, a masking, and a heat treatment. This results in a total increase in the number of fabrication process steps of about 12%. If this additional technology can result in even a small decrease in the area required for a given circuit, in this case if the die size could be reduced by as little as 10% on an edge with no effect on die-sort yield, then the total cost of producing the microcircuit would be decreased. Since the addition of some of the other technological capabilities can be expected to decrease die size appreciably for a given circuit function, the increased processing complexity which will be incorporated in integrated circuitry production as time goes on will further decrease the production costs.

Of course, one way to assure high die-sort yield is to perform no die-sort testing, assuring that the yield is always unity. Since the total cost of the microcircuit is generated by multiplying all the costs up through final test by the final yield, however, it is more important that Y_f be high than that Y_d be maximized, since Y_f also multiplies the assembly costs. Accordingly, it usually pays to perform tests that can be done easily on the unpackaged dice at the die-sort operation if these result in appreciable yield loss. Since the planar techniques employed in making the microcircuits under consideration result in structures which show negligible degradation during the usual assembly operations, the die-sort test is an efficient cost-reducing operation.

It cannot be overemphasized, however, that the most important way of achieving low-cost microcircuits is to design the circuits themselves for low cost. One must limit the component values as much as possible to those achievable in small areas. He must accept large tolerances on individual components, and must avoid such luxuries as high-voltage requirements wherever possible.

To illustrate the importance of these considerations, curve 1 of Fig. 5-60 shows the voltage at 10 μa current through the isolation diodes from a production run of micrologic elements, measured at room temperature. One sees that 70% of the isolation regions have voltages of at least 100 volts for this particular current. Hence, for individual components, one could specify 100-volt ratings and suffer only a 30% yield loss. However, if he is interested in a microcircuit involving several of these isolation regions, the situation can be altered appreciably. The curves labeled 2 to 4 and 8 in Fig. 5-60 are generated from curve 1, for 2, 4, and 8 isolation regions, respectively, assuming that each is chosen at random from the distribution of curve 1. For eight such regions, one sees from curve 8 that only 10% of the groups would have all their voltage ratings in excess of 100 volts, while 50% would exceed 20 volts and 70% exceed 10 volts. Accordingly, a specification on a microcircuit involving eight isolation regions requiring 100 volts isolation capabilities would result in a sevenfold decrease in yield over the identical circuit requiring only 10-volt ratings. Although this, in some respects, is a pessimistic example

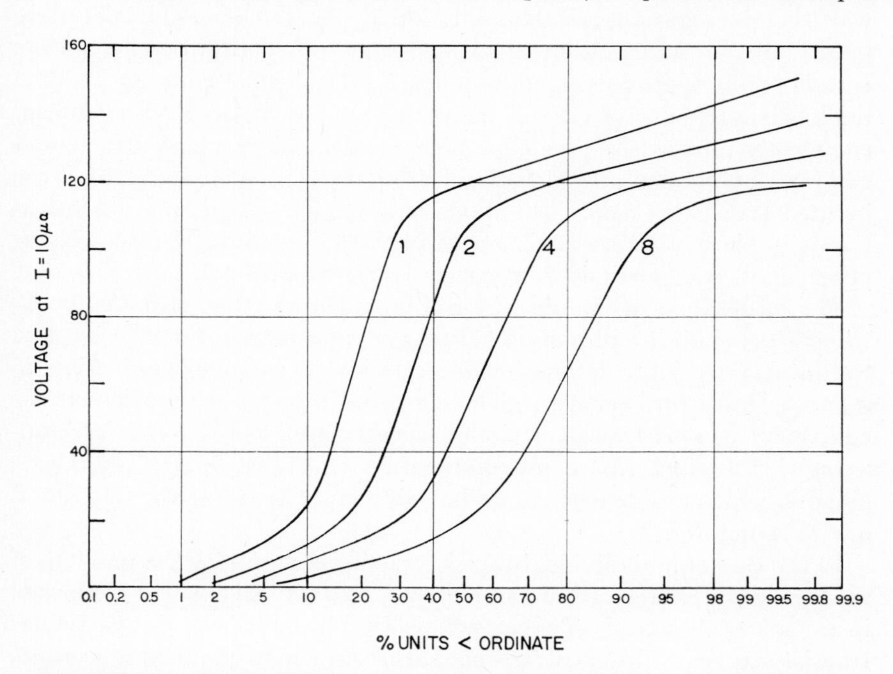

FIG. 5-60. Distribution of isolation breakdown voltage for 1, 2, 4, and 8 isolation regions, respectively. The curves with several regions were generated from the curve for one region, assuming a random distribution.

since no allowance is made for the possibility that there may be area correlations so that the high-leakage regions are not completely randomly distributed, this correlation is relatively small.

The qualitative conclusion that an unrealistically high-voltage rating results in an extremely severe cost penalty is certainly correct. It has been shown, in addition, that the planar protected junctions of the type shown in Sec. 5-3 are very stable. Even when they show excessive leakage initially, they show no tendency to drift. Accordingly, for operation at 10 volts, structures good to 100 volts are not necessarily more reliable than those good only to 10 volts initially.

Specifications should be established realistically on the basis of need. Unrealistic specifications are much more expensive for microcircuits than they are for individual components. Adjusting circuit design toward low voltage is, in addition, the correct direction in which to go to achieve greater reliability from other considerations as well.

In order to evaluate the promise of the integrated-circuitry approach with respect to cost, it is useful to compare this directly with that for transistors. A transistor in an isolated region, similar to one shown in Fig. 5-27, including all contacts made to the top surface, requires an area of about 100 mils^2, assuming that one restricts himself to 1-mil line widths and separations and 0.5-mil total indexing tolerances for successive masks—the tolerances allowed in micrologic. A conventional separate transistor on the other hand requires a 20-mil-square die as about a practical minimum for ease of handling. If one allows a 10-mil border completely around the edge of a microcircuitry die for lead attachment and for separation of one microcircuit from the next, one can get 64 of the isolated transistors mentioned above in a $\frac{1}{10}$-in. square area. Alternatively, in place of a transistor, one can have a 4-kilohm resistor. On the other hand, one gets only 25 separate transistors from an area this size.

The microcircuit assembly cost will be much less than that for the 25 transistors, since it will undoubtedly have considerably fewer leads; and the package which contains it will certainly be less expensive than 25 separate transistor packages. Therefore, one can expect the microcircuit equivalent to a 64-transistor circuit to cost much less than the 25 transistors if the yields are at all comparable. Fortunately, for the microcircuits, yields are defined somewhat differently than for the individual device structures.

In the case of an individual transistor, one is faced with the problem of specifying the parameters of this device, so that it will operate when used in any one of thousands of circuits, which can be built from this particular transistor type. This requires some rather arbitrary numbers concerning the various device parameters, to define a good unit.

In the case of the transistor incorporated into a microcircuit, however, the definition of yield becomes more closely related to its particular operation. This transistor is good if, when used with *exactly* those other components in the circuit, it will perform the *specific* function required of the circuit itself. This is a much less stringent requirement than the former, and results in yields of entire microcircuits being of the same order as those for individual components. Indeed, this order-of-magnitude equivalence of yields is necessary for the production of microcircuits to be feasible. Fortunately, this is the case. Actually the example considered above is a worse case, in the respect that all the transistors in the microcircuit were considered as requiring separate isolation regions.

In actual practice, in such a complex circuit, there would usually be several instances where common regions could be used, resulting in the possibility of even more transistors in the given area. Also, the inclusion of additional technologies, for example the addition of thin-film resistors or of some of the epitaxial isolation techniques, can further increase the amount of circuitry per unit area. Even at the circuit-complexity level of the micrologic gate—three transistors, four resistors, and six lead bonds—the cost of producing the integrated circuit is approximately equal to the cost of producing the individual components in equivalent quantities.

As the complexity is increased, microcircuits are favored more and more strongly, until one reaches the point where the yield because of the complexity falls below a production-worthy value. The point at which this occurs will continue to push rapidly in the direction of increasing complexity. As the technology advances, the size of circuit function that is practical to integrate will increase rapidly for a variety of reasons. The problems associating yield and area will be better understood, and the yield improvements which will result will allow the use of larger areas.

More circuitry will be possible in a given area through the use of finer-scale structures. For example, changing from the 1-mil line width with ½-mil indexing tolerance to ½-mil line width with ¼-mil indexing tolerance is probably within the present state of the art. This decreases the area of a given circuit by a factor of four. Additionally other structures, such as some of those described in Sec. 5-3 and others still to be conceived, will increase the available electronic function per unit area.

Another contribution to the cost of microcircuitry that must be considered is the tooling cost required to make a new circuit. If one considers that a standard production technology exists, within which one can make a large variety of circuits by changing only the set of masks employed and the electrical testing, the tooling cost to be considered is that of masks and test equipment.

Two approaches to the mask problem have been considered. One involves making a complete set of optimum masks for each new circuit.

The other considers the standard item to be a wafer containing a variety of individual component structures such as transistors and resistors, rather than an available technology, and makes only one mask—the intraconnection mask—which is superimposed on this "master wafer," to intraconnect the already existing components into the new circuit.[52]

This single-mask technique offers the advantage of rapid turnaround time, because it circumvents a delay of up to several weeks in the early fabrication steps inherent in a smoothly flowing production line with adequate in-process inventory. A single mask requires less time to prepare than a complete set of masks and can be made for a few hundred dollars. This single-mask master-wafer technique, however, obviously makes extremely nonoptimum use of the available area. Accordingly, it is suitable only for very small quantities of microcircuitry, probably less than 100, and where short turnaround time is important.

An optimum set of masks appears the more economical approach for any microcircuit requiring more than several hundred finished structures and for any very complex circuit. A set of masks optimized for a particular circuit probably costs two to four times as much as an individual mask. Since even in this case the mask amortized over the first several hundred circuits contributes only a small fraction of their cost, it is not an overriding consideration, except for extremely small-usage items where the desirability of integration must be especially carefully considered anyhow.

Testing costs are a variable that is difficult to estimate in general, since they are so dependent upon detailed specifications. It is possible to conceive of quite flexible microcircuitry test equipment having an assortment of available forcing functions and measuring schemes that can be programmed for a specific circuit. Testing a microcircuit with such a tester should be less expensive than testing the equivalent number of separate components.

At least for the present, the third contribution to the cost of producing microcircuits, the circuit-engineering expense, is not likely to decrease, since the circuit design engineer, experienced at designing with standard components, must familiarize himself with an entirely new set of boundary conditions. After this initial educational period, which can certainly be compressed through maximum early consultation with the integrated-circuit manufacturer, the circuit-engineering expense should be no higher than with conventional electronics. In fact, the use of the integrated technology will simplify the job of component selection. An important part of the circuit designer's job will be the consideration of appropriate functional tests to assure proper circuit performance.

Although detailed analysis of system cost is beyond the scope of this chapter, some general arguments can be made. Consider as an example the logic portion of a computer. Beyond the components and other parts cost, the cost to complete the system can be divided into engineering, assembly, and checkout. Engineering costs with microelectronics should be greatly reduced because only the layout of the logic need be considered. Because of substantially fewer parts and, accordingly, simpler layout, assembly cost is also decreased. Beyond that, checkout is simpler and less expensive, since there are fewer available test points and fewer possibilities for error. Each of the circuits has been completely evaluated before insertion. It has been estimated that the cost of designing, assembling, and checking out a first-prototype logic section of a digital computer using micrologic would be only 20% of the cost of one that using standard components.[53] Early experience confirms this prediction at least qualitatively.[54]

In considering the impact of semiconductor integrated circuitry upon system reliability, one must appreciate that each integrated circuit replaces many components and interconnections, compared with a system made from conventional components. In order to improve over-all system reliability, it is necessary only that the integrated circuit exceed the reliability of the total of the features which it replaces. Not enough experience with systems employing integrated circuits has been accumulated to date to demonstrate that the anticipated improvement in reliability has been realized, although what experience does exist is not in contradiction with the prediction. In order to estimate the magnitude of the improvement to be expected, it is useful to look at the failure mechanisms for these circuits, and to compare them with devices of established reliability. The most obvious comparison to make is with silicon planar transistors, for which a large amount of data has been accumulated.

The semiconductor integrated circuits employ identical *pn* junction structures with the planar devices and are dependent upon the same silicon dioxide insulating layer for passivation. Even the feature of running the metallization over the silicon dioxide is used in some transistor types and has been the subject of extensive reliability evaluations. The lead bonding and packaging for microcircuits can be identical with that for individual transistors. For example, the package usually employed for micrologic elements is identical with the transistor package, except for the number of leads. Not all packages proposed for microcircuits are directly comparable with conventional transistor packaging. Those which differ must be considered independently.

The principal additional failure modes which one introduces as possibilities in such integrated circuitry as micrologic are those associated with the long metal intraconnections. The possibility of the occurrence of breaks in these intraconnections does not exist to the same degree in conventional transistors. Of course, the introduction of new technologies such as film resistors increases the number of failure modes. No single change in technology, however, introduces more possibilities of new failure modes than a change in the basic packaging scheme.

On ultrareliable transistors the residual failure mode is generally mechanical rather than any drift in the semiconductor structure itself. This has been confirmed both for large-scale life testing, such as that to which the planar transistors for the Minuteman missile system were subjected, and for accelerated testing, wherein a relatively small quantity of units are stressed until failure. For the Minuteman devices, failure rates less than 10^{-8} per hr were established. The residual mode was the opening of lead bonds. One sees no a priori reason why microcircuit reliability should differ significantly from that of the transistors, providing the packaging is equivalent, since there are no other important additional failure modes expected.

Considerable life-test data have been accumulated on micrologic elements. More data are avilable on these than on any other integrated circuits. Both operating-life and high-temperature storage data exist. Over 6,400,000 unit-hours of operating-life test at 125°C ambient, the maximum operation temperature for these structures, has been accumulated, representing some 2,070 elements. These were production units, mostly gates, although several flip-flops were also included.

Most of the units were rejects from the ordinary final test specifications for such problems as high isolation leakage or low available node current, resulting from high collector load-resistance values. Accordingly, if borderline units represent an increased tendency toward failure, this life test should be a worse case. One failure was observed in these tests. This resulted from a lead bond opening on one device. A group of these units were monitored in detail for parameter drift. No measurable drift was observed.

The storage-life data showed no failures at 150°C storage in 266,000 unit-hours. At 300°C storage, however, 16 failures were observed in 840,000 unit-hours, resulting in an estimated mean time to failure of 52,500 hr. All 16 failures were open circuits or high resistance in the lead connection. This mean time to failure at 300°C storage can be compared directly with a mean time to failure of 206,000 hr under identical storage conditions, measured for the Minuteman planar transistors. The ratio of a factor of four in mean time to failure is somewhat greater than the ratio of lead bonds of 5:2.

Whether this difference is significant is difficult to say, since the two products represent a different degree of production maturity. It does establish, however, that the microcircuit demonstrates stability about comparable to that of an individual component. Accordingly, a system composed of a given number of microcircuits should be about as reliable as a system containing that number of components; or, stated another way, a given system made from microcircuits should demonstrate an appreciably improved reliability compared to the equivalent system made from conventional components, since fewer components would perform the same function.

5-7. SUMMARY AND CONCLUSIONS

The structures and technology described above are reasonably representative of the present state of semiconductor integrated circuitry. This concept of functional electronics, wherein completed electronic functions are produced within and upon a single monolithic block of semiconductor material, is rapidly assuming a major role in advanced electronic systems. Its promise of improved reliability at decreased cost is being realized. Just as the performance capabilities of transistors are seldom limiting the performance of systems made by conventional electronics today, the microcircuits will soon deliver as high performance as is useful for the entire system. With decreased cost and improved reliability, larger and more complex systems than ever before practical can be considered. The amount of circuit function to be put economically in a single functional electronic block will increase rapidly. This, in turn, will further contribute to cost and reliability improvements in systems of a given size. While we have used digital computers and logic circuits as examples, because the advantages are most dramatic in these cases where large numbers of repetitive circuits are employed, the advantages will carry over into all of electronics. We can fully expect to find integrated semiconductor functional electronic blocks in consumer products when the technology has assumed a higher degree of maturity.

Much of the data and information in this chapter was previously unpublished. It represents work done by many contributors at Fairchild Camera and Instrument Corporation. While to single out a few is probably unfair to many others, I would like to give credit to D. Farina and N. Gault for supplying many of the data, and to V. H. Grinich and R. N. Noyce for critically reading the manuscript. I would like to thank Mrs. Ruth Ann Cameron and William Jimenez for their work on the figures, and Mrs. Helen Bonfadini for extracting and preparing the manuscript.

REFERENCES

[1] Molecular Electronics Receiver, *Semicond. Prod.*, vol. 5, p. 46, April, 1962.

[2] J. A. Hoerni, IRE Electron Devices Meeting, Washington, D.C., October, 1961; Fairchild Semiconductor, Division of Fairchild Camera & Instrument Corp., Mountain View, Calif., Technical Article TP-14, 1960.

[3] D. L. Klein, G. A. Kolb, L. A. Pompliano, and M. V. Sullivan, Electrochemical Society 119th Meeting, Indianapolis, Ind., May 1–3, 1961.

[4] See H. Reiss and C. S. Fuller, in N. B. Hannay (ed.), "Semiconductors," Reinhold Publishing Corporation, New York, 1959, for an extensive bibliography.

[5] See the first reference cited in Reiss and Fuller, above.

[6] R. L. Batdorf and F. M. Smits, *J. Appl. Phys.*, vol. 30, p. 260, 1959.

[7] C. S. Fuller and J. A. Ditzenberger, *J. Appl. Phys.*, vol. 27, p. 544, 1956.
R. C. Miller and A. Savage, *ibid.*, p. 1430.
C. T. Sah, H. Sello, and D. A. Tremere, *J. Phys. Chem. Solids*, vol. 11, p. 288, 1959.

[8] F. A. Trumbore, *Bell System Tech. J.*, vol. 39, p. 205, 1960.

[9] L. A. D'Asaro, *Solid-State Electron.*, vol. 1, p. 3, 1960.

[10] A. D. Kurtz and C. L. Gravel, *J. Appl. Phys.*, vol. 29, p. 1456, 1958.

[11] C. J. Frosch and L. Derrick, *J. Electrochem. Soc.*, vol. 104, p. 547, 1957.

[12] M. M. Atalla, E. Tannenbaum, and E. J. Scheibner, *Bell System Tech. J.*, vol. 38, p. 749, 1959.

[13] J. E. Sandor, *Electrochem. Soc. Electron. Div. Abstr.*, vol. 11, p. 228, May, 1962.

[14] M. M. Atalla, in H. Gatos (ed.), *Proc. Conf. Elemental and Compound Semicond.*, Interscience Publishers, Inc., New York, 1960.
J. R. Ligenza, *J. Electrochem. Soc.*, vol. 109, p. 73, 1962.

[15] C. T. Sah, H. Sello, and D. A. Tremere, *J. Phys. Chem. Solids*, vol. 2, p. 228, 1959.

[16] L. B. Valdes, *Proc. IRE*, vol. 42, p. 420, 1954.

[17] W. W. Happ and W. Shockley, *Bull. Am. Phys. Soc.*, ser. 2, vol. 1, p. 382, 1956.
B. McDonald and A. Goetzberger, *J. Electrochem. Soc.*, vol. 109, p. 141, 1962.

[18] W. R. Wilcox and T. J. LaChapelle, *Electrochem. Soc. Electron. Div. Abstr.*, vol. 11, p. 194, May, 1962.

[19] P. Baruch, C. Constantin, J. C. Pfister, and R. Saintesprit, *Discussions Farraday Soc.*, vol. 31, p. 76, 1961.

[20] A. P. Hale and B. D. James, IRE Electron Devices Meeting, Washington, D.C., October, 1960.

[21] J. C. Marinace, *IBM J. Res. Develop.*, vol. 4, p. 248, 1960.

[22] H. C. Theurer, *J. Electrochem. Soc.*, vol. 107, p. 29, 1960.

[23] S. K. Tung, in J. B. Schroeder (ed.), "Metallurgy of Semiconductor Materials," vol. 15, p. 87, Interscience Publishers, Inc., New York, 1962.

[24] J. Corrigan, in J. B. Schroeder (ed.), "Metallurgy of Semiconductor Materials," vol. 15, p. 103, Interscience Publishers, Inc., New York, 1962.

[25] B. E. Deal, *J. Electrochem. Soc.*, vol. 109, p. 514, 1962.

[26] R. Glang and E. S. Wajda, in J. B. Schroeder (ed.), "Metallurgy of Semiconductor Materials," vol. 15, p. 27, Interscience Publishers, Inc., New York, 1962.

[27] H. Basseches, S. K. Tung, R. C. Manz, and C. O. Thomas, in J. B. Schroeder (ed.), "Metallurgy of Semiconductor Materials," vol. 15, p. 69, Interscience Publishers, Inc., New York, 1962.

[28] R. M. Warner, *Electronics*, vol. 35, p. 49, 1962.

[29] W. C. Dash, *J. Appl. Phys.*, vol. 33, p. 2395, 1962.

[30] W. G. Spitzer and M. Tannenbaum, *J. Appl. Phys.*, vol. 32, p. 744, 1961.

[31] C. C. Allen and E. G. Bylander, in J. B. Schroeder (ed.), "Metallurgy of Semiconductor Materials," vol. 15, p. 113, Interscience Publishers, Inc., New York, 1962.

[32] Kodak Photosensitive Resists for Industry, Eastman Kodak Company Publication P-7, Rochester, N.Y., 1962.

[33] H. Kressel, H. Veloric, A. Blicher, and D. Rauscher, *Electrochem. Soc. Electron. Div. Abstr.*, vol. 11, p. 190, May, 1962.

[34] J. L. Langdon, W. E. Mutter, R. P. Pecorarv, and K. K. Schuegraf, IRE Electron Devices Meeting, Washington, D.C., October, 1961.

[35] Solid Circuit Type SN510-515, Bulletins DL-S 612015-8 Texas Instruments, Inc., Dallas, Tex., October, 1961.

[36] A. Evans, Research on Semiconductor Single Crystal Circuit Development, Electronic Technology Laboratory, Contract AF33(616)6600, Wright Air Development Division, Air Research and Development Command, Technical Note 61-25, U.S. Air Force, Wright-Patterson Air Force Base, Ohio, August, 1961.

[37] R. E. Crippen, Private Communication.

[38] W. M. Lafky, Private Communication.

[39] R. E. Craig and D. E. Farina, Private Communication.

[40] Compatible Techniques for Integrated Circuitry, 2d Quarterly Report, U.S. Air Force Contract AF33(616)8276, prepared by Motorola Semiconductor Products, Inc., for U.S. Air Force Aeronautical Systems Division, Wright-Patterson Air Force Base, Ohio, 1961, p. 33.

[41] W. J. Ostrander and C. W. Lewis, in Luther E. Preuss (ed.), *Natl. Symp. Vacuum Technol. Trans.*, vol. 8, 1961.

M. Beckeman and R. E. Thun, *ibid.*

[42] G. W. A. Dummer and J. W. Granville, "Miniature and Microminiature Electronics," p. 247, John Wiley & Sons, Inc., New York, 1961.

[43] Reference 40, p. 19.

[44] W. Shockley, *Proc. IRE,* vol. 40, p. 1365, 1952.

[45] C. T. Sah, *Proc. IRE,* vol. 49, p. 1623, 1961.

[46] L. A. D'Asaro, *IRE WESCON Conv. Record,* 1959, part 3, p. 37.

[47] W. F. deBoice and J. F. Bowker, *Electronics,* vol. 35, p. 37, 1962.

[48] Texas Instruments, Inc., Contract AF33(616)8339, Wright-Patterson Air Force Base, Ohio.

[49] J. Kilby, IRE 4.10 Committee Meeting, Seattle, Wash., August, 1962.

[50] J. A. Narud, W. C. Seelbach, and N. Miller, *Computer Design,* vol. 1, p. 26, 1962.

[51] Micrologic, Fairchild Semiconductor, Division of Fairchild Camera & Instrument Corp., Mountain View, Calif., August, 1961.

[52] *Electron. Equipment Eng.,* vol. 10, p. 14, September, 1962.

[53] J. Nall, R. Anderson, D. Farina, R. Norman, and J. Campbell, AIEE Microsystems Conference, Los Angeles Section, May, 1961.

[54] B. Miller, *Aviation Week and Space Technol.,* May 21, 1962.

CHAPTER 6

FUNCTIONAL DEVICES

By Ian M. Ross and Eugene D. Reed

> . . . the aim of electronics is not simply to reproduce physically the elegance of classical circuit mathematics—rather, it is to perform desired electronic system functions as directly and as simply as possible from the basic structure of matter.*

6-1. THE TYRANNY OF NUMBERS

The rapid and continuous growth of the electronics industry at large has been accompanied by a concurrent growth in both the complexity and the size of individual electronic equipments. This development is the direct response of a forward-looking and progressive industry, sensitive to the needs of this era, the need to perform more complicated systems functions, the need to communicate more and over longer distances, and the need to process more information. In the early 1950s, an advanced computer contained somewhat over a thousand active devices. By the end of the decade this number had risen to the order of 100,000, and considering that there are, in general, about five passive devices associated with each active device, machines, therefore, already exist with a device count in excess of one-half million. There are equipments now in development with counts in the many millions. Even in the average home, there can today be found several thousand electronic parts.

Much of the ability of the electronics industry to meet this challenge, and in doing so to quicken still further the pace of this trend, has been due to the availability of ever cheaper and more reliable devices. Semiconductor and solid-state devices, both active and passive, have been improving in performance steadily, and at the same time dropping in cost at an impressive rate. System designers have thus found it technically as well as economically feasible to conceive new systems containing ever larger numbers of component parts. It is clear, however, that this trend

* J. A. Morton, New Directions in Component Development, keynote address, Electronic Components Conference, May 6, 1959.

cannot continue forever. While the demand for larger or more complex electronic systems will surely continue to grow, the approach to meet the demands of these new systems using traditional building blocks, in even greater numbers, is certain to encounter basic difficulties and ultimate limitations.

It is imperative, therefore, to explore all possible means of easing the problems growing out of this "tyranny of numbers." These problems are manifold and serious. They start with the sheer economics of fabricating the individual component parts; extend to the cost of assembling, mounting, and interconnecting them into reliable systems; and include the problems of housing and transporting of large systems and, finally, the problems in the logistics of maintaining the operation of complex systems in the field. Unless these problems of the tyranny of numbers are attacked directly and effectively, the electronics industry cannot continue to respond indefinitely to the increased demands placed on it.

Microminiaturization per se, since it deals only with the problems of size and weight, is clearly not an answer to the numbers problem. On the other hand, any measure which is successful in bringing about a reduction in the number of component parts will, in general, satisfy the objectives of microminiaturization.

There are, however, important activities in progress today in electronics which do, consciously or otherwise, attack some of these problems. One such major activity is the continuing effort to develop discrete devices of lower cost, higher reliability, and superior performance. Where successful, this approach alleviates many of the adverse effects of large numbers by reducing the initial cost and the cost of maintenance. It also attacks the numbers problem directly in two ways: (1) Improved reliability permits reduced redundancy, and (2) improved device performance often reduces appreciably the number of devices required to fulfill a given system function.

For example, in computer logic, a measure of effectiveness is the number of bits processed per unit time. Since the number of logic circuits required is, in general, inversely proportional to the speed of the active devices involved, it follows that higher speed in these devices will effectively lower the number needed. In a similar manner, superior stability in a resistor may eliminate the need for compensating feedback circuits, and thus also is effective in reducing the device count.

In the application of improved discrete devices, considerable effort has been applied to decreasing the cost and increasing the reliability of interconnections through standardization of interconnection patterns. Micromodule is a good example. Although effective in alleviating some of the difficulties of the numbers problem, such schemes do not attack the basic problem, namely, the numbers themselves.

A more recent and far-reaching approach is the widespread activity on integrated circuits. A broad range of technologies and techniques are brought together here in an attempt to fabricate groupings of interconnected elements on a single substrate. Included are the deposited metal films which can produce networks of interconnected resistors and capacitors. Similarly, deposited films of magnetic or superconducting materials can be used to create interconnected arrays of memory or logic elements. In addition, there are semiconductor integrated circuits where a combination of various technologies including deposition and diffusion are used to produce simple circuits containing interconnected resistors, capacitors, diodes, and transistors.

The integrated-circuit approach attacks the tyranny of numbers in two ways. First, since many connections are made simultaneously in a single process step, there results a corresponding reduction in the number of connections which would otherwise have to be made individually. Secondly, since a group of elements now forms a single entity, there results a great reduction in the number of physically separable pieces which must be fabricated and handled. The integrated-circuit approach therefore results in distinct improvements in all the problems enumerated earlier. Nevertheless, it too fails to come to grips with the heart of the problem.

One feature common to all the attacks on the tyranny of numbers outlined so far is their reliance on classical circuit concepts and traditional circuit elements for performing needed functions. Even after these circuits have been optimized and the number of elements pared down to an irreducible minimum, and when groupings of these elements have been integrated on single substrates, the total number of circuit elements still remains embarrassingly large. No approach, in fact, which is based on classical-circuit theory can make a real inroad into the basic problem of numbers.

Classical-circuit theory and circuit elements, after all, are merely convenient bridges between the needed system functions and the basic properties of matter upon which, in the final analysis, the operation of all systems must depend. Substantial inroads can therefore only be made by abandoning entirely the classical-circuit concepts and by invoking basic properties of matter to perform the desired system functions as directly and simply as possible. These functions will then be achieved without compounding and multiplying the numbers of component parts. Devices conceived in this manner have been called *functional devices.*

A useful criterion for a functional device is that, while it performs a function that would otherwise need a conventional circuit, it is not possible to identify within it the elements of the conventional circuit. It then follows that there is no useful equivalent circuit for a functional device.

The functional-device concept promises a direct solution to some of the most pressing problems of electronics now and in the future. Because it is not tied to the exhaustively exploited concepts of classical-circuit design, it offers boundless opportunities. Not only does it accomplish existing functions much more elegantly, but it also may pave the way toward the achievement of entirely new functions.

6-2. FUNCTIONAL DEVICES

There already exist a large number of devices, which by the definitions given above, should be classed as functional devices. Some were invented many years ago, well before the problem of numbers became so pressing, and certainly before the basic concepts of functional devices had been formulated. Nevertheless, these devices were the outcome of using physical phenomena to satisfy systems needs, and resulted in a level of simplicity or performance previously unobtainable. Some of the more recent examples, on the other hand, were the direct result of deliberate attempts to develop functional devices for specific needs.

The examples to be discussed in this section were chosen to illustrate the principles of functional devices and the advantages to be gained from their use. It is not the intent to cover in detail the theories of operation nor the technologies which are peculiar to the individual devices. Functional devices are, in fact, not limited to any single technology or even groups of interrelated technologies.

The first group of examples to be described illustrates how the principles of piezoelectricity have been combined with the properties of the propagation of elastic stress waves in insulators and semiconductors to produce a variety of functional devices. The second group of examples illustrates some of the ways in which the principles of transistor action in multijunction semiconductor structures have been exploited.

One of the oldest but still most pertinent examples of a functional device is the *piezoelectric quartz-crystal resonator*. It came into being long before the concept of the functional device had been formulated; yet it remains an excellent illustration of four principal features of this approach.

1. The direct utilization and combination of basic properties of matter, i.e., the piezoelectric effect and mechanical resonance.
2. Lack of correspondence between classical-circuit equivalent and the functional device; i.e., neither inductance nor capacitance can be isolated or identified within the quartz resonator.
3. Reduction of component count, i.e., replacement of LC network by a single entity.
4. Superior performance, i.e., higher Q and greater frequency stability.

It is interesting to observe that piezoelectricity has only very recently been combined with semiconduction to give rise to a new functional device, *the elastic-wave amplifier*. This device may well turn out to be the forerunner of a whole family of related functional devices. A brief description of the ultrasonic amplifier will follow the next example, *the ultrasonic delay line* which, in the technical as well as chronologic sense, forms the bridge between the quartz crystal and the ultrasonic amplifier.

Delay lines find applications in radar applications such as moving-target indication, video integration, and chirp techniques; they are also used for digital storage in computers and for time marking and pulse coding. In all these applications, the ultrasonic delay line provides time delay of electric signals by converting them to mechanical stresses (Fig. 6-1) using a suitable transducer, propagating these stresses as elastic waves through a prescribed path in the delay medium, and, finally, reconverting them to an electric signal at the output transducer. Delays up to a few milliseconds can be achieved at frequencies from 10 to 60 Mc. The transducers, which are either piezoelectric crystals or piezoelectric ceramics, are most efficient when used near their resonant frequency. Hence, delay lines are bandpass devices and the signals must be applied in the form of r-f pulses with a carrier frequency close to the transducer resonant frequency. Bandwidth of 30 to 40% may be obtained with losses ranging from a few to 50 db. One of the earliest delay media used was mercury. This has now been entirely superseded by solid delay media made of fused quartz, aluminum tapes, or glass tapes.

The property chiefly responsible for the performance of the ultrasonic delay line is that stress waves propagate through the appropriate media 100,000 times slower than the velocity of light, hence making this means of providing delay many orders of magnitude more efficient than electromagnetic delay lines.

Another exploited attribute of the ultrasonic delay line is that by choosing a suitable mode of operation, delay can be made to vary with frequency in a prescribed manner. *Dispersive delay lines* employing this principle are utilized in chirp radar to collapse a long frequency-modulated pulse into a narrow pulse of high amplitude, and vice versa (Fig. 6-2).

Lumped-element electric networks have been designed to provide delay characteristics comparable to those typical for ultrasonic delay lines. These networks, however, require hundreds of elements and many adjustments; they occupy large volumes, are expensive, exhibit high attenuation, and, in many cases, produce delay characteristics with unavoidable ripples. An ultrasonic delay line, on the other hand, may consist simply of an aluminum strip about 1 in. wide, about $\frac{1}{10}$ in. thick,

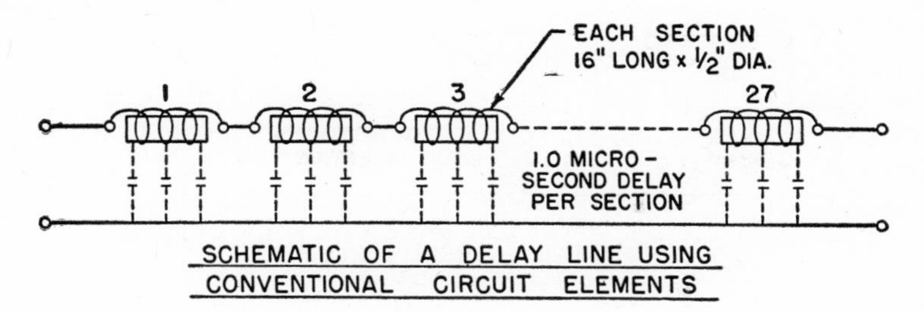

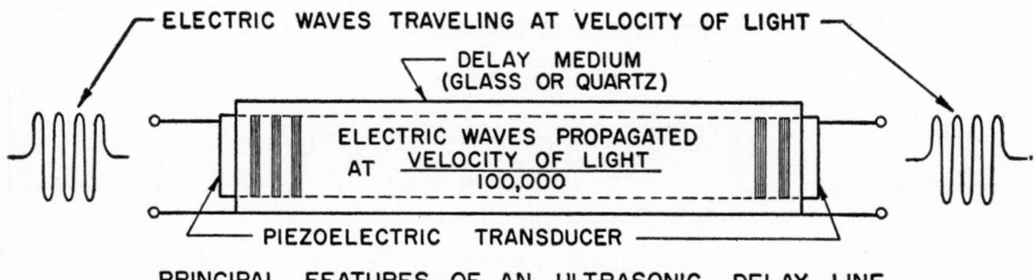

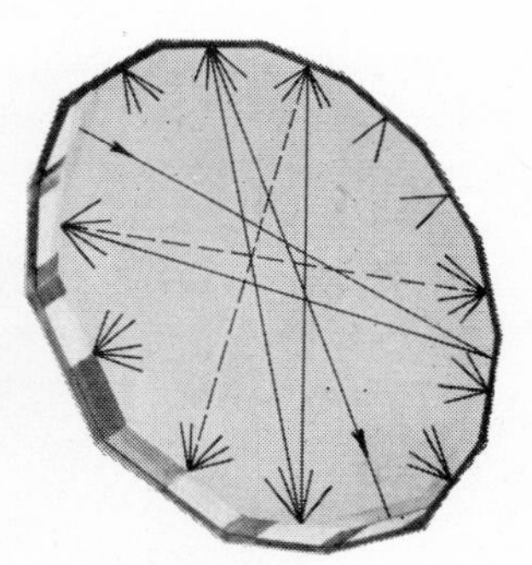

GLASS DELAY LINE LONGITUDINAL WAVES	GLASS DELAY LINE SHEAR WAVES	QUARTZ DELAY LINE SHEAR WAVES
DELAY = 12 MICROSECONDS	DELAY = 16 MICROSECONDS	DELAY = 707 MICROSECONDS
CENTER FREQ.= 15 MEGACYCLES	CENTER FREQ.= 15 MEGACYCLES	CENTER FREQ.= 20 MEGACYCLES
BANDWIDTH = 4.5 MEGACYCLES	BANDWIDTH = 6.5 MEGACYCLES	BANDWIDTH = 12 MEGACYCLES
LOSS = 8 DB	LOSS = 8 DB	LOSS = 33 DB

FIG. 6-1. Delay lines using conventional circuit elements are 50 to 150 times as long as ultrasonic delay lines for the same delay. For comparable electrical performance they are generally limited to short delays of a few microseconds. Glass having a low temperature coefficient of delay is used for ultrasonic delay lines under 100 μsec. Quartz, since its loss is $\frac{1}{50}$ that of glass, is used in lines having delays up to 4,000 μsec. Such delay lines consist of a quartz polygon and make use of multiple internal reflections.

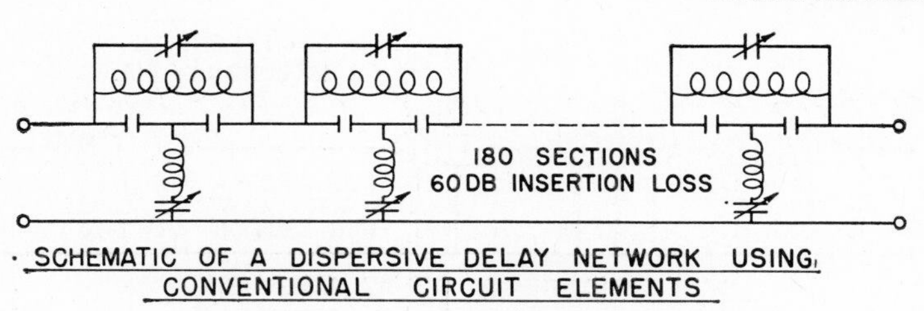

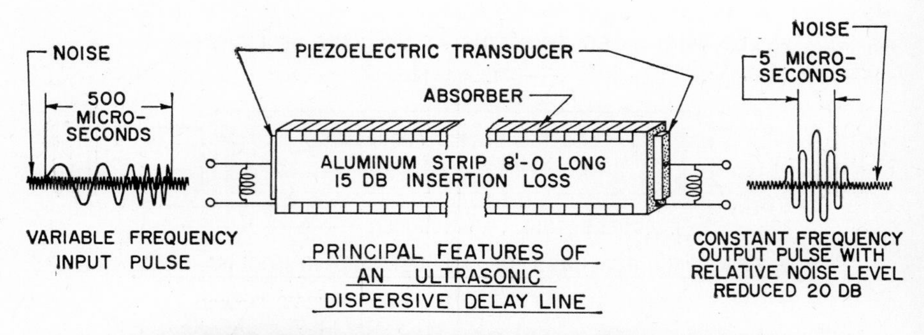

FIG. 6-2. A several-hundred-microsecond dispersive delay line, if made of conventional-circuit elements, would require hundreds of components. A drastic reduction in this component count coupled with superior electrical performance is provided by the ultrasonic delay line.

15 ft long, and coiled within a diameter of 12 in. Four such lines interleaved within the same space provide a function which, when executed with lumped-element networks, would require two 6-ft relay racks containing over 5,000 active and passive devices.

A newcomer to the family of functional devices and potentially, perhaps, the most important one is the *elastic-wave amplifier* (Fig. 6-3). It promises to provide, in a single entity, amplification in the 100-Mc range which can now only be obtained in a multistage amplifier. The properties of matter exploited are piezoelectricity and semiconduction, both simultaneously present in a semiconductor. The active material used in this device to date is cadmium sulfide.

Direct amplification of high-frequency sound waves is possible in cadmium sulfide because an elastic wave traveling through a piezoelectric material produces a sinusoidal longitudinal electric field which travels along with the wave. Since the material is also conductive, this sinusoidal electric field will cause electron bunching in a correspondingly periodic manner. By means of an applied d-c electric field, the electron bunches are made to drift in the same direction as—but somewhat faster than—the elastic waves. Interaction between the bunched electrons and the piezoelectric field results in a slowing of the electrons with consequent loss in kinetic energy and a corresponding gain in energy by the electric field. This gain mechanism is similar to that giving rise to the amplification of electromagnetic waves in traveling-wave tubes.

In the original experiments, a crystal of CdS was used with quartz transducers at both ends. As much as 38 db of gain was observed at 45 Mc in a crystal a little over ¼ in. long. Although still in the research stage, this new device or, rather, the new principles exploited here—namely, the combination of piezoelectricity and semiconduction—promise to lead to a new class of functional solid-state devices such as amplifiers, oscillators, isolators, and improved high-frequency delay lines.

A frequently used example of a semiconductor functional device is the *pnpn diode*. Figure 6-4 shows a schematic diagram of a *pnpn* diode together with the *VI* characteristic of the device. It is seen that, in one direction of current flow, the characteristic consists of a region of high impedance at low current and a region of low impedance at high current,

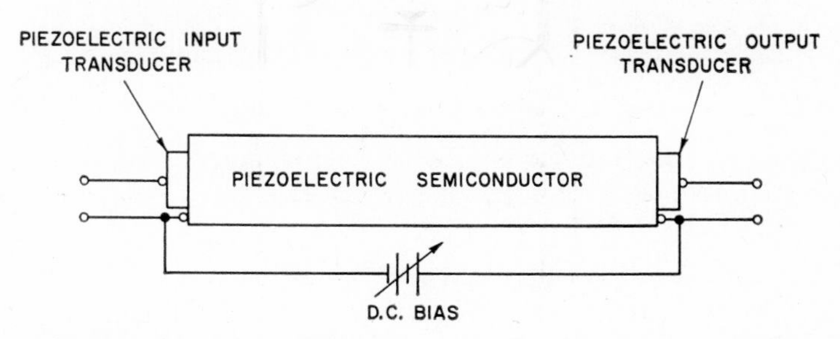

FIG. 6-3. The elastic-wave amplifier uses a piezoelectric semiconductor as active material. Gain is obtained when the velocity of the electric waves is greater than that of the elastic waves. This velocity ratio is controlled by the d-c bias voltage. Gain of 38 db at 45 Mc has been obtained in a 7-mm length of cadmium sulfide.

these two being separated by a region of negative resistance. By choice of a proper load line, the device can be made to operate stably in either the high-impedance or the low-impedance regions. Such bistable performance can be used to enable the achievement of many functions commonly found in electronics. Typical of these are the temporary memory functions that are the essence of registers, counters, etc. The need for a device that could perform such functions was indeed recognized and formulated some years before the invention of the device.

Figure 6-4 also shows the conventional circuit that would be needed to perform the same function that is performed by a *pnpn* diode. It is seen that the *pnpn* diode fits the definition of a functional device, in that nowhere within it can the elements of the circuit be identified. In the circuit, a resistor shunting the emitter to the base of one of the transistors

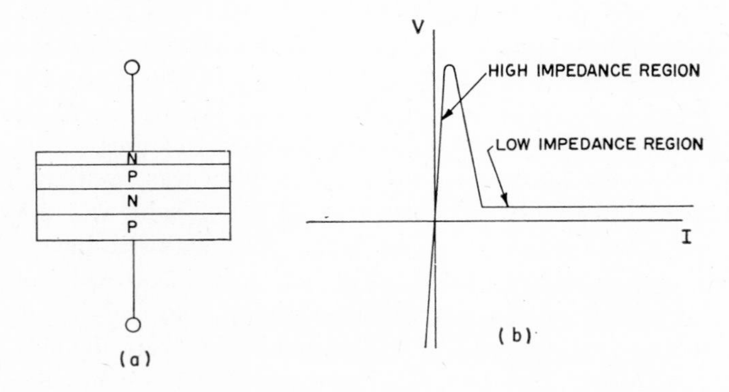

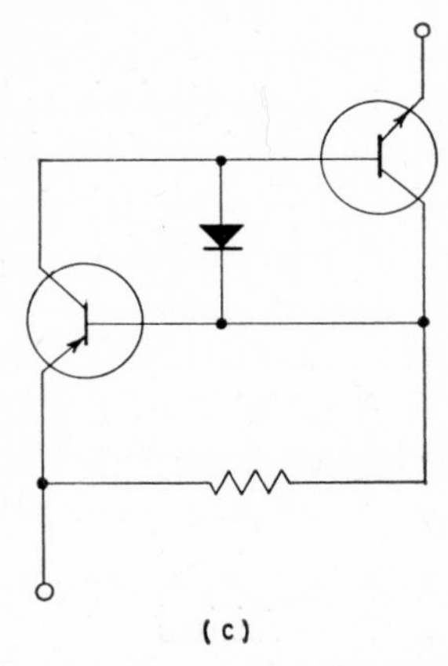

FIG. 6-4. *pnpn* diodes: (*a*) Schematic diagram of the device showing four conductivity regions and two leads; (*b*) *VI* characteristics showing the high- and low-impedance regions in one direction of current; (*c*) the diagram of the circuit which a *pnpn* diode replaces.

is used to reduce the effective gain at low current of that transistor. In the diode, the equivalent function is achieved making use of the recombination of electrons and holes in the neighborhood of the *pn* junctions.

Although the *pnpn* diode is a simple and well-established device, it is truly a functional device. The reduction in the number of needed components that can result from its use is great. The *pnpn* diode is made in one piece of silicon, contains three simple junctions, and has a total of two terminals. The circuit it replaces, on the other hand, contains four components, with a total of ten terminals that need to be interconnected in order to provide the required function. Finally, these diodes have demonstrated switching times of a few nanoseconds and are, therefore, comparable in performance to the very best that can be achieved with conventional circuits.

As suggested previously, one useful application of *pnpn* diodes is in counters. In one special form of counter called the *stepping switch*, a more complex functional device based on the *pnpn* diode has already been demonstrated. A stepping switch typically consists of a row of identical circuits or devices in each of which can be stored a binary 1 or a binary 0. This array of circuits or devices is interconnected in such a way that all except one of them is in the 0 state. It is further arranged that, upon the receipt of a suitable stimulus, the binary 1 steps from circuit to circuit or from device to device in a predetermined manner. It is this stepping action from which the counter derives its name.

A study of a stepping switch reveals that it contains two primary functions. The first is storing binary information at discrete points in space, and this can be performed, of course, by *pnpn* diodes. The second function is transmitting the information upon receipt of a suitable stimulus from point to point in a predetermined direction.

Figure 6-5 is a schematic diagram of the *stepping element** which is an extension of a *pnpn* diode, and is specifically designed to perform the two primary functions of a stepping switch.† The stepping element has two contacts, C and D, in addition to the usual two contacts of a *pnpn* diode. The device can be triggered from the low-current to the high-current condition by applying a signal to either one of these additional leads. However, when the element is in the high-current condition, a potential appears at lead D that is capable of triggering another similar device, while the potential appearing at lead C is not capable of triggering a similar device.

* L. A. D'Asaro, A Stepping Transistor Element, 1959 *IRE WESCON Conv. Record*, , part 3, pp. 37–42, 1959.

† The predecessor of this device is a gas-discharge stepping tube described by M. A. Townsend in Construction of Cold Cathode Counting and Stepping Tubes, *Elec. Eng.*, vol. 69, pp. 810–813, September, 1950.

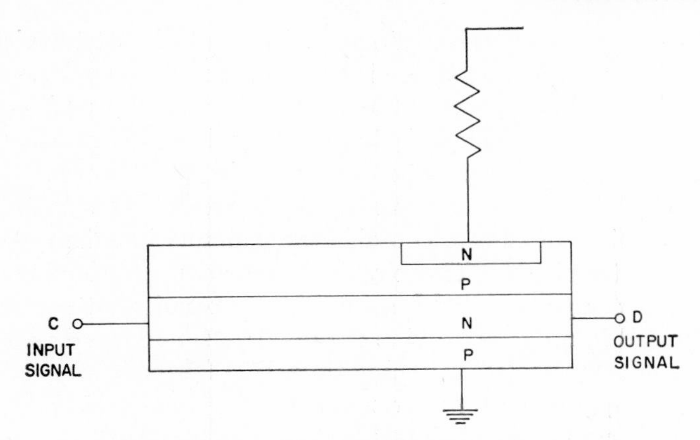

FIG. 6-5. The semiconductor stepping element.

This structure, therefore, yields the possibility of transferring the high-current state or binary 1 in a predetermined direction. A number of these elements can then be connected together to create a stepping switch. These connections can be made by leads or, more elegantly, through the semiconductor itself, thus creating one functional device capable of performing the functions of a complete stepping switch. Figure 6-6 is a photograph of such a structure, while Fig. 6-7 shows a similar

FIG. 6-6. A photograph of a sample stepping switch in which stepping elements have been fabricated on a single piece of material and in the form of a circle to give continuous operation. The outside diameter in the picture is 40 mils.

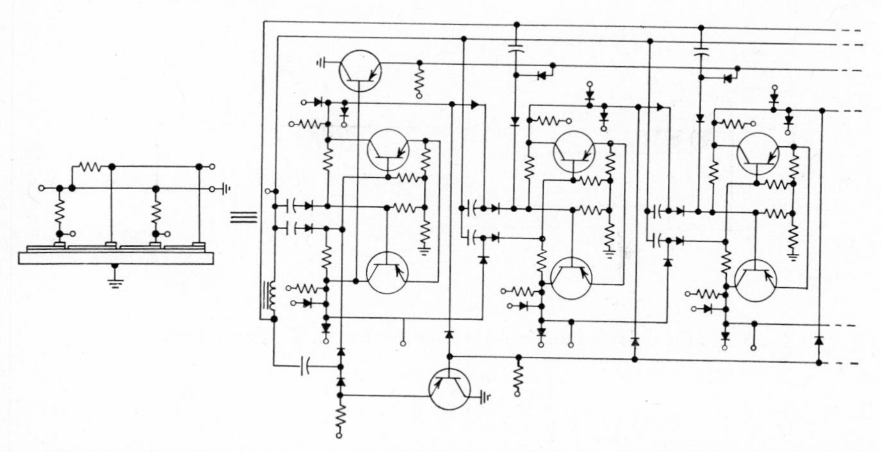

FIG. 6-7. Comparison of a stepping transistor with the circuit which performs a similar function. A component count shows that there is a reduction of about 15 to 1 in going to the functional device.

structure and compares it to the conventional circuit which performs approximately the same function. A detailed count shows that there is a reduction by a factor of about 14 times in the number of individual components or piece parts which are required to perform the desired function.

Another area of computer technology in which there has been an effort to develop functional devices is that associated with registers. A register typically consists of a row of identical circuits or devices, in each of which can be stored a binary 1 or 0. The *pnpn* diode can again provide this binary storage.

A version of register is the *shift register* in which, upon command, the information stored in the row of devices or circuits is made to shift one step to the left or right. Study of the shift register again reveals that it contains two prime functions. First is the storing of binary information at discrete points in space, and this can be performed by the *pnpn* diodes. The second function is transmitting this information from point to point but with delay.

Thus, some form of delay line is needed. There are a number of physical phenomena which can form the basis of a functional delay line. The ultrasonic version has already been discussed. For this application, a semiconductor functional device has been proposed.* As shown schematically in Fig. 6-8, the structure is essentially that of a *pnpn* with an extended base region. When the *pnpn* is switched into the high-current condition, electrons are injected into the *p*-type base. These can

* J. T. Wallmark, Design Considerations for Integrated Devices, *Proc. IRE*, vol. 48, pp. 293–300, March, 1960.

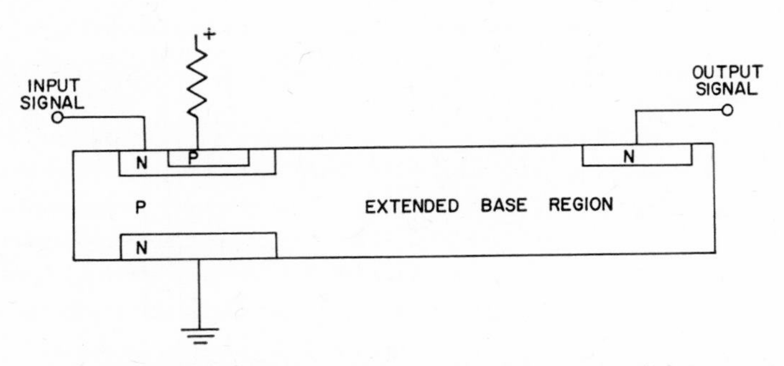

FIG. 6-8. A functional device for use in shift registers.

then travel along the extended base and be collected after the transit delay at the remote contact.

Thus, by a single extension of one region of a *pnpn* device, the inventor has added the needed delay function to the already existing bistable behavior. As with the stepping element, many of these shift-register elements can be connected together to form a complete shift register; and again the most elegant method is to use the semiconductor itself as the means of interconnection, and thus produce a functional device performing the complete function of a shift register. The savings to be achieved by such a procedure are equal to about a factor of 16 to 1 in the number of separate components or piece parts needed.

6-3. THE FUNCTIONAL-DEVICE APPROACH

Having demonstrated the power of the functional-device approach in solving one of the most pressing problems in the future of electronics, the numbers problem, it remains to explore the means by which such an approach can be implemented. At first sight, the demand for functional devices may be interpreted as a demand for major invention, and major invention cannot be programmed. It could, therefore, be argued that there is no methodical procedure for implementing the functional-device approach. However, it will be shown that the two essential ingredients in a functional-device program are not basically new to the field of electronics.

The first requirement in a functional-device program is to define the needed functions and their properties. The analogous requirement in a conventional-device program is to define the needed circuit elements and their properties. In the course of the evolution of conventional circuits, methods of analysis and synthesis were developed, and numerous measures of effectiveness were defined, to permit the relation of the terminal properties of devices to their performance in circuits. The concept and use of four-pole parameters and the concept of gain-band product are typical examples of such analytical tools and measures. In a functional-device approach, the equivalent analytical tools and measures of effectiveness must similarly be generated to permit relating the terminal properties of functional devices to the performance of these devices in a system or subsystem. Such an endeavor is not new in concept, only in scope.

The second requirement in a functional-device program is the identification of those properties of matter that best enable the realization of the desired system functions. This phase embraces the inventive process. The inventive process enters a conventional-device program in a closely analogous manner. The rapid emergence of new devices during the last decade or so is a vivid demonstration of the outpouring of useful invention that can be stimulated, motivated, and recognized when it occurs, by a proper definition of desired objectives. Typical examples are ferrite devices, varactors, and masers. There is every reason to expect similar progress in the development of functional devices, provided the proper questions are asked. In other words, the inventive talent of the device engineer, though not programmable, can be effectively stimulated and channeled.

The major change in the organization of electronics effort in response to a shift to the functional-device approach is a shift in the interface between the systems engineer and the device engineer. This shift is illustrated in Fig. 6-9.

In the classical-circuit approach, the interface separates the circuit element from the circuit or network in which it is used. In the functional-device approach, the interface has moved in the direction toward the complete system, and now separates the device from the subsystem.

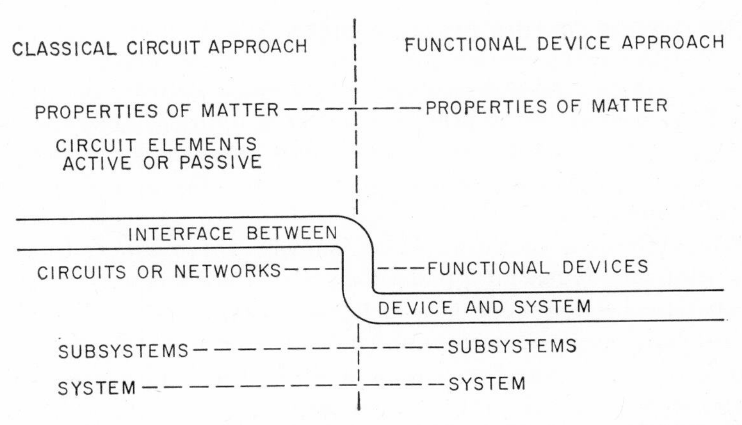

FIG. 6-9. Shift in the interface between the systems engineer and the device engineer.

As can also be seen from Fig. 6-9, one complete step in the progression from the properties of matter to the achievement of a working system has been eliminated.

Close cooperation between the device engineer and the systems engineer will be even more important in the era of functional devices than it has been before, there now being more at stake. Since the individual functional device, by definition, forms a larger portion of the over-all system than does the circuit element, it represents a larger fraction of the total investment. Decisions made at the interface, therefore, will be of even greater economic significance than they are today.

To achieve the interaction needed in functional-device programs, some reeducation will be required for both the device engineer and the system designer. Both will have to change their basic viewpoints. In the past, these two, between them, had to determine the best match between feasible components and feasible circuits to perform the desired system function. In the era of functional devices, they will have to determine the best match between feasible functions and the over-all system requirement.

The system designer then must start at a new level of synthesis, specifying his needs only in terms of basic system functions with properly weighted objectives. This, in turn, will stimulate his imagination and effort to higher levels of sophistication in system organization and logical design. At the same time, the range of possible solutions open to the device designer will be greatly broadened in terms of such functional requirements. For example, the opportunities for bold new ideas emerging is much greater in response to the request "Give us signal delay," than to the request "Develop network components for us to fit our delay line design."

6-4. THE IMPORT OF FUNCTIONAL DEVICES

Microelectronics programs, in general, promise substantial and rapid alleviation of many of the problems resulting from the tyranny of numbers. However, because they are based on traditional concepts of circuit design, they are limited in scope and, therefore, are doomed to reach the point of diminishing returns. This, together with the fact that they do not attack the root of the problem, namely, the numbers themselves, means that they represent a palliative rather than a cure.

There have been previous periods in the history of electronics when the conventional approaches were running dry. It was only when a basically different approach was taken or a major invention occurred that an upsurge of new growth occurred. A good example is the state of the electronics industry in the middle thirties, then limited by having grid-controlled tubes as the only active devices. Miniaturization and integration had been exploited, as evidenced by the emergence of miniature tubes and combinations of tubes sharing the same enclosure. Yet, it was only with the advent of microwave-tube technology that an upsurge of communications was made possible, and with the advent of solid-state devices that, similarly, an upsurge in information processing could occur.

A functional-device program does represent a fundamentally new approach and goes directly to the root of the numbers problem. Before such a program can be fully effective, new analytical tools will have to be developed to permit the proper formulation of the functions to be performed and the determination of the terminal properties of the devices needed to perform these functions; inventions will be required in order to realize the functional devices themselves. The functional-device approach is, therefore, a long-range program, with potential benefits so profound, that it cannot be ignored.

INDEX